CYBER TERRORISM AND INFORMATION WARFARE

III. Critical Infrastructure Protection Issues

Yonah Alexander
and
Michael S. Swetnam

Seventh Volume, Second Series
Terrorism: Documents of International & Local Control

Yonah Alexander
Donald J. Musch
General Editors

Oceana Publications, Inc. / Dobbs Ferry, New York

You may order this or any other Oceana publications by visiting Oceana's website at http://www.oceanalaw.com

Terrorism: Documents of International and Local Control, Second Series

9780379214154

9780379214185

Library of Congress Catalog Card Number: 99-75945

Manufactured in the United States of America on acid-free paper.

TABLE OF CONTENTS

CHAPTER 1

LEGAL FOUNDATIONS

STUDIES AND CONCLUSIONS

Report to the President's Commission on Critical Infrastructure Protection 1997

This report was submitted to the President's Commission on Critical Infrastructure Protection, and informed its deliberations and recommendations. This report represents the opinions and conclusions solely of its developers.

CONTENTS

ACKNOWLEDGMENTS

The *Legal Foundations* series of reports of the President's Commission on Critical Infrastructure Protection (PCCIP) resulted from the concerted efforts and hard work of several individuals. The Commission gratefully acknowledges Commissioner Stevan D. Mitchell and Assistant General Counsel Elizabeth A. Banker for their leadership and important contributions in developing the *Legal Foundations* series of reports. Their research, writing and analytical contributions were essential to the success of the effort.

The Commission also acknowledges Lee M. Zeichner, Esq. of LegalNet Works Incorporated and his staff, for conceptualizing and maintaining the Legal issues database and for providing tireless research support. Finally, the Commission acknowledges the contributions of Senior Consultant Paul Byron Pattak for his deft editing of this compilation.

PREFACE

Executive Order 13010 established the President's Commission on Critical Infrastructure Protection (PCCIP) and tasked it with assessing the vulnerabilities of, and threats to, eight named critical infrastructures and developing a national strategy for protecting those infrastructures from physical and cyber threats. The Executive Order also required that the PCCIP consider the legal and policy issues raised by efforts to protect the critical infrastructures and propose statutory and regulatory changes necessary to effect any subsequent PCCIP recommendations.

To respond to the legal challenges posed by efforts to protect critical infrastructures, the PCCIP undertook a variety of activities to formulate options and to facilitate eventual implementation of PCCIP recommendations by the Federal government and the private sector. The PCCIP recognized that the process of infrastructure assurance would require cultural and legal change over time. Thus, these activities were undertaken with the expectation that many would continue past the life of the PCCIP itself.

The *Legal Foundations* series of reports attempts to identify and describe many of the legal issues associated with the process of infrastructure assurance. The reports were used by the PCCIP to inform its deliberations. The series consists of 12 reports:

1. *Legal Foundations*
2. *The Federal Legal Landscape*
3. *The Regulatory Landscape*
4. *Legal Authorities Database*
5. *Infrastructure Protection Solutions Catalog*
6. *Major Federal Legislation*
7. *Adequacy of Criminal Law and Procedure (Cyber)*
8. *Adequacy of Criminal Law and Procedure (Physical)*
9. *Privacy and the Employer-Employee Relationship*
10. *Legal Impediments to Information Sharing*
11. *Federal Government Model Performance*
12. *Approaches to Cyber Intrusion Response*

and two special studies:

- *Information Sharing Models*
- *Private Intrusion Response*

Legal Foundations: Studies and Conclusions is the overall summary report. It describes the other reports, the methodologies used by the researchers to prepare them, and summarizes the possible approaches and conclusions that were presented to the PCCIP for its consideration. The series has been sequenced to allow interested readers to study in detail a specific area of interest. However, to fully appreciate the scope of the topics studied and their potential interaction, a review of the entire series is recommended.

PART ONE
LEGAL ACTIVITIES

With Executive Order 13010, the President's Commission on Critical Infrastructure Protection (PCCIP) was tasked to assess the vulnerabilities of, and threats to, eight named critical infrastructures, and also to develop a national strategy for protecting those infrastructures from physical and cyber threats. The Executive Order also required the Commission to consider the legal and policy issues raised by efforts to protect the critical infrastructures and to propose statutory and regulatory changes necessary to effect potential Commission recommendations. The results of these efforts are documented in the 12 reports that constitute the *Legal Foundations* series.

The first part of this summary report describes the legal survey work done to better inform the PCCIP about the complex webs of legal and regulatory authorities surrounding each of the critical infrastructures, and documents the methodology used to identify legal issues and generate material for potential recommendations. It also contains a summary of the substantive legal work that further informed the Commission's efforts.[1]

The first step characterized the legal and regulatory "landscapes"—providing overviews of the legal and regulatory climates surrounding each of the critical infrastructures. This initial landscaping project, and the difficulties inherent in adequately capturing legal authorities relating to the protection of critical infrastructures, stirred the undertaking of several other activities and the creation of a number of products. The following sections capture the results of these efforts.

The Federal Legal Landscape

The landscape process began by surveying attorneys from the offices of General Counsel from the ten agencies represented on the Commission: the Departments of the Commerce, Defense, Energy, Justice, Transportation, and Treasury; and the Central Intelligence Agency, Federal Bureau of Investigation, Federal Emergency Management Agency, and the National Security Agency. The General Counsel attorneys were briefed on the goals and objectives of the PCCIP. They, in turn, provided the Commission with accounts of their agencies' legal authorities and mechanisms relating to infrastructure assurance.

---

1 The legal materials presented have been compiled only from publicly available materials. The *Legal Foundations* series of reports contains statements of the law and legal analysis. While there are classified materials that are relevant to the legal issues associated with infrastructure assurance, they have not been included within the scope of these materials.

To aid those efforts, the Commission provided the survey matrix shown in Figure 1 below. The matrix breaks down "infrastructure assurance" into the components of the three-part definition from the report of the Critical Infrastructure Working Group (CIWG), a definition that has continued to serve well through the Commission's efforts. Infrastructure assurance is defined as ". . . the surety of readiness, reliability and continuity of infrastructures such that they are: (1) less vulnerable to disruptions or attack; (2) harmed to a lesser degree in the event of a disruption or attack; and (3) can be readily reconstituted to reestablish vital capabilities." The matrix encourages further analysis by specific critical infrastructure and purpose of the authority (i.e., legal authority or enforcement mechanism, physical or cyber threats). The General Counsel attorneys were also encouraged to identify plans and projects "under construction" that appeared to relate to assurance efforts and to suggest other agencies with stakes in infrastructure assurance efforts.

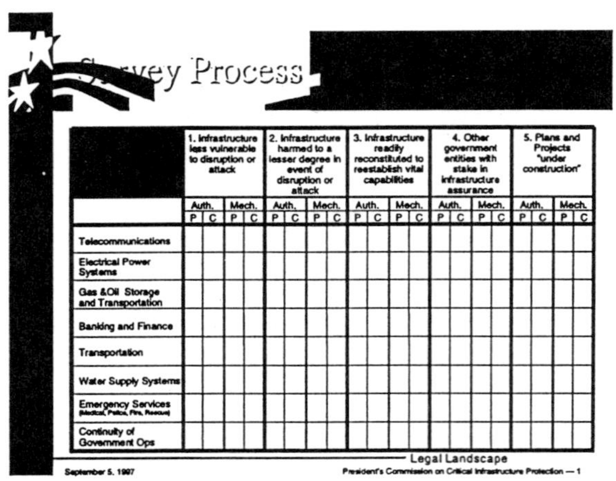

	1. Infrastructure less vulnerable to disruption or attack				2. Infrastructure harmed to a lesser degree in event of disruption or attack				3. Infrastructure readily reconstituted to reestablish vital capabilities				4. Other government entities with stake in infrastructure assurance				5. Plans and Projects "under construction"			
	Auth.		Mech.		Auth.		Mech.		Auth.		Mech.		Auth.		Mech.		Auth.		Mech.	
	P	C	P	C	P	C	P	C	P	C	P	C	P	C	P	C	P	C	P	C
Telecommunications																				
Electrical Power Systems																				
Gas &Oil Storage and Transportation																				
Banking and Finance																				
Transportation																				
Water Supply Systems																				
Emergency Services (Medical, Police, Fire, Rescue)																				
Continuity of Government Ops																				

Figure 1: Infrastructure Assurance Authorities Survey Matrix

From the preliminary results of the survey effort, it became apparent that to achieve a reasonably complete picture of the Federal government's authorities and mechanisms that could be used to achieve infrastructure assurance objectives, a broader outreach effort would be needed.

* * * *

[Figure 2 Omitted]

The Regulatory Landscape

In addition to *The Federal Legal Landscape*, which focuses primarily on broad authorities possessed by the Federal government, the Commission felt that a study of the current and past regulatory climates of the critical infrastructures would be very helpful. Historically, the critical infrastructures differ in the degree to which they have been regulated, and the reasons for their continued regulation. Some were regulated predominantly by concern for broad access to service, while others were regulated over matters relating to market power, public safety, or consumer confidence. With restructuring occurring across some of the critical infrastructures—results of efforts to open markets and increase competition—they now also differ in the extent to which they are being restructured. Potential impacts on security and reliability were areas of concern and study. A study of the regulatory landscape, titled *The Regulatory Landscape* was undertaken to provide a better understanding of the way regulatory tools and mechanisms—Federal, state, local or private—are being, or could be used to promote infrastructure assurance objectives.

Legal Authorities Database

The Legal Authorities Database is an outgrowth of various "landscaping" projects. The Commission recognized early on that, given methodological and time constraints, the landscaping efforts could not reasonably be expected to capture, for example, the full range of international bodies, treaties, and conventions relevant to the infrastructures or infrastructure assurance. Nor would these landscaping efforts adequately account for or explain "authorities" that—although not promulgated by governments per se—nonetheless can be influential in effect. We refer here to the myriad policies, guidelines, and standards (such as auditing standards) that carry force of law for private sector and public sector owners and operators of critical infrastructures. Neither the legal nor the regulatory landscape, separately or together, would provide a complete panorama of legal authorities associated with infrastructure assurance, nor would they provide an adequately broad picture of relevant cross-infrastructure issues.

The legal and the regulatory landscapes do a good job of capturing many legal authorities relating to infrastructure assurance, which as a result, may be used to *promote* or *enhance* infrastructure assurance. Some current laws, regulations, policies, or guidelines, however, may actually operate in ways that are *antithetical* to achieving infrastructure assurance objectives. Identifying, reviewing, understanding and revising those authorities in appropriate ways may prove to be as valuable as creating new ones.

To fill this void, the Commission recruited a team of attorneys to cull through the United States Code, the Code of Federal Regulations, Executive Orders, treaties, select state legislation and other relevant authorities with "infrastructure assurance" in mind. The authorities found to bear on infrastructure assurance—positively and negatively—were then incorporated into a database. The database was originally built around key words and concepts relevant to assurance efforts. These included topics such as indications and warning, information sharing, education and awareness, information security, emergency response and others. (Although the database, now with over 14,000 entries, has grown beyond these topics, these keywords and concepts continue to serve as useful search terms to guide research efforts.) The database represents a source of data which could facilitate implementation of PCCIP recommendations by Federal, state and local governments and by the private sector.

Figure 3: Legal Authorities Database CD Cover

Infrastructure Protection Solutions Catalog

The *Infrastructure Protection Solutions Catalog* is a compendium of prior recommendations made by informed individuals, study commissions, and other groups who have weighed in on issues related to the Commission's work. This document contributed not only to the identification of possible legal issues, but also served as a validation that a number of broad solution areas were adequately explored.

PART TWO
LEGAL ISSUES

The Commission was asked to identify legal and policy issues raised by efforts to protect the critical infrastructures, and to make recommendations for reform. Studies were undertaken to identify relevant issue areas.

Originally, some 16 discrete issue areas were identified, but later consolidated to seven.[2] These issue areas focus not only on using existing authori-

2 Of these, liability was addressed as an economic issue by the Commission to insure that it would be considered not only from a legal perspective, but from a broader economic perspective as a tool for social change.

ties to promote infrastructure assurance objectives (e.g., through greater inclusion of cyber concerns into existing statutes), but also on the need to remove existing legal impediments to infrastructure assurance-related actions. Results captured in the following studies in the *Legal Foundations* series reflect the significant background research completed to support the Commission's work.

Major Federal Legislation

This study explores major areas of Federal legislation, such as the Defense Production Act of 1950, the Stafford Act/Federal Response Plan, the Computer Security Act of 1987, the War Powers Resolution and related authorities, and the Nunn-Lugar-Domenici legislation. It reviews whether these evolving bodies of authority appropriately take into account threats to the critical infrastructures. It also addresses whether these authorities could be considered for modification to better address physical and cyber threats, or whether separate legislation would be preferable for this purpose.

Adequacy of Criminal Law and Procedure (Cyber)

This study explores the adequacy of Federal and state substantive criminal law to deter and punish those who commit computer-based attacks on critical infrastructures. It also evaluates procedural law, and whether it may place undue impediments on law enforcement's ability to conduct investigations.

Adequacy of Criminal Law and Procedure (Physical)

This study explores the adequacy of criminal law to respond to and deter physical attacks on critical infrastructure.

Privacy Laws and the Employer-Employee Relationship

This study looks at laws that prohibit or restrict infrastructure owners and operators from making use of various devices and processes routinely used by the Federal government in conducting background investigations and issuing security clearances to employees in highly sensitive positions. It explores the balance between concerns over security and privacy.

Legal Impediments to Information Sharing

This study attempts to forecast the various legal impediments that might inhibit the flow of infrastructure assurance-related information between and among government bodies and the private sector. It suggests ways in which an information sharing mechanism might be structured to minimize these legal impediments.

Federal Government Model Performance

This study explores five discrete areas through which the Federal government can unilaterally alter its own behavior in order to encourage the private sector, state or local governments to act consistently with infrastructure assurance objectives. They include performance measurement; publication of infrastructure assurance-related data; identification and dissemination of "best practices," procurement reform; and certification programs.

Approaches to Cyber Intrusion Response

This study assumes that the number and severity of computer intrusions will continue to grow, and that a conventional law enforcement response might not be sufficiently robust to maximize deterrence. It explores viable alternatives to traditional criminal enforcement mechanisms (investigative and prosecutorial) to serve as a supplemental deterrent to unauthorized computer intrusions.

PART THREE
CONCLUSIONS

An overarching and unsurprising conclusion from the *Legal Foundations* series of studies is that law has failed to keep pace with technology. Some laws capable of promoting infrastructure assurance objectives are not as clear or effective as they could be. Still others can operate in ways that may be adverse to security concerns. Sorting them all out, of course, will be a lengthy and massive undertaking, involving efforts at local, state, Federal, and international levels. The studies within the *Legal Foundations* series represent a jump-start to a possible process of reform.

The conclusions presented in this paper reflect a principled approach to achieving infrastructure assurance. They reflect efforts to assess and understand the legal and regulatory landscapes of the critical infrastructures; the work of prior bodies (and the courses of action they recommended); and research on pressing legal issues that emerged from supporting the Commission's efforts. Much of the legal work presented throughout this series of studies examines the operation of the laws and regulations within and across the critical infrastructures at a micro-level. This following section collects observations and conclusions, so as to place them into a larger perspective—as a series of attainable steps to lay the legal foundation for the processes and mechanisms that should be considered in efforts to achieve continued assurance of the critical infrastructures.

The Process of Infrastructure Assurance

"Infrastructure Assurance" is not merely an end state, but a continuous process. It is a process of continuing to improve capabilities in five general areas: (1) policy formulation; (2) prevention and mitigation; (3) operational warning; (4) incident management; and (5) consequence management.

Responsibilities in these areas are performed today to varying degrees within each of the critical infrastructures by the Federal government, state and local governments, trade associations, and by individual owners and operators of critical infrastructures. "Legal Activities" described in *The Federal Legal Landscape* and *The Regulatory Landscape* revealed shortcomings of current efforts. First, current efforts are largely uncoordinated. Infrastructure regulators (Federal and state), law enforcement, intelligence, emergency preparedness and management, defense and other communities formulate policies that control the security and reliability of the critical infrastructures, but they do so by-and-large independently. Second, vital information is not currently shared between and among infrastructure sectors, and between and among government in ways that would feed a more complete understanding of the risks facing the critical infrastructures. Policy and information are developed and held in pockets which do not come together to be analyzed and de-conflicted.[3]

Roles of the respective parties in infrastructure assurance should periodically be reevaluated and redefined to reflect the new environment in which the critical infrastructures will operate. Recommendations for organizing a national infrastructure assurance effort will, by necessity, reflect the need for new roles—greater private sector participation in policy formulation, effective information sharing between and among government and the private sector, and greater sharing of resources for research and development. Many of the conclusions presented here are intended to enable these roles to be undertaken by those most suited to the task.

Legal Foundations Conclusions

The conclusions drawn from *Legal Foundations* are designed to jump-start the gradual process of reform. They are set forth in three parts. Existing laws can be made to better serve the government in an effort to take the lead and serve as a model of standards and practices for the private sector. Other areas of law, with careful attention, can enable infrastructure owners and operators to take precautions proportionate to the threat. Still other areas of law can be molded to enable a greater degree of government-in-

---

3 Not only is policy uncoordinated and vital information unshared under the current construct, but the focus within few of these functional areas is on "infrastructure assurance." Regulators are primarily concerned with safety and reliability; owners and operators with competitiveness and growth; law enforcement with prosecuting criminals; intelligence with gauging threats; defense on preparing for war—these are the traditional roles of these constituencies. However, in the face of new threats and vulnerabilities, new technologies, and increasing irrelevance of jurisdictional boundaries, some of these roles may merit continued re-examination and realignment.

dustry partnership in areas such as information sharing. Many of these proposals call for further study of areas deserving additional attention.

Enabling the Federal Government to Take the Lead

Achieving increased protection depends on voluntary cooperation. In order to attract it, government must demonstrate its own commitment to the protection of the critical infrastructures and lead by example. The first set of conclusions presents several ways the Federal government may better fulfill its traditional governmental responsibilities through performance-based measures, modest legislative initiatives, and by enhancing deterrence to criminal activities that threaten critical infrastructures.

Federal Government Model Performance

Model performance means measures that can be undertaken unilaterally by the Federal government, involving minimal alterations to structures, expenditures or regulations, that will enhance assurance for government and carry over positive effects for owners and operators of the critical infrastructures. Techniques such as the greater use of performance measures, publishing of comparative data, certifications, and revision of government procedures for procurement were identified and studied. (*See Federal Government Model Performance.*)

Publication of Infrastructure Assurance-Related Data

The Federal government can make use of information already in its possession to better assess threats and vulnerabilities, and to encourage others to do likewise. The publication of comparative data within an industry, such as on-time arrival and departure statistics for airlines, may serve to promote particular concerns, such as information security and reliability, without resorting to regulation. Much of this information may already be collected and may prove useful to consumers in light of increased competition and choice between providers of critical infrastructure services. Accordingly, a conclusion is that:

- The Administration could direct FCC's Network Reliability and Interoperability Council to initiate a pilot study as to the feasibility of publishing comparative infrastructure assurance-related data for the telecommunications industry. This data may be related to reliability, security or other issues identified during the course of the pilot study. The study should focus on whether publication is likely to achieve infrastructure assurance objectives; the types of data to collect and publish; whether current data collection efforts are sufficient or new ones are needed; and other possible impacts, positive and negative, of publication. The NRIC study should be followed by similar studies in other critical infrastructure sectors.

Procurement Reform

The Federal government could, where feasible, incorporate infrastructure assurance concerns into large pending procurements such as the re-competition of FTS 2000. Also, because of easy availability of waivers and other gaps in procurement policies and regulations, assurance objectives might not be adequately addressed by current procurement processes. Accordingly, a conclusion is that:

- An interagency task force could be convened to identify large pending procurements related to infrastructure assurance issues, study whether infrastructure assurance objectives are being adequately taken into account, how they may be adapted, and based on the lessons learned, suggest revisions for the procurement process generally.

Standards and Certifications

The Federal government could serve as a model for the private sector with respect to its own information security standards and compliance with those standards. Standards can provide a foundation for government-sponsored certification programs to signify compliance with security-related objectives. Some government-sponsored certification programs (for example, EnergySTAR), do not require large bureaucracies to oversee implementation and enforcement, but make available marketing incentives to encourage private sector participation. These programs place the burden of compliance principally on those who opt to use the certification. They can be inexpensive for the government to administer, and can be enforced principally through civil remedies rather than large regulatory bureaucracies. Accordingly, a conclusion is that:

- Lead Agencies can consider the creation and use of certification programs which are inexpensive to administer and enforce, and which provide incentives for adoption of standards relating to information security and information technology services and products.

Performance Measurement

The Federal government could better assess its progress in implementing these and other infrastructure assurance-related goals. Effective tools for doing so have been put in place through the "Reinventing Government" initiative. The Government Performance and Results Act (GPRA) currently requires five-year strategic plans and performance measures for major functions and operations of federal agencies. These performance measures are reviewed by OMB and Congress as part of the budget process. The Information Technology Management Reform Act (ITMRA) also requires the setting of performance measures related to the use of information technology. The required performance measures, however, do not specifically include information security. Accordingly, a couple of conclusions emerged:

- Federal agencies can be directed to include assigned roles and functions relating to infrastructure assurance specifically within their strategic planning and performance measurement framework; and
- The Information Technology Management Reform Act can be amended to specifically require agency Chief Information Officers to develop performance measures for the security of their information systems and to submit evaluations to OMB as required by the statute.

Revisiting Federal Legislation in Light of Infrastructure Assurance Objectives

Many areas of Federal legislation that enable prevention and mitigation, reconstitution, response and recovery to incidents involving the critical infrastructures were written before the emergence of a recognizable cyber threat. It is not clear whether many of these authorities would apply, and should apply, in the event of a major cyber-related event. Until the dynamics of such an event are better understood, legislative change would be premature. However, key issues were identified and conclusions reached in order to incorporate infrastructure assurance issues within these legislative frameworks. (*See Major Federal Legislation.*)

The Defense Production Act

The Defense Production Act (DPA) is an important mechanism for security of the nation's industrial and technological base. Its authorities, including priorities in contracts, financial incentives, and voluntary agreements, may assist in the reconstitution or recovery of a critical infrastructure made necessary by a physical or cyber event. The DPA authorities and triggering mechanisms, and efforts underway at modernization were reviewed to determine whether these powers might be available and adequate in light of emerging threats, vulnerabilities, and related challenges. Accordingly, some conclusions emerged:

- That the Administration and Congress could review the DPA in light of infrastructure assurance objectives; more specifically, that:

  1. Congress could consider amending the DPA Declaration of Policy to include a finding of how critical infrastructures are essential to national security;

  2. Lead agencies associated with the critical infrastructures could study the energy provision for priorities in contracts as a potential model for other critical infrastructures in order to ensure continued operation and clear authority for use in reconstituting such infrastructures after an event;

3. Funding for the DPA Fund and financial incentives could be supported and appropriated and that such funds are made available for research and development related to the critical infrastructures; and

4. The Administration could direct Federal agencies with authorities pertaining to the critical infrastructures to review the DPA's authorities and work with industry to make available such authorities when needed to respond to a critical infrastructure incident.

The Stafford Act/Federal Response Plan

The Stafford Act and the Federal Response Plan set out the parameters of the Federal response to major disasters as declared by the President. The Federal Emergency Management Agency's authority to prevent, mitigate, and respond to physical or cyber incidents affecting the operation of the critical infrastructures may require some degree of clarification or modification. Accordingly, the analysis led to a conclusion that:

- The Administration should undertake a study of Stafford Act and Federal Response Plan mechanisms to determine their applicability and suitability to cyber-induced disasters as well as their current implementation with regard to prevention and mitigation. Such a study would also take into account other disaster recovery authorities and their potential impacts on infrastructure assurance goals, as well as the desirability of providing direct assistance to infrastructure owners and operators.

The Deterrence-Related Authorities

Authorities such as the War Powers Resolution and the U.N. Charter set out conditions for use of force against nations. Many of these authorities were written before the emergence of information warfare and cyber threats. As these threats evolve, the mechanisms and policies that enable defense of our nation should be reviewed and updated to ensure they enable an adequate and effective response, and instill adequate deterrence where appropriate. Accordingly, a conclusion is that:

- The Administration, including the National Security Council and other appropriate agencies, could ensure that the U.S. strategy for responding to an information warfare attack addresses the potential legal issues associated with current definitions of "attack" contained in important domestic and international legislation.

Nunn-Lugar-Domenici

The Nunn-Lugar-Domenici legislation creates a partnership involving many Federal agencies and state and local governments to increase responsive capability to weapon of mass destruction incidents across the United

States. It focuses on providing training, access to equipment, and information to local first responders. While these programs are in their initial phases, already state and local police, fire, and medical officials are requesting an expanded effort in this area. More resources for training and equipment may be required based on updated assessments of threats, and may also require that the effort be expanded in scope to address other infrastructure-related events. Accordingly, a conclusion can be drawn that:

- Congress could consider whether the current Nunn-Lugar-Domenici program should be expanded to incorporate other critical infrastructure issues, including attacks on infrastructures by means other than Weapons of Mass Destruction, as well as the need for training, awareness, and information sharing efforts directed at state and local responders on the potential impact of disruptions of critical infrastructures, particularly information and communications, on emergency response efforts.

Enhancing Criminal Deterrence for Acts Targeting Critical Infrastructures

Deterrence plays an important preventative role against attacks on critical infrastructures. Deterrence through the criminal law should be built not only through Federal investigative and prosecutive capabilities, but also on state and local, and international capabilities. It should be built to prevent physical and cyber threats to infrastructures.

Physical Threats

Over 30 major Federal criminal laws may be invoked in response to physical attacks on critical infrastructures, and the statutes confer adequate investigative jurisdiction and authority in the vast majority of instances. However, several potential shortfalls remain relating to the ability of this complicated patchwork of laws to deter crimes against critical infrastructures. The Federal Sentencing Guidelines, for example, provide for lighter sentencing for attacks against property, however serious. More severe sentences are reserved for attacks against people and attacks using traditional, violent means-such as explosives and firearms. Thus, the Sentencing Guidelines may not adequately take into account the severity of consequential damages arising from attacks on critical infrastructures. For example, damage resulting from the "downstream" effects of a denial of service attack may result in the possibility of disproportionately light sentences for some forms of attack on critical infrastructure. (*See Adequacy of Criminal Law and Procedure (Physical).*) Accordingly, conclusions can be drawn that:

- The U.S. Sentencing Commission should expand its Sentencing Guidelines to include greater flexibility to address actual and consequential damages, including "downstream" damage to property or loss of service that might result from attacks on critical infrastructures; and

- The Sentencing Commission should consider expanding coverage of Guidelines to make available stricter penalties for uses of biological and chemical weapons not resulting in death.

Two potential deficiencies were identified with respect to purely *intrastate* attacks against critical infrastructures-even when such attacks result in severe damage. In these instances, in order legally to assume jurisdiction over an investigation or prosecution, the Federal government must demonstrate that the incident affects interstate commerce. This may be a particularly difficult determination to make at the earliest stages of an investigation, before the scope of an attack or its effects are well known. Accordingly, a conclusion can be drawn that:

- Congress should consider making a finding that certain critical infrastructures are "instrumentalities of interstate commerce" to help ensure the immediate availability of Federal law enforcement authorities to respond to crimes against those infrastructures and subject those convicted to stiffer Federal penalties. Congress should consider which of the critical infrastructures should be included in such a finding and establish an appropriate nexus between those infrastructures and interstate commerce for inclusion in appropriate legislation.

Legislation that offers reward monies for information leading to the capture of terrorists was reviewed. Under these legal authorities, Congress authorizes the Attorney General and the Secretary of State to administer rewards and payment-for-information programs. It was found that these laws effectively supplement other Federal criminal legislation to protect critical infrastructures, but further research revealed that these programs might be more effectively administered if included as line items in participating agencies' budgets. Accordingly, a conclusion can be drawn that:

- Monetary reward programs for information leading to capture and arrest of criminals should be included as a line-item in participating Federal agencies' budgets to ensure proper funding and implementation of such programs.

Cyber Threats

Efforts are ongoing, within many states, to keep current effective computer crime legislation. Many have produced statutes complete with civil remedies, restitution, forfeiture and special sentencing provisions that may provide needed deterrence in the area, particularly with respect to particular classes of offenders such as juveniles. Prosecution and rehabilitation of juveniles has been predominantly a state and local responsibility. Some states have adopted novel approaches specifically tailored to the unique nature of the crime and the maturity of the offenders. Other states and the federal government can learn from state efforts in this area, and can consider modifications to their laws to address what may be a growing problem. (*See Adequacy of Criminal Law and Procedure (Cyber).*) Accordingly, a conclusion to pursue would be that:

- The Department of Justice sponsor a comprehensive study aimed at compiling demographics of computer crime offenders, comparing the effectiveness of various state approaches to computer crime and discovering effective ways of deterring and responding to computer crime and abuse by juveniles.

The recent approach adopted by the U.S. Sentencing Commission in revising Sentencing Guidelines that apply to violations of the Computer Fraud and Abuse Act was recognized during these studies. The Sentencing Commission expanded applicable definitions of "harm" and "loss" to include interruptions in service; disruptions or delays in delivery of vital services endangering lives; invasions of privacy; and the cost to the victim of damage assessment, restoration of service and data, and loss of business revenue due to interruption of service. It may be advisable to expand this approach to sentencing and sentencing enhancement for possible application in conjunction with the full range of charges that might be implicated by abuse of computers and computer networks, including but not limited to violations of the Wire Fraud statute, the Wiretap statute, the Electronic Communications Privacy Act, or Fraud Relating to Access Devices. Accordingly, a conclusion is that:

- The Sentencing Guidelines Commission consider expanding the application of its broader reformulation of harm and loss (in Guidelines Section 2B1.1, as it applies to violations of the Computer Fraud and Abuse Act and theft of trade secrets) to other forms of electronic crime and crimes relating to information and information technology.

The Department of Justice is currently exploring ways to ease undue administrative burdens on Federal law enforcement officers investigating various forms of computer and high technology crimes that cross Federal jurisdictional boundaries. Law enforcement investigators are currently required to obtain court orders or warrants in many jurisdictions through which a communication might pass or even where electronic information is physically stored. A nationwide capability might allow these procedures to be conducted across jurisdictional boundaries with the authorization of only one Federal judge. These studies bring a recognition of a need for capabilities that, without altering existing processes, would allow for court orders obtained in one jurisdiction to be used in multiple jurisdictions. The Administration and Congress should also be sensitive to such needs, given the value of a rapid law enforcement response. (This conclusion is based on the understanding that this would not constitute an expansion of law enforcement's current ability to perform these functions, but that modifications would merely increase the speed and efficiency by which existing law enforcement capabilities are administered). Accordingly, a conclusion emerges that:

- The Administration could endorse and promote efforts currently underway to develop specific procedural changes to assist law enforcement in the investigation of computer crimes, including modification of existing procedures to create effective cross-jurisdictional trace and search warrant capabilities, and that Congress expeditiously consider enactment of such legislation.

The U.S. has been a forerunner among other nations in efforts to clarify and improve current law enforcement procedures pertaining to computer crime. The Administration and Congress could continue to recognize and support these international efforts. This may require, for example, careful comparison of existing standards and procedures for searching and intercepting electronic data during an investigation, as well as support of internationally focused efforts to assure that other countries have the legal and technical capability to conduct and assist in computer crime investigations. Second, the United States could continue with efforts to establish a strong network of international law enforcement agencies and telecommunications carriers to draw on in investigations. The creation and maintenance of such a network can facilitate prompt trace and identification of foreign sources of intrusions. Third, the United States could continue to promote efforts to enhance international cooperation in computer crime investigations, whether through new international arrangements or traditional mutual legal assistance treaties. The U.S. could further ensure that existing and future efforts cover the broadest range of potential sources of attack (i.e. where possible, agreements should cover more than members of recognized multilateral organizations). Accordingly, conclusions are drawn that:

- The efforts of U.S. delegations to several international bodies could be acknowledged and fully supported as contributing to infrastructure assurance. Potential areas for continued or expanded efforts in the international area should be highlighted. Specifically:

  1. The U.S. should continue to act as an international leader in efforts to clarify and improve current procedures for investigating computer crime;

  2. The U.S. could work to create a network of international law enforcement agencies and telecommunications carriers to draw on in the course of international investigations of computer crimes; and

  3. The U.S. could continue to promote efforts to enhance international cooperation in computer crime investigations through treaties and

other cooperative arrangements both through existing international fora and with countries not represented in those organizations.

Legal Impediments to Vulnerability Assessments

Existing laws may create unnecessary legal impediments to the performance of vulnerability assessments on federal computer systems. The Computer Fraud and Abuse Act criminalizes a wide variety of misconduct premised on unauthorized access to government (and private) computer systems. The legislation is silent, however, as to what might constitute valid authorization for security teams to attempt penetrations without running afoul of the criminal law. Legislative change does not appear to be required, but agencies could clarify attendant procedures to facilitate sound vulnerability assessment practices. Accordingly, a conclusion is that:

- Agency Chief Information Officers could establish procedures for obtaining expedient and valid authorization to allow vulnerability assessments to be performed on government computer systems. This would require a clear designation by government agencies regarding who may authorize access to their computer systems for this purpose.

Enabling Private Sector Response

Federal authorities that could be strengthened or expanded to allow the Federal government to more adequately accomplish infrastructure assurance objectives were reviewed for the *Legal Foundations* studies. Potential legal impediments that might prevent owners and operators from taking appropriate action to safeguard portions of critical infrastructure within their control and responsibility were also considered. The conclusions contained in this section focus on providing owners and operators greater ability to take protective action.

Additional Deterrence against Cyber Crimes

Deterring cyber-related acts targeting critical infrastructures requires an understanding of the unique nature of this sort of crime. The disparate motivations of the offenders, the continual growth of technologies for committing such crimes, and the decreasing importance of geographic and legal boundaries complicate the application of criminal law to these acts. Many states, the U.S. Congress, the Administration and international bodies have been paying considerable attention to computer crime over the past five years. The result has been improvements in the substantive criminal law, sentencing provisions, investigative procedures and international cooperation. But there is room for more progress. New solution options should be considered to address investigative difficulties and complexities arising in international investigations. Traditional procedures have proven less than effective in responding to the jurisdictional challenges connected with computer crime cases. (*See Approaches to Cyber Intrusion Response.*)

Clearly the capability to do harm and the availability of powerful intrusion tools will continue to grow. Accordingly, there is a need for more powerful ways of deterring criminal behavior. We also have reaffirmed, however, that victims are limited in their ability to detect unauthorized intrusions and in their willingness to report intrusions, once detected, to law enforcement. This has led to a question of the degree to which the threat of a criminal prosecution currently serves to deter would-be computer criminals, and has prompted analysis of ways to increase the effectiveness—hence the deterrent value—of a law enforcement response. Further complicating deterrence is the ability of intruders to span several traditional legal jurisdictions, which exacerbates the difficulties facing local and state response in this cyber-dimension.

These jurisdictional issues may place a tremendous burden on a centralized law enforcement response. Given the highly resource-intensive and time consuming nature of computer intrusion investigations, this centralized response could quickly be taxed by increases in computer intrusion activity, increases in detection, or even increases in reporting. (Consider, for example, the decentralized mechanisms that developed in response to the growth of automobile transportation, and the corresponding need for enforcement of traffic and parking laws. Similarities are apparent: pervasiveness of rights of way, differing rules governing bodies in motion and at rest). This led to an examination of ways to supplement investigative and prosecutive capabilities to increase deterrence.

Supplementing Investigative Capabilities

The growing use of private responsive capabilities has fueled a rapidly expanding industry of cyber-security specialists and cyber investigative services. In addition, state-licensed private investigators are taking on additional responsibilities in the cyber world. While these services fulfill vital needs of many victims for confidentiality and control, potentially valuable information useful to assessing the scope and nature of the threat is lost in the process. Further, these sensitive duties are being performed by individuals whose qualifications, methods and accountability are unproven. Accordingly, a conclusion can be reached that:

- Congress consider encouragement of increased sharing of information relevant to the scope and nature of the threat, through new ways of facilitating the natural growth of private sector cyber-investigative capabilities.

One possibility might be to consider nationwide licensing of private security specialists under conditions mutually beneficial to the growth of the industry and the development of a predictive threat warning capability.

Professional licensing could facilitate responsible growth of the profession by ensuring that trained, qualified and fully insured personnel are available to perform these sensitive duties. Professional licensing could also accelerate the establishment of much-needed industry standards and best practices, and can even serve the interests of privacy by making practitioners more accountable. The licensing body might also facilitate the passing of relevant information in the interests of government and private parties, while meeting clients' needs for confidentiality and control over routine investigations.

The expected growth in incidents requiring investigative response, and formation of effective partnerships between responders may free law enforcement resources for more efficient and targeted use. The availability of private professionals would not interfere with law enforcement's ability to investigate any incidents that come to or are brought to its attention. With private professionals providing a supplemental response, law enforcement might be freed from the weight of preliminary work associated with distinguishing between nuisance intrusions and real threats, such as dangerous intrusions into federal government systems or incidents involving economic espionage.

It is foreseeable that law enforcement reporting might actually increase, as professional practitioners, reluctant to exceed their legal authority and risk sanction or loss of license, would be inclined to encourage clients to make proper referrals to law enforcement when available private response options have been exhausted.

Supplementing Prosecutive Capabilities

A logical complement to expanding investigative capabilities would be to expand the ability of intrusion victims to proceed in civil actions. Civil remedies are currently available for violations of the Federal Computer Fraud and Abuse Act, through they have rarely been exercised. The interests of deterrence might be served, however, by making civil penalties enforceable internationally through appropriate multilateral mechanisms and agreements. Accordingly, a conclusion from this study is that:

- The President could seek to expand the availability of civil remedies for computer-related violations through appropriate international bodies, multilateral and bilateral agreements.

Countering the "Insider" Threat

Enabling an effective and reliable private intrusion response and an international civil remedy for computer crime are just two ways to help owners and operators to take greater action to protect the critical infrastructures. Another important area relates to the insider threat.

Insiders arguably pose the most immediate and credible threat to the nation's critical infrastructures. The Federal government is able to guard against misdeeds by insiders to some degree by virtue of its authority to conduct background investigations, issue security clearances, and conduct periodic reinvestigations. The private employers who are the owners and operators of the critical infrastructures often do not have the same ability to screen applicants for certain highly critical positions, or to re-investigate current employees prior to placement in such positions.

Employee Screening

In the vast majority of instances, an individual employees' interests are likely to outweigh the need for the enhanced security available through rigorous employee screening. But there may be some positions of such sensitivity within the critical infrastructures that security concerns could call for the availability of heightened screening procedures. Reconsideration of existing laws that hinder an employer from obtaining, with the consent of the applicant or employee, certain job-related background information may need to be reconsidered in light of infrastructure assurance concerns.[4]

Currently, in some states, private employers are unable to get access to criminal history information, are prohibited from requesting or using criminal, financial or employment information, and must avoid tort liability for sharing unfavorable employment history. These restrictions have come about as a result of state and Federal concerns over privacy, fair employment, or rehabilitation of criminals, among others. (*See Privacy Laws and the Employer-Employee Relationship.*) Accordingly, the conclusions which emerge are:

- The Attorney General could convene a group of professionals from law, state and federal legislatures, labor and management organizations and the privacy community to explore existing laws governing employer-employee relations in light of infrastructure assurance objectives. This group could also recommend measures to achieve an appropriate balance between the needs for critical infrastructure owner and operators to conduct appropriate employee background investigations with individual employees' interests in personal privacy; and

- State legislatures could consider adopting "consent" as a baseline for allowing employers to request background information from employ-

---

4 In such instances, it is *not* the intention here to mandate such procedures, or even to advocate their use, but to increase the availability of such procedures to certain owners and operators who feel compelled to invoke them. Additionally, any information so obtained would be have to be used in accord with fair information practices.

ees and potential employees for sensitive positions within critical infrastructures, subject to fair information practices.

The group convened by the Attorney General will be in the best position to identify the potential impediments to adequate employee screening for highly sensitive positions. It can also determine which of the critical infrastructures may be in need of such heightened protections and which positions within those infrastructures should be subject to such screening procedures. These recommendations may take the form of model legislation, guidelines, or even preemptive Federal legislation. While this group is studying this issue, the states may wish to examine their laws in light of infrastructure assurance concerns that we have raised and to determine whether they adequately provide for consent as a baseline.

Polygraph Examinations

Another area that would benefit from re-examination in light of infrastructure assurance concerns and the potential risks from insiders is the Employee Polygraph Protection Act (EPPA). Federal law currently prohibits private employers from requiring or even asking employees to submit to polygraph examinations except under very narrow and carefully described circumstances involving an investigation into employee wrongdoing. A second narrow exception exists for employers in certain security professions-so that security companies, such as those that provide armored car services or alarm system installation services to specially enumerated businesses, can periodically polygraph their employees (again, subject to carefully defined conditions and circumstances). But the same rationale that militates in favor of having this exception available for providers of *physical security services* should apply with equal force to providers of *information security services.* This approach is entirely consistent with the Commission's orientation to remain vigilant to both physical and cyber security issues. Accordingly, a conclusion can be drawn that:

- Congress narrowly expand existing exemptions to the Employee Polygraph Protection Act (EPPA) to include within the scope of its exemptions those who are in the business of providing information security services.

Note that this conclusion neither advocates nor endorses the use of polygraph examinations by employers. It merely seeks for Federal law to extend equal treatment to both physical and cyber-security specialists. There are probably other many other areas of the law that would benefit from being examined and considered for similar treatment.

Enabling Government-Industry Partnership

Infrastructure assurance attempts to bring together diverse interests of the Federal and state and local governments, owners and operators of the infrastructures, and users of infrastructure services. Some of the functions that need to be performed for infrastructure assurance are best performed by those who own and operate the critical infrastructures. Others should be carried out by the government. However, even the most productive efforts of the private sector alone and the government alone will leave a significant void where only joint action will accomplish desired goals. The most prominent example is information sharing. An information sharing mechanism that depends solely on government information or private sector information will only represent part of the picture.

Legal Impediments to Information Sharing

The success of an information sharing mechanism for infrastructure assurance will in large part depend on the creation of a trusted environment where participants, both government and private sector, are encouraged to share sensitive information on a voluntary basis. Several legal impediments currently exist that may prevent or discourage such participation. Impediments to government and private sector participation include apprehension over potential antitrust liability, tort liability, national security concerns, classification of information, legal processes compelling public disclosure, and concerns over the protection of proprietary and trade secret information. These impediments are implicated to varying degrees depending on the precise structure of an information sharing mechanism. However, some important observations surfaced concerning these issues generally as a result of the study. (*See Legal Impediments to Information Sharing.*)

Antitrust Liability

Potential contributors of information from outside of government have often expressed reluctance to participate in the sharing of specific threat and vulnerability information because of impediments they perceive to arise from antitrust and unfair business practice laws. Accordingly, a conclusion drawn is that:

- The Department of Justice could offer limited assurances to the private sector that participation in recognized information sharing processes would not, at least under certain conditions, run afoul of such laws, and consider providing appropriate guidelines to inform such participation.

Tort Liability

The creation of and participation in an operational threat-warning capability may raise liability concerns over failure to share certain information that might have otherwise prevented harm to a critical infrastructure. These failures could carry liability consequences for public and private participants. Determinations such as these, however, are highly dependent on the respective roles and responsibilities of those engaged in the information sharing and threat warning process, and how those roles and responsibilities are specifically delineated. For this reason, any complete study of these operational liability issues must necessarily be performed concurrent with the design of the information sharing capability itself. Accordingly:

- The Federal government could undertake a detailed study of the liability issues which may arise surrounding both government and private sector participation in an infrastructure assurance-related information sharing process.

National Security

With increases to the globalization of business and communications come heightened concerns over the sharing of sensitive information. Currently, many Federal agencies have specific guidelines controlling interaction with foreign corporations or corporate entities with significant foreign ownership. These guidelines, regulations and authorities reflect a variety of disparate approaches. Accordingly:

- The National Security Council could study whether the Federal government should standardize its approaches to and guidelines for sharing information with foreign corporations in light of potential national security benefits and increased government efficiency.

Meanwhile, to facilitate sharing of information for infrastructure assurance purposes, it is vital that a public-private information sharing mechanism establish and adopt appropriate guidelines for sharing information with foreign corporations. Accordingly:

- In the short term, the National Office could set guidelines for the sharing of infrastructure threat and vulnerability information with foreign corporations.

Confidential Information

The Freedom of Information Act and other related laws control the conditions under which information in the possession of the federal government can be made available to the public. Potential participants in an information sharing mechanism may require some degree of assurance that the sensitive information they contribute will remain confidential if shared with the Federal government. Accordingly:

- Appropriate protection of specific private-sector information could be required by, for example, inclusion of a b(3) FOIA exemption in enabling legislation.

Classified Information

Classification issues may arise due to the need to continue to protect classified information in the possession of the government—the contents of which might contribute substantially to a threat warning capability. Accordingly:

- The need for classification of certain information, or certain bodies of aggregated information, and the impact that classification would have on the dissemination process should be considered by the Federal government offices responsible for critical infrastructure assurance.

Trade Secret And Proprietary Information

If appropriate mechanisms for protecting proprietary and trade secret information are not incorporated as part of an information sharing mechanism for infrastructure assurance, private sector participants will be reluctant to share sensitive information. Accordingly:

- The National Office could require appropriate protection of information containing trade secrets or other forms of proprietary information.

State Government Participation

Many of the legal impediments to information sharing identified at the Federal level exist at the state level as well. These include issues regarding liability for participation in information dissemination and the effects of state laws and regulations requiring public access to government information. The great diversity between state laws further complicates any efforts to facilitate and maximize state government participation in information sharing. Accordingly:

- A study group could research and identify legal impediments to information sharing at the state level and propose solutions and draft model legislation to maximize state government participation in information sharing.

PART FOUR
INFRASTRUCTURE ASSURANCE, LAW AND CULTURE

The conclusions emerging from these studies are intended to help government and the owners and operators of the critical infrastructures do what is needed to achieve enhanced infrastructure assurance. They suggest actions by the Federal government, the private sector, and, where appropriate, by the government and private sector in partnership.

This work was intended to lay a foundation for a gradual process of legal reform. The tools identified through the preparation of the twelve studies and two special studies, and the roles suggested by the conclusions may prompt action by necessary parties. Those who are most closely and intimately involved in these areas, and ultimately in the best position to effect change, to reexamine the laws, regulations and policies under which they operate will find these studies most useful.

A technical definition of "infrastructure assurance" has guided this legal inquiry. Under this definition, infrastructure assurance refers to the "surety of readiness, reliability and continuity of infrastructures such that they are (1) less vulnerable to disruptions or attack; (2) harmed to a lesser degree in the event of a disruption or attack, and (3) can be readily reconstituted to reestablish vital capabilities." This definition has served well for compiling relevant authorities, parsing issues, and developing recommendations. Such a technical definition may not be as effective a platform for implementation of concepts and achievement of objectives. Infrastructure assurance is a dynamic set of processes that include individual and shared responsibilities across government and private sectors and, thus, demand implementation through a long-term series of gradual reforms, behavioral reforms, and corresponding legal reforms. In short, it requires cultural change, which results only from changes in behavior.

Viewed in terms of cultural change, "infrastructure assurance" will see businesses respond by including emerging threats and vulnerabilities in their risk management methodology; governments respond by organizing themselves in such ways as to be more able to detect and, in appropriate circumstances, respond to emerging threats and vulnerabilities; and even private citizens, in their homes and offices, modify the way that they practice computer and information security.

The legal community has a responsibility to remain vigilant to these dynamic changes, and to insure that laws facilitate and do not unduly impede it. The conclusions set forth here are the beginnings of a gradual process, one that must take hold at the local, state, Federal and international levels.

CHAPTER 2

INFORMATION-SHARING MODELS

A "LEGAL FOUNDATIONS" SPECIAL STUDY

Report to the
President's Commission
on Critical Infrastructure Protection
1997

This report was prepared for the President's Commission on Critical Infrastructure Protection. The report represents the opinions and conclusions solely of its developer, Legal Networks, Incorporated. The Commission has not made a judgment on, nor should the publication of this document be taken to represent an endorsement by the Commission for the content of the material contained herein.

CONTENTS

INTRODUCTION

The following discussion describes and analyzes various approaches to information sharing. Each of the approaches, or models, reflects methods adopted by Federal and state government entities and, in many cases, private companies and institutions.

This discussion is divided into two parts. The first part is a discussion of five basic models of information acquisition and dissemination. Where appropriate and useful, bullet points showing "stovepipes" and "filters" are inserted into the text.

The term *stovepipe* is widely used within the intelligence community and refers to systems of vertically aligned information systems that do not communicate with one another. Information stovepipes control the flow of information relevant to infrastructure assurance, and deliver that information to a broadly diverse (and usually functionally disconnected) grouped of public and private sector users.

Filters control the channeling of information concerning infrastructure assurance. Typically there are two types: jurisdictional and legal. Filters limit the types of information which may be collected and used. Within government, for example, jurisdictional and legal filters channel, or block, the flow of information based on "need to know", Congressional grants of authority, or presidential tasking, such as in Executive Orders.

The second part of this paper is a more thorough discussion of key cross-model issues, such as national security, business proprietary matters, antitrust, and liability for disclosure/failure to disclose. These applications illustrate regimes that both thwart and encourage information sharing across critical infrastructures.

The *Legal Foundations* study *Information-Sharing Models* is the companion volume to the *Legal Foundations* study *Legal Impediments to Information Sharing*. The two documents should be studied together to fully understand the issue of information sharing. *Information-Sharing Models* details the current approaches to information sharing and sets the stage for further understanding the concepts described in *Legal Impediments to Information Sharing*.

PART ONE
MODELS OF INFORMATION ACQUISITION AND DISSEMINATION[1]

Information Clearinghouse Model

Information clearinghouses exist in a variety of forms. Some are administered by Federal agencies, under authorization from statute, regulation or executive order, while others are administered by private entities such as trade associations. There are clearinghouses that are exclusively for information sharing between entities at different government levels, such as between:

- Federal entities,
- Federal and state entities,
- Private entities,
- Federal and private entities, and
- All levels of government—including the municipal level—and the private sector.

The information clearinghouse model includes different methods for gathering information. Some provide for voluntary participation and provision of information, while others are mandatory, requiring that specific public and private entities provide information.

There are also differences in the methods for dissemination of information. Some clearinghouses serve as repositories of information where requesting parties can find information, while others are required actively to distribute information. Additionally, some clearinghouses provide for wider access to information by non-participating parties than others.

Despite these differences, the hallmarks of an information clearinghouse are centralized collection and organization and widespread dissemination of information to participating parties.

This is distinct from other models which, while providing for the collection and dissemination of information, more severely limit the release of information to parties providing the information.

---

1 These information-sharing models reflect the research of Mitchell Baxter, Esq. and Lee M. Zeichner, Esq. of LegalNetworks, Incorporated, a Washington DC consulting company.

Applications of The Clearinghouse Model

Department of Commerce Information Clearinghouse

This clearinghouse was established by Congress through 15 United States Code Sections 1151-1157. The stated purpose of the statute is to "make the results of technological research and development more readily available to industry and business, and to the general public," through a "central clearinghouse for technical information which is useful to industry and business." *15 U.S.C. § 1151.* Congress directs the Secretary of Commerce to "establish and maintain . . . a clearinghouse for the collection and dissemination of scientific, technical, and engineering information." *15 U.S.C. § 1152.*

Congress further directs the Commerce Department to "search for, collect . . . and catalog . . . and to make such information available to industry and business, to State and local governments, to other agencies of the Federal government, and to the general public." *15 U.S.C. § 1151(a).* Additionally, the Secretary is directed to effect . . . the removal of restrictions on the dissemination of scientific and technical data where considerations of national security permit the release of such data to industry and business." *15 U.S.C. § 1152(c).*

This information clearinghouse is administered by a Federal agency, the Department of Commerce; information is collected on a voluntary basis, and made available to all interested parties, and on an equal basis. The only restrictions placed on dissemination of information relate to national security matters and other confidential matters such as business proprietary information and trade secrets.

Information is created and communicated between:
Federal government to Federal government,
State government entities, and the private sector.

Information Clearinghouse: Collect and Disseminate Information to All Interested Institutions and Parties. Prevent Dissemination of Classified Information and Trade Secrets, or Information Designated by Private Entities as Business Proprietary Information.

- **STOVEPIPE: Department must refer all relevant technical information to the heads of other government agencies and the military where such information would be useful.** ***15 U.S.C. § 1154.***

FILTERS: National Security and business proprietary/trade secrets

---

National Crime Information Center (NCIC)

Congress instructs the Attorney General to "acquire, preserve, and exchange identification records and information" relating to criminal identification and other information which would assist in identifying related matters. *28 U.S.C. § 534.* Regulations written pursuant to this statute create a detailed information sharing system in response to the Congressional mandate. According to 28 Code of Federal Regulations 20.31, "the Federal Bureau of Investigation shall operate the . . . computerized information system . . . to link local, state and Federal criminal justice agencies for the purpose of exchanging NCIC-related information, . . . which includes information in the Computerized Criminal History (CCH) File, a cooperative Federal-State program for the interstate exchange of criminal history record information."

Information is created and communicated between:
Federal government to Federal government,
State government entities, and the private sector.

Information Clearinghouse: Collect and Disseminate Information to Authorized Parties in Government and the Private Sector.

STOVEPIPE: Congress instructs the Attorney General to exchange such information with authorized officials of the Federal Government, the States, cities, penal institutions and other institutions—which include entities as diverse as railroad police departments and private college and university police departments. *28 U.S.C. § 534(a)(4).*

National Driver Register

This information clearinghouse, under the supervision of the Secretary of Transportation, was established by Act of Congress, under 49 United States Code Chapter 303. The purpose is "to assist chief driver licensing officials of participating States in exchanging information about the motor vehicle driving records of individuals." *49 U.S.C. § 30302(a).* Participation in the program is voluntary, and States may opt in or out at their own discretion. *49 U.S.C. § 30303, 23 C.F.R. 1327.5.*

Access to the information is limited to:

- "Chief Driver Licensing Officials" of participating States
- The Chairman of the National Transportation Safety Board
- The Administrator of the Federal Highway Administration in furtherance of accident investigation [2]

---

2 The National Highway Traffic Safety Administration is administering the fully-elec-

- The individual himself or herself [3]

The individual may also request that the information be disclosed to third parties, such as the Federal Aviation Administration, if the individual is seeking an airman's certificate, or to the Secretary of Transportation, if the individual is seeking employment as a locomotive operator. *See e.g., 49 C.F.R. Pt. 240* (requiring each person seeking certification as a locomotive operator to request that a check of the NDR be conducted, and the results forwarded to his employer.)

Significantly, only these entities may obtain information from the clearinghouse. Thus, information in the clearinghouse is specifically exempted from FOIA disclosure. *49 U.S.C. § 30305(c).* Further, insurance companies must use other routes to gain access to this data—such as through state legal regimes and clearinghouses. There are criminal penalties for "unauthorized disclosure and receipt" of clearinghouse data. *49 U.S.C. § 30307(a).*

Note that participation in this information clearinghouse is voluntary, while dissemination of the information is restricted. Contrast this structure with that of the following one, which, while apparently similar, contains significant differences.

Information is created and communicated between:
State government and voluntarily sent to the Federal government.
Federal government and provided to designated State government officials and specified Federal government agencies.
Private persons that request information from Federal government Clearinghouse.
Note: Criminal sanctions for unauthorized access and use of information.

Information Clearinghouse: Collect and Disseminate Information to Authorized Parties in Government and the Private Sector.

- **STOVEPIPE: The National Driver Register channels information to qualifying State and Federal departments and agencies.** *23 C.F.R. Pt 1327.*
- **FILTER: Access is limited to chief licensing officials of States, and select officials of the NTSB, Federal Highway Administration, and the individual.** *49 U.S.C. § 30305*

---

tronic register system. *23 C.F.R. 1325.1-1325.4.*

3 49 *U.S.C. § 30305.*

National Motor Vehicle Title Information System

This information clearinghouse is also established by Act of Congress, pursuant to 49 U.S.C. § Chapter 305. Unlike the National Driver Registry, which is administered by the Department of Transportation, this information clearinghouse is supervised by the Attorney General. The purpose of the regime is to provide individuals and entities instant and reliable access to information maintained by the States related to automobile titling. *49 U.S.C. § 30502.* Information includes validity of titling documents, whether an automobile bearing a particular vehicle identification number has ever been junked, or salvaged, and odometer disclosures. *Id.*

Congress requires all States to participate in this information clearinghouse—in stark contrast to Congress' mandate in the National Driver Register, which offers States a choice. *49 U.S.C. § 30503.* Additionally, there are reporting requirements which obligate private sector owner/operators—such as junk yard owners, salvage yard operators, and insurance carriers—to provide information to the information clearinghouse. [4]

Access to information is much broader than the National Driver Registry. States, law enforcement officials, prospective automobile purchasers, and insurance companies all have access to the information. *49 U.S.C. § 30502(e).*

Information is created and communicated between:

State government and sent to the Federal government; participation is mandatory. Federal government and provided to State government officials and Federal government agencies. Note: Private persons and corporations may obtain information from the Federal government clearinghouse.

Information Clearinghouse: Collect and Disseminate Information to Parties in Federal, State and Local Governments; Private Sector May Also Obtain Information.

STOVEPIPE: Information Clearinghouse channels information to all levels of government and the private sector.

---

4 Owner/operators must file monthly inventory reports on junked and salvaged vehicles. *See 49 U.S.C. § 30504.*

Environmental Protection Agency Solid Waste Information System

The Administrator of the EPA is directed by 42 U.S.C. § 6983 to collect and disseminate information relating to methods, costs and management practices for the collection, disposal, financing and reduction of solid waste. The statute further calls for the establishment of a central reference library, and the development of model accounting systems and codes. *Id.*

Pursuant to Federal legislation, Congress directed the EPA to implement information programs "for the rapid dissemination of information on solid waste management . . . including the results of any relevant research, investigations . . . studies, or other information which may be useful," and "educational programs to promote citizen understanding of the need for environmentally sound . . . practices." *42 U.S.C. § 6983(e).*

There are close to 100 information clearinghouse models at the Federal, state, and local government level.

Advisory Committee/Commission Model

Federal and State government committees and commissions are a common form of information collection and dissemination in the United States. Some are established to study one issue or set of issues, and then disband (*e.g.* the President's Commission on Critical Infrastructure Protection (PCCIP)), while others are established and charged with a continuing mission. There are commissions that deal only with information sharing within a particular Federal department or agency, some within the entire Federal government, and some which are concerned with relationships between Federal departments and agencies and their State counterparts.

The distinguishing feature of this model is a centralized body which gathers information and works in a collaborative manner to coordinate collection, analysis, and dissemination of the information; these entities also resolve specific issues, which, at least in part, define the committee or agency's mission or purpose.

Applications of the Advisory Committee/Commission Model

Interagency Security Committee (ISC)

Established by Executive Order 12977, the ISC is intended to:

> *"enhance the quality and effectiveness of security in and protection of buildings and facilities in the United States occupied by Federal employees for nonmilitary activities, and to provide a permanent body to address continuing government-wide security for Federal facilities. Executive Order 12977."*

The committee is comprised of representatives of all Federal departments, plus representatives from several Federal law enforcement agencies.

In addition to establishing security policies and strategies, the Committee is charged with:

1. encouraging agencies with security responsibilities to share security-related intelligence in a timely and cooperative manner *(E.O. 12977 Sec 4(A))*, and

2. assisting in "developing and maintaining a centralized security database of all Federal facilities." *Id. Sec. 4(E).*

The Executive Order mandates compliance—that is, all relevant agencies must share the information. The President also requires all relevant agencies to comply and cooperate with the Committee's policies and recommendation, except where the Director of Central Intelligence determines that compliance would jeopardize intelligence sources and methods. *Id. at Sec. 6(b).*

Information is created and communicated between:
Federal government to Federal government.

Note: Federal government agencies required to provide information to a committee, which collects, analyzes, and disseminates the information within the Federal government.

- **STOVEPIPE: Security-related intelligence is to be shared between all members of the committee and ultimately, after analysis, to specific government agencies and entities.**
- **FILTER: Classified, National Security Information.**

Advisory Committee for Trade Policy and Negotiation

Some committees collect specific and specialized information from the private sector; the information is then used, in conjunction with information collected by the Federal government to address an issue, or series of issues. As part of the Trade Act of 1974, 19 U.S.C. § 2111 et seq., Congress provides that "the President shall seek information and advice from representative elements of the private sector and the non-Federal governmental sector with respect to . . . negotiating objectives, bargaining positions . . . and the operation of any trade agreement," as well as other trade policy matters. *19 U.S.C. § 2155(a).*

Congress further directs the President to establish an advisory committee, to provide overall policy advice on such matters. *19 U.S.C. § 2155(b).* The committee is to be made up of representatives of non-Federal government, labor, industry, agriculture, small business, service industries, retailers and consumer interests. The U.S. Trade Representative is directed to make available to the committee such information and resources as it requires. *19 U.S.C. § 2155(c).*[5]

Information is created and communicated between:
Private Sector to Federal Government.

Note: Federal Government Required to Collect Specialized Information from Cross-Section of Private Sector; Information used by Executive Branch.

- **STOVEPIPE: Information in separate, private sector stovepipes transferred to Federal government. (*e.g.*, specialized knowledge on an industry's trade-related knowledge provided to USTR; USTR assists in information collection).**

Advisory Committee on Intergovernmental Relations

This committee is established by 42 U.S.C. § 4272, in order to "strengthen the ability of the United States federal system of government to meet the problems of an increasingly complex society by promoting greater cooperation, understanding and coordination of activities between the separate levels of government." *5. C.F.R. 1701.3*

Two of the primary objectives are

1. "[b]ringing together representatives of the Federal, State and local governments for the consideration of common problems, and

2. [p]roviding a forum for discussing the administration and coordination of Federal grant and other programs requiring intergovernmental cooperation." *Id.*

Information resources include a wide range of private sector and Federal, state, and local government representatives. The committee is composed of 26 members, six appointed by the President, (three from the executive branch and three private citizens), three Senators appointed by the President of the Senate, and three Members of Congress appointed by the Speaker of the House of Representatives. The remaining four Governors,

---

5 The Act additionally provides for the establishment of individual general policy committees, at the President's option. *Id.*

three members of State legislative bodies, four majors, and three elected county officials are all appointed by the President. *5 C.F.R. 1701.4.*

To ensure that the committee does not become a political vehicle, designators to the committee are required to appoint bipartisan membership. *5 C.F.R. 1701.5.*

Information is created and communicated between:

Private Sector and Federal, State, And Local Government
Note: Shared And Analyzed, And Ultimately Offered To All Levels Of Government And The Private Sector.

A Plethora of Advisory Committees - Federal Advisory Committee Act, S 2, 5 U.S.C. § A. App. 2:

(a) The Congress finds that there are numerous committees, boards, commissions, councils, and similar groups which have been established to advise officers and agencies in the executive branch of the Federal Government and that they are frequently a useful and beneficial means of furnishing expert advice, ideas, and diverse opinions to the Federal Government.

(b) The Congress further finds and declares that—

(1) the need for many existing advisory committees has not been adequately reviewed;

(2) new advisory committees should be established only when they are determined to be essential and their number should be kept to the minimum necessary;

(3) advisory committees should be terminated when they are no longer carrying out the purposes for which they were established;

Mandatory Exchange or Provision of Information Model

Under this model, one entity is required, through a variety of legal mechanisms, to provide information to another entity. The distinguishing characteristic of this model is that the collecting entity is generally under no obligation to disseminate the information—except to related or supervisory agents.

In some cases, there is a statutory requirement that one party give information of a certain type to another party. In other cases, an agency or department is given subpoena power to compel production of information, testimonial or documentary, from another entity or individual.

Federal Subpoena Authority

Typically, Federal departments and agencies are given subpoena power to gather information in furtherance of their statutory missions. Several common examples include:

- **The Federal Communications Commission (FCC)** has "the power to require by subpoena the attendance and testimony of witnesses and the production of all books, papers, schedules of charges, contracts, agreements, and documents relating to any matter under investigation." 47 U.S.C. § 409(e). The regulations relating to the issuance of subpoenas by the FCC, which outline in great detail how the legal mechanism operates, are found at 47 C.F.R. 1.331.
- **The U.S. International Trade Commission** [6] (ITC) is given extensive powers to gain access to information pertinent to its investigations. The Tariff Act of 1930 provides that the Commission has "the right to copy any document, paper or record, pertinent to the subject matter under investigation, in the possession of any person, firm . . . or association engaged in the production, importation or distribution of any article under investigation . . . may summon witnesses . . . may require any [party] to produce books or papers . . . and . . . to furnish in writing in such detail and in such form as the Commission may prescribe, information in their possession pertaining to such investigation." 19 U.S.C. § 1333.
- To facilitate the collection of relevant information from private entities, including non-US industries and corporations, the ITC has extensive rules governing the collection of Business Proprietary Information. *See* 19 C.F.R. 207.7 (1997) Under this regime, parties may submit sensitive business information (trade secrets) to the Commission and attorneys involved in the case. These regulations carry severe sanctions for lawyers, and other consultants, who divulge business proprietary information—even if by mistake. Penalties include disbarment (19 C.F.R. 207.7(d), (e)) and referrals to the US attorney for further prosecution and publishing violators names in the Federal Register.

---

6 Trade-related investigations often involve the ITC and the International Trade Administration ("ITA"), under the Department of Commerce. The ITA has similar rules and regulations for collection of business proprietary information—including sanctions for the release of that information. *See* 19 C.F.R. 353 et seq.

- **The Secretary of Transportation** is given similar power to subpoena witnesses and records related to a proceeding or investigation pursuant to 49 U.S.C. § 502 (d), and 49 U.S.C. § 13301.

Private Subpoena Authority

The Federal Rules of Civil Procedure provide a mechanism for private litigants to compel the production of witness testimony, and documentary evidence, and written answers to interrogatories. The general discovery provision is Rule 26; the authority to issue subpoenas is found in Rule 45.

Reports, Disclosures and Notification

In some instances, a party is given a statutory mandate to provide certain information to another party, either with or without the need for a request.

- In the statute providing for the powers and duties of the Inspector General of the Department of Energy, Congress requires that "the Inspector General shall report expeditiously to the Attorney General whenever the Inspector General has reasonable grounds to believe there has been a violation of Federal criminal law. *42 U.S.C. § 7138(j).*
- Under the Tariff Act of 1930, at 19 U.S.C. § 1332(g), Congress specifies that the U.S. International Trade Commission "shall put at the disposal of the President . . . the Committee on Ways and Means of the House of Representatives, and the Committee on Finance of the Senate, whenever requested, all information at its command, and shall make such investigations and reports as may be requested by the President or by either of said committees or by either branch of Congress."
- Executive Order 10450, which concerns security requirements for government employment, mandates that "[w]whenever there is developed or received by any department or agency indicating that the retention of employment of any officer or employee of the Government may not be clearly consistent with the interests of the national security, such information shall be forwarded to the head of the employing department or agency."

Cooperation with Other Agencies

Listed below are two of many examples in which Congress or the President, via Executive Order, mandates information sharing:

- Congress, at 19 U.S.C. § 1334, requires the ITC to "act in conjunction and cooperation with the Treasury Department, the Department of Commerce, the Federal Trade Commission, or any other departments, or independent establishments of the Government, and that such [entities] shall cooperate fully with the Commission . . . and, when directed by the President, shall furnish to the Commission, on its request, all . . . information in their possession relating to any of the subjects of investigation by the Commission.

- Executive Order 12333, which concerns national security coordination, mandates that the heads of all Executive Branch departments and agencies shall give the Director of Central Intelligence access to all information relevant to national intelligence needs.

Voluntary Cooperation or Exchange of Information Model

Under this model, one entity exchanges information even in the absence of a Congressional or Presidential mandate. The exchange can take place pursuant to statutory *authorization* to communicate, through an organizational structure that facilitates such information exchange, or on an informal basis. The statutory exchanges tend to involve those between Federal agencies, or Federal and State agencies. Private exchanges tend to occur through organizations such as trade associations, or through informal channels, such as a private industry or through individual, private sector contacts.

Applications of the Voluntary Cooperation or Exchange of Information Model

Emergency Federal Law Enforcement Assistance (EFLEA)

This law was enacted to give States a means to request Federal law enforcement assistance in the event of an emergency:

> *"an uncommon situation which requires law enforcement, which is or threatens to become of serious or epidemic proportions, and with respect to which State and local resources are inadequate to protect the lives and property of citizens or to enforce the criminal law." 42 U.S.C. § 10502(3).*

Assistance means funds, equipment, training, *intelligence information,* and personnel. *See 42 U.S.C. § 10502(1).*

Under EFLEA, Congress offers a mechanism for States to apply to the Attorney General for such information. The Attorney General then may approve or disapprove the application, based on certain criteria specified in the statute, such as the nature of the emergency, availability of resources, cost, and various federalism issues. *42 U.S.C. § 10501.*

Attorney General Seeking State or Local Assistance

When the Attorney General or President determines that an immigration emergency exists, and State or local assistance is required, Department of Justice regulations provide a mechanism for such a request and exchange of information. *See 28 C.F.R. 65.84.* Under the regulations, the Attorney

General is authorized to negotiate the terms and conditions of the assistance with the State or local government, and funds can be provided by a reimbursement agreement, grant or cooperative agreement.

In exigent circumstances, the Attorney General is authorized to seek State or local assistance and agree to provide funding without a written agreement. *28 C.F.R. 65.84(c).*

Secretary of Transportation Cooperating with States

In carrying out his responsibilities under the general authority enabling statute, 49 U.S.C. § 502(c), the Secretary of Transportation may:

(1) confer and hold joint hearings with State authorities;

(2) cooperate with and *use the services, records,* and facilities of State authorities; and

(3) make cooperative agreements with a State to enforce the safety laws and regulations of a State and the United States related to highway transportation.

These arrangements include the all information held by the State government relating to safety laws and highway transportation.

Exchange of Information through Organizations

Most industries and private sector interest areas have trade associations and other organizations established for the purpose of representing their interests as a whole before Congress and the individual State legislatures, and Federal and State executive agencies and Courts. These organizations typically provide for the sharing and exchange of information, such as through newsletters, journals, and other publications, trade practice groups and conferences.

In their role as lobbyists and advocates, these organizations also serve as a major conduit for information flowing from the private sector to Federal, State and local governments.

Informal Exchanges of Information

In the private sector, an individual or group at one company may simply contact a counterpart at a different company to discuss a matter relevant to both companies, or even to all companies in a particular industry segment.

There are many reasons that private companies would need to exchange information, as opposed to using a third-party intermediaries. If a security breach or attempt is discovered, for example, it is important that one company be able to immediately warn other companies facing the same risk. It also allows companies which may have incomplete information about a security threat or criminal activity to combine information resources.

Cyber-related breaches offer the most obvious example. Where such breaches raise public relations issues, companies are especially sensitive to government information exchange requests, for fear of loss of control and publicity. Under these circumstances, information exchanges between company Systems Administrators and Operators may solve problems for the future and offer both companies an opportunity to pool resources without adverse publicity.

Note, however, that these private exchanges of information—including those discussed in numbers 3 and 4 above—may give rise to Federal and state antitrust violations. We discuss these antitrust issues more fully under the Cross-Model Issues, discussed below.

Summary

Voluntary Cooperation or Exchange of Information Model

Federal, State, and local governments exchange information on a voluntary basis.
Private sector similarly exchanges information with other private entities.

Threat and Hazard Detection and Notification Model

This model is, in one sense, a combination of the information clearinghouse and mandatory notification models, with the additional requirement that the collector of the information must then disseminate the information through specific channels or to specific parties; these parties may are granted the right to receive the information.

There are several levels of notification, ranging from simple and unsophisticated models (*e.g.*, sending notices to be posted on a wall) to more complex, expensive, and far reaching (*e.g.*, issuing general public announcements or extensive notification campaigns which require contacting all individual members of a particular group).

Applications of the Threat and Hazard Detection and Notification Model

Emergency Planning and Right-to-Know

This multi-tiered structure of State and local planning districts and emergency response committees is established by 42 U.S.C. § Chapter 116. Under this system, each local entity is required to evaluate potential hazards, available resources, and appropriate responses to various emergencies. *42 U.S.C. § 11003.* A national response team is established to review each local committee's plans, as well as to issue guidance documents to assist in the preparation of emergency plans. *42 U.S.C. § 11003(e), (f), (g).*

Additionally, the operator of any facility determined to be relevant to the plan, by virtue of its production, use or storage of hazardous materials, is required to designate a representative to participate in the planning process, and to provide information as requested by the committee. *42 U.S.C. § 11003(d).*

In the event of the release of "an extremely hazardous substance referred to in [the Act], "from a facility, the operator of the facility is required to provide immediate notice to the community and State emergency planning officials. *42 U.S.C. § 11004.*

- STOVEPIPE: Facility operator required to provide information to committee, and to notify planning officials in the event of a release.
- FILTER: Trade secrets are protected from public disclosure. 42 U.S.C. § 11042.

Natural Oil and Hazardous Substances Pollution Contingency Plan

This plan, codified at 40 C.F.R. 300, shows a fully developed example of this type of model. It provides for the establishment of a National Response Team, *40 C.F.R. 300.110,* and Regional Response Team, *15 C.F.R. 300.115,* which are to coordinate and share information in the preparation of contingency plans. *40 C.F.R. 300.205.*

Industry groups and academic organizations are encouraged to participate in planning, committing resources, and sharing technical and scientific information; the Scientific Support Coordinator is authorized to act as a liaison between the On Scene Coordinator and the Regional Project Manager, and all interested organizations. *40 C.F.R. 300.185.*

In essence, this is an information clearinghouse within the threat and hazard detection and notification model.

There are notification requirements and a permanent infrastructure is established and continuously manned for handling emergency responses. *40 C.F.R. 300.125.* A chain of command is established, and provision is made for down-line and lateral communication of information.

Finally, recognizing that "it is imperative to give the public prompt, accurate information on the nature of the incident and the actions underway to mitigate the damage," Section 300.155 calls for the establishment of a Joint Information Center to coordinate the collection and dissemination of information through appropriate media and other channels. [Again, a clearinghouse is established within this model.]

The following two examples highlight the differences in notification required, both in the form and the breadth of notification required.

FDA Consumer Right-to-Know About Pesticide Chemical Residue

As part of a larger statute requiring the establishment of standards regarding tolerance of pesticide residues, which is found at 21 U.S.C. § 346a, there is a provision regarding public disclosure and notification of the findings of such research. *21 U.S.C. § 346a(o).* This section provides that the Administrator of the Food and Drug Administration, in consultation with the Secretaries of Agriculture and Health and Human Services, "shall publish in a format understandable to a lay person, and distribute to large retail grocers for public display (in a manner determined by the grocer), . . . a discussion of the risks and benefits of pesticide chemical residues in or on food," and other specific information about chemical pesticide residues which may be useful to consumers. *Id.*

- FILTER: Confidential business information is protected from disclosure. *21 U.S.C. § 346a(I).*

This provision, which calls for a narrow scope of notification, should be contrasted with the following one, which calls for a much broader scope of notification.

Consumer Product Safety Commission Remedies Respecting Banned Hazardous Substances

15 U.S.C. § 1274(a) provides that, "[i]f any article or substance . . . is defined as a banned hazardous substance, . . . and the Commission determines . . . that notification is required to adequately protect the public . . . the Commission may order the manufacturer or any distributor or dealer . . . to take any one or more of the following actions:

(1) To give public notice that the article or substance is a banned hazardous substance.

(2) To mail such notice to each person who is a manufacturer, distributor, or dealer of such article or substance.

(3) To mail such notice to every person to whom the person giving the notice knows such article or substance was delivered or sold."

Additionally, the Commission has the power to order repairs and changes to be made, or for recalls of the article or substance. *15 U.S.C. § 1274(b).*

Note: The latter two examples are relevant to infrastructure assurance in that they go to the questions of the form of notification of hazard utilized, and the breadth of "audience" the notification is required to reach. The FDA provision calls for posters to be mailed to large grocers, omitting smaller grocery chains, in addition to "mom-and-pop" grocery stores, while the CPSC provision requires public announcements ***plus*** *individual notification of* ***every*** *party who may be affected.*

This issue is more fully discussed in the Cross-Model Issues Section of this discussion, under the subheading "Liability for Failure to Disclose or Inform."

PART TWO
CROSS-MODEL ISSUES

There are a number of issues which are relevant to many, if not all of the models. These issues will be discussed in this section, with reference to particular models as necessary. In one sense, the issues may be referred to as "stovepipes" and "filters."[7]

---

7 *See* definitions in introduction of this discussion.

Stovepipes & Filters—
Limiting the Flow of Information to the Necessary Parties

- Stovepipes - In all of these issues, as well in all of the models, the "stovepipe" issues involve the following inquiries:
    (1) Who gets access to information?
    (2) Who provides access to information?
    (3) Who must be informed?
    (4) Who may be informed?
- Filters - In all of these issues, as well as in all of the models, the "filter" issues raise the following inquiries.
    (1) What mechanisms cause information to be withheld?
    (2) Who is responsible for withholding information?
    (3) What liability is there for filter "breaches?"

National Security

This discussion will be limited to key statutes and executive orders which highlight how national security rules affect the flow of information within and outside of government channels.

- STOVEPIPES AND FILTERS: The entire national security infrastructure is a study in stovepipes and filters: who gets access and what information is withheld from disclosure.

The National Security Infrastructure

The general statute for the establishment and structure of the national security infrastructure is 50 U.S.C. § Chapter 15. Coordination for national security is discussed beginning at *50 U.S.C. § 402,* which establishes the National Security Council and the Central Intelligence Agency. Protection of the operational files of the CIA is provided for at *50 U.S.C. § 431,* which exempts from disclosure under *5 U.S.C. § 552a* those files which document intelligence or counterintelligence operations or arrangements, means of collection, or sources.

50 U.S.C. § 435 establishes the requirement that no employee of the executive branch, except as permitted by the President, may be given access to classified information without an appropriate background check.

Classifying, Safeguarding and Declassifying Information

Executive Order 12958 "prescribes a uniform system for classifying, safeguarding and declassifying national security information." The Implementing Directive for the Order is found at *28 C.F.R. 2001*. Each department and agency of the Federal Government has promulgated rules pursuant to this Order. For example, the Department of Transportation rules are found at 49 C.F.R. Part 8, the Department of Justice rules are found at 28 C.F.R. Part 17, and FERC rules are found at 18 C.F.R. Part 3a.

Information Sharing

Executive Order 12333 provides that the agencies within the intelligence community, in addition to participating in intelligence activities, shall "ensure the establishment by the Intelligence Community of common security and access standards for managing and handling foreign intelligence systems, information, and products."

Liability for Disclosure of Classified Information

There are numerous statutes sanctioning—with criminal penalties—the disclosure of classified information. For example, 18 U.S.C. § 798, which is part of the chapter relating to espionage and censorship, provides that "[w]hosoever knowingly and willingly communicates . . . to an unauthorized person, or publishes, or uses in any manner prejudicial to the safety or interests of the United States . . . any classified information" relating to codes, ciphers or communication intelligence activities, is subject to a fine, prison term, and forfeiture of assets derived as a result of, and used in the facilitation of the violation.

"Foreign" Person—Corporate Access to Confidential Information

A major area of concern is how to determine to which entities it is in the national interest to disclose sensitive information. Many key infrastructure-related companies may not be entirely American-owned. What exactly is a "foreign company? The answer to this question is not entirely clear.

Even Congress has expressed uncertainty over the concept, as evidenced by a hearing held on November 1, 1989 by the House of Representatives Subcommittee on Science, Research and Technology and the Subcommittee of International Scientific Cooperation. The subject of the hearing was *"What Is A U.S. Company?"* The purpose of the hearing was to identify the guidelines under which foreign owned and U.S. based multinational corporations should be allowed to participate in Federally-funded research projects, consortia, or joint ventures.

Various agencies have defined "foreign" differently for their own purposes. For example, the Department of Energy has established a set of sophisticated guidelines for determining the level of foreign interest, ownership, control, or influence over a government contractor. These guidelines, which are codified at 48 C.F.R. 952.204-73, contain a series of questions that draw on similar rules in several legal regimes, including customs, tax, and securities regulation.

Department of Energy Listing of Foreign Interest and Ownership Questions

Several Questions include:

- Does a foreign interest own or have a beneficial ownership interest in five percent or more of your organization's voting securities?
- Identify the class of shares issued which are owned by foreign interests, listed by country.
- Does your organization own ten percent or more of any foreign interest? Supply the particulars.
- Does any foreign interest have management positions such as directors, officers, or executive personnel in your organization? Provide particulars.
- Does any foreign interest control or influence, or is any foreign interest in any position to control or influence the election, appointment, or tenure of any of your directors, officers, or executive personnel? Identify and discuss.
- Does your organization have any contracts, binding agreements, understandings, or arrangements with foreign interests that cumulatively represent ten percent or more of your organization's gross income? Provide particulars on any intellectual property ties, trade secrets, licensing, patent agreements, cartel or industry ties . . .
- Is your organization indebted to foreign interests? If include debentures, discuss whether the securities are convertible, etc.
- Does your company derive any income from Communist countries included in Groups Q, S, W, Y, and Z in Supplement No. 1 in 15 C.F.R. part 770 (*e.g.*, Cuba, Syria, North Korea.).

These questions are presented to contractors to determine whether such foreign interest rises to such a level as to pose a security concern and must be certified.

The Internal Revenue Service has its own classification of "controlled foreign Corporations" for tax purposes, which is codified at 26 U.S.C. § 957. This provision considers a controlled foreign corporation one in which more than 50 percent of the stock is owned by U.S. shareholders.

Access to Classified Information by Non-United States Citizens

Executive Order 12968, Section 2.6 provides that "where there are compelling reasons in furtherance of an agency mission, immigrant alien and foreign national employees who possess a special expertise may, in the discretion of the agency, be granted limited access to classified information only for specific programs, projects, contracts . . . for which there is a need for access." The access is limited to that level which the U.S. Government "has determined may be released to the country of which the subject is currently a citizen, and such limited access may be approved only if the prior 10 years of the subject's life can be appropriately investigated."

Private Sector and other Confidential Matters

This issue cuts across nearly all of the models, as it is necessary to ensure that information which private companies wish to keep secret remains secret. Without built-in guarantees of such confidentiality, no business would engage in information sharing of any kind with either a government or private entity, unless forced to do so. Even then, the amount, veracity, and breadth of information would likely be limited.

Protective Measures for Trade Secrets and Proprietary Information

Confidential materials are protected through a variety of mechanisms in the public and private sectors. These include such measures as statutes mandating that information received by government agencies in the execution of their statutory missions be held in confidence, restricting access to such information, and providing penalties for the disclosure of such information. There are also judicial (JPOs) and administrative protective orders (APOs) which permit confidential information to be provided to certain parties, such as attorneys, while the information is withheld from other parties, such as their clients.

Examples include:

- The Tariff Act of 1930, at 19 U.S.C. § 1677f(b) provides that information submitted to the ITC which is designated as proprietary by the person submitting it, shall not be disclosed to any person outside of those Commission persons who are directly involved in the relevant investigation, or to an officer of the U.S. Customs Service who is conducting an investigation regarding fraud under the Act, without the consent of the person submitting it. The Act further provides for limited disclosure of certain proprietary information under protective order. *19 U.S.C. § § 1677f(c).*

- By statute codified at 49 U.S.C. § 1114, the NTSB may disclose trade secrets only—

  (1) to another . . . instrumentality of the U.S. Government when requested for official use

  (2) to a Committee of Congress . . . having jurisdiction . . . when requested;

  (3) in a judicial proceeding under a court order that preserves the confidentiality of the information without impairing the proceeding; and

  (4) to the public to protect health and safety after giving notice to any interested person . . . and an opportunity for that person to comment . . . if the delay . . . would not be detrimental to health and safety.

Balancing Information Disclosure with other Public Policy Considerations

An interesting feature of this statute is the provision for the protection of the voluntary submission of information to the NTSB. The provision prohibits the release of such information "if the Board finds that the disclosure of the information would inhibit the voluntary provision of that type of information." *49 U.S.C. § 1114(b)(3).*

A similar provision, valuing disclosure of information to protective bodies over the release and use of such information by third parties, can be found at 49 U.S.C. § 504(f), which states that "[n]o part of a report of an accident occurring in operations of a motor carrier . . . and required by the Secretary [of Transportation], and no part of an investigation of the accident made by the Secretary, may be admitted into evidence or used in a civil action for damages related to a matter mentioned in the report or investigation."

- The U.S. criminal code has a chapter devoted to the protection of trade secrets, 18 U.S.C. § Chapter 90, which has two primary provisions. The first, which is under the heading "Economic Espionage," at 18 U.S.C. § 1831, prohibits the theft of trade secrets with the intention of benefiting any foreign entity. Individuals violating this section are subject to be fined up to $500,000 and imprisoned for up to 15 years; organizations face fines up to $10,000,000.
- The second provision, 18 U.S.C. § 1832, punishes the theft of a trade secret with the intention of benefiting anyone other than the owner of the trade secret with a fine and imprisonment of up to 10 years. Organizations face fines of up to $5,000,000. The statute also provides for

judicial orders to preserve confidentiality of trade secrets in prosecutions under the statute. *18 U.S.C. § 1835.*

- There are provisions in the Federal Criminal and Civil Codes for protection of confidential matters which may be uncovered in court proceedings. Federal Rule of Criminal Procedure 6(e) prohibits disclosure by grand jurors and grand jury employees of any matter occurring before the grand jury. Exceptions are made for disclosure to government attorneys, government personnel as deemed necessary by a government attorney, and for a few other disclosures such as other grand juries, or state law enforcement officials under limited circumstances.
- Federal Rules of Civil Procedure 26 and 45, which concern discovery and subpoenas, both contain provisions for the issuance of protective orders to ensure the confidentiality of sensitive information.

Antitrust Liability

There are several antitrust issues which arise in the context of infrastructure information sharing. The mere act of information sharing about security-related matters should not raise questions of antitrust or unfair trade liability, which generally arise in the contexts of price fixing and market splitting. Such questions do arise, however, in the context of exclusion of one or more companies from access to information. In this event, there is a possibility that an excluded company could claim that it is the victim of a boycott, or that it is being denied access to an "essential facility." [8]

While much informal information sharing takes place, many companies remain hesitant to provide even crucial security information to competitors, for fear of antitrust liability.

While such a claim may be brought, it is not clear how likely it would be that such a claim could be successfully prosecuted, civilly or criminally. The standard for establishment of an "essential facility" is high, requiring that the excluded party have no other means for gaining access to, or duplicating the denied facility.

One question that must be factored into this equation is the potential likelihood of this occurring. It would appear to be i the best interests of all companies in a particular industry to either share no information with any other companies, or to cooperate with all companies desirous of ensuring industry-wide security. Nevertheless, the possibility does exist that two or more companies, privately or through the establishment of a restrictive

---

8 An "essential facility" is well defined in antitrust law and includes situations where a person or company denies access to a facility that is necessary to conduct business. Information can also be considered an "essential facility."

trade organization, might attempt to drive other competitors out of the market by withholding key information.

Liability for Failure To Disclose or Inform

Duty to Provide Information

1. What is the duty of the Federal government to disclose information to owner/operators of a critical infrastructure?

2. What is the duty of an owner/operator of a critical infrastructure to disclose information to

- *Federal government?*
- *A different owner/operator? A competitor?*

3. Duties arise by operation of law or by contract:

- *Absent a well-defined duty, there is typically no liability for failure to disclose information.*
- *Duties may be imposed at the Federal level by the President (Executive Order), Congress (statute), the Courts (court decision);*
- *Typically, each State (through the legislature or the courts) addresses*
- *whether private entities have an affirmative duty to disclose information.*
- *A duty may arise between private parties by contract (e.g., Sprint promises to notify AT&T of any cable break) or by tort law.*
- *It is well-recognized tort law that one who undertakes to warn the public of a danger and thereby induces reliance must perform his good samaritan task in a careful matter. That is, if you choose to inform, you bear the consequences of informing incorrectly or incompletely.*

Congress, state legislatures and courts have all addressed general duties to act. For critical infrastructure assurance and information-sharing purposes, the affirmative duty to disclose information raises two issues. The first involves liability of the Federal government's failure to disclose information in its possession to a private party. The second involves the liability of a private party for failure to disclose or report information which the party is under a statutory, regulatory, or, potentially, a common law duty to disclose.

The *Federal Torts Claim Act* creates potential liability for the Federal government in dealing with private parties. The possibility exists that a party who has sustained a loss that could have been prevented had certain information been relayed to him will attempt to hold the entity responsible for intentionally or negligently withholding that information. Such a claim could be brought under the Tort Claims Act, which provides that the United States shall be liable in the same manner and to the same extent as a private individual under like circumstances, with many exceptions, including those for pre-judgment interest and punitive damages. *28 U.S.C. § 2674.*

There is an additional exception to U.S. liability which is relevant here: 28 U.S.C. § 268(a) exempts "any claim based upon an act or omission of an employee of the Government, exercising due care, in the execution of a statute or regulation." The question to be addressed is whether this would bar a claim against the U.S. from an aggrieved party who was omitted from notification of critical security information.

Additionally, many statutes, regulations and executive orders providing for disclosure of certain information have provisions stating that they do not create any right or claim against the US. In these cases, there would be no duty to disclose information to an owner/operator of a critical infrastructure. [9]

For example, E.O. 12812, which provides for the declassification of materials regarding POWs and MIAs, states in Sec. 3: "This order is not intended to create any right or benefit, substantive or procedural, enforceable by a party against the United States, its agencies or instrumentalities, its officers or employees, or any other person."

E.O. 12951, regarding the release of imagery acquired by space-based reconnaissance systems, contains the same disclaimer.

At this point it is also worth considering the different *notification requirements* in the discussion of the Threat and Hazard Notification Model, *Sections 3 & 4 of Model V,* above. One example (FDA Pesticide Residue) simply calls for the government to send posters to major grocers, excluding smaller grocers (and, more importantly, that sector of the American public which shops at these smaller stores) while the other (CPSC Notification of Banned Hazardous Substances) mandates individual notification by mail of every person who may be affected.

---

9 The Department of Health and Human Services, by regulation, limits liability to acts or omissions by government employees only within limited parameters. "With respect to covered individuals, only acts and omissions within the scope of their employment (or contract for services) are covered. If a covered individual is providing services which are not on behalf of the covered entity, such as on a volunteer basis or on behalf of a third-party (except as described in paragraph (d) of this section), whether for pay or otherwise, acts and omissions which are related to such services are not covered." *42 C.F.R. 6.6 (1997)*

The scope of disclosure required by statute or regulation is relevant. The determining factor in liability under the Tort Claims Act is whether the government agent(s) were negligent in the execution of their duty. If a statute calls for notification of large telecommunications companies, then smaller telecommunications companies may not have a cause of action.

Privacy

Privacy issues are discussed in a separate paper, but should be mentioned here, as they must be taken into account, and applied to each model and cross-model issue as a filter.

CHAPTER 3

PRIVATE INTRUSION RESPONSE
A "LEGAL FOUNDATIONS" SPECIAL STUDY

Report to the
President's Commission
on Critical Infrastructure Protection
1997

This report was submitted to the President's Commission on Critical Infrastructure Protection, and informed its deliberations and recommendations. This report represents the opinions and conclusions solely of its developers.

Introduction

Private Intrusion Response originated among the studies conducted to prepare reports in the *Legal Foundations* series to inform the deliberations and recommendations of the President's Commission on Critical Infrastructure Protection (PCCIP). It was subsequently submitted for publication to the *Harvard Journal of Law & Technology* by Stevan D. Mitchell, Esq. the PCCIP Commissioner representing the Department of Justice) and Elizabeth A. Banker, Esq., PCCIP Assistant General Counsel.

The views expressed in *Private Intrusion Response* are those of the authors and do not necessarily reflect those of the President's Commission on Critical Infrastructure Protection (PCCIP) report the U.S. Department of Justice.

For more information or to obtain a copy of *Private Intrusion Response,* please contact the Harvard Journal of Law & Technology, (http://jolt.law.harvard.edu/).

CONTENTS

I. Introduction

Law enforcement is improving its ability to respond to and deter computer-based intrusions. A private response is developing on a parallel track, in ways that may be more responsive than current government efforts to the private sector's need for confidentiality and control over sensitive investigations. The private response is taking shape in an environment where liability and standards of conduct are largely undefined. The simultaneous growth of divergent governmental and private responses hinders the nation's ability to estimate the size and scope of the threat of computer intrusions, share information about vulnerabilities, and lay a foundation for an effective threat warning capability. A professional licensing scheme for certain classes of computer security specialists may provide a basis for compromise, cooperation, and enhanced deterrence.

Misuse of computer systems appears to be the modus operandi of an increasingly broad spectrum of actors, including those without authorization to enter a system and those who exceed their valid authorization. They range from recreational hackers seeking a challenge, to disgruntled employees out for revenge, to those pursuing financial gain through theft of trade secrets and proprietary data, and even terrorists or nation-states seeking to further foreign policy or military objectives. Protecting vulnerable systems is increasingly vital given our increasing dependence on them for information, communication, and commerce. As a result, we are beginning to see dramatic changes within the computer security industry and in the mechanisms put in place by governments to provide responses to computer misconduct. But as both the frequency and reporting of incidents increase, the resources made available to prevent, investigate,(2) and respond to the consequences of incidents—particularly those incursions that appear to originate from external sources—will become increasingly taxed.

The most visible responsive resources currently reside within the federal law enforcement community. A federal governmental response has developed for a number of reasons. Geographic dispersal of networks and the need to investigate incidents that cross state boundaries implicate federal jurisdiction. The expense of training, equipment, and conducting computer investigations often price computer crime expertise out of the range of state and local police resources.(3)

The federal government has begun to equip itself to address an expected increase in the volume of computer intrusions,(4) raising basic questions about the ability of federal law enforcement to effectively and efficiently resolve large numbers of incidents. Even assuming the availability of adequate resources—an assumption we question—not all cases brought to the attention of law enforcement will be investigated.(5) Still others may not be prosecuted.(6) In addition, the response typically pursued by law enforcement is geared toward identifying, apprehending, and prosecuting the intruder. Such a response is clearly a critical element in deterring such activity, but is not necessarily consistent with the business objectives of a corporation that has been the victim of a serious incident.

Businesses have a primary need to repair damage and restore service to customers, a process often complicated by an ongoing criminal investigation. While some businesses may also be interested in pursuing criminal prosecution, other business considerations, such as the need to control costs and maintain customer confidence in the reliability of service and in the security and confidentiality of transactions and records, may militate against initiating a public response. The result to date has been a low rate of reporting intrusion incidents to law enforcement.(7)

Whether viewed as alternatives to the traditional criminal law enforcement response or as supplements to such a response, there is are several existing and emerging avenues for responding to intrusion activity. Statutory civil remedies are in place at the state and federal levels, though they are seldom pursued.(8) Contract and tort remedies have been proposed as potential vehicles for settling disputes between private parties for unauthorized use of systems.(9) However, all of these alternatives are dependent on identifying the source of the intrusion—identification that must be sought through an investigatory process. Identifying the source of an unauthorized intrusion can be costly and time-consuming, causing businesses carefully to weigh the respective benefits of initiating a public response, a private response, or no response at all.

Due to the sensitive nature of the work performed by security firms and their customers' desire for confidentiality, there are few published discussions of the services related to tracking the sources of intrusions. Many of our insights were gained through confidential interviews with members of the security community. Many similar conclusions were drawn as an outcome of the 1996 Security in Cyberspace Hearings.(10) In addition, our observations receive anecdotal support from advertisements for computer security services available on the Internet.

Richer options are becoming available. Some computer security experts have begun to provide services to clients that can ensure confidentiality and control over their systems, while providing the necessary security measures and intrusion detection and response capabilities. Given the sensitive nature of the work involved in provision of computer security services, and the legal complexities of conducting private investigations into intrusion incidents, the potential benefits of creating an oversight mechanism for computer security experts engaged in such activities seem compelling. In addition, a licensing scheme could be administered in such a way as to provide mutual benefits to private security experts, government, and owners of compromised systems.

Computer security practitioners could benefit from working within the parameters of a more clearly defined legal and liability climate and from the marketing advantages that a license may afford, including enhanced public trust. The government could benefit from receiving limited information about incidents that are currently investigated and resolved without its knowledge. System owners could benefit by having available a broader array of intrusion response options. The public could benefit from having more of these sensitive operations performed by licensed professionals. Several security practitioners agree that merely raising the prospect of such an approach would contribute in important ways to awareness of the problem and would begin an important public policy dialogue.

Given current uncertainty over the size and scope of the future threat,(11) and of the ability of technological solutions and existing government structures to provide an adequate response, further consideration of the idea seems warranted. Vehicles to provide effective forms of public and private response are becoming increasingly entrenched and institutionalized. It may now be an opportune time to consider the extent to which limited and carefully defined cooperation may be mutually beneficial. We have produced this article to provide a perspective on the disadvantages of increasingly uncoordinated and potentially inadequate means of response, and to encourage further exploration of the range of available and emerging options.

II. Convergent Trends, Diverging Responses

It is unclear to what extent routine computer-based intrusions will proliferate. It is similarly unclear to what extent traditional forms of governmental response will be adequate to address and deter this behavior. It is not necessary, however, to project accurately huge growth in these areas in order to begin thinking about next steps. Technology will continue to provide more effective ways of not only preventing, but also detecting unauthorized intrusions and unauthorized use. Reluctance to report anomalous activity may decrease as events become more commonplace. These trends will place a progressively greater burden on a growing federal law enforcement response. While alternative and effective forms of response are developing within the private sector, developments are taking place under conditions that make it difficult for the government adequately to assess the scope of the problem or to develop a predictive threat-warning capability. Perhaps there is a way for private experts to function in a way that also advances important societal and governmental interests.

A. Potential Growth of Computer-Related Misconduct

Not much is known about the future size and scope of computer-related misconduct.(12) Statistics and surveys compiled to date are of limited utility, but do reflect trends that would support qualified assumptions. It is reasonable to expect steady increases in the number of people with the technical capability to commit computer-related misconduct, and the continued widespread availability of increasingly harmful and easy-to-use "hacker" tools. It is also reasonable to assume public and private institutions' growing dependence on information technology, and growing interdependence on the information and processes that are generated by and shared among them. Taken together, these trends could mean an increase in the number of unauthorized intrusion incidents, and an increase in the severity of the potential effects of any single intrusion incident.

Congress has begun to ask difficult questions about the ability of law enforcement and the defense community, as currently configured, to serve adequately national interests in this area.(13) There has been little or no discussion, though, about contributions that could be made through the formalization of the resources that are already trained and equipped to work within the private sector. Because it makes good administrative sense to consider extant resources before building anew, these potentially powerful capabilities should also be included within the framework for discussion.

We may be at a stage where the proliferation of personal computers and computer networks resembles the birth and expansion of automobile transportation, although at an accelerated pace. We hold out high hopes for security measures, but if they do not fulfill their promise, the next few years could see unlawful instances of computer-related misconduct—if unchecked or undeterred—become as common as traffic infractions. Traffic infractions, however, are not all handled through conventional "criminal" channels. Consider the overlapping and decentralized criminal and administrative enforcement mechanisms that developed in response to the growth of automobile transportation. Similarities are apparent:(14) Respective rights of way are pervasive across our borders. Jurisdiction is often shared. Different rules govern bodies in motion (electronic interceptions and "moving violations") and at rest (stored electronic communications and parking infractions). Before the problem becomes this large, we need to assess existing divisions of response authority.

Because we do not know very much about the future scope of the problem, we cannot now know very much about the effectiveness of any particular set of solutions. Technology may lead to promising ways of preventing minor intrusions and detecting major ones, thus making enforcement truly manageable. Law enforcement resources and capabilities will continue to grow, perhaps obviating the need to consider alternatives. Civil enforcement regimes are likely to assume a more prominent role as well. But certain trends create cause for concern: Computer crimes are more difficult to detect than other forms of crime. A large percentage appear to go undetected, with others detected only long after having been committed. Even when detected, equities often militate in favor of not reporting incidents. And many of the same factors that make detection so difficult also make responding to an incident an inordinately time and resource-intensive undertaking. All of this has contributed to the rapid growth of a private-sector response capability. But this capability is one that arose to prevent and respond to discrete incidents, and is thus highly decentralized, creating accompanying concerns.

It may now be the best time to begin a public dialogue about possible ways of supplementing existing response capabilities. Perhaps a coordinated effort to clarify the roles of the public and private sectors with respect to investigation and responsive legal action is warranted, and could be done in a manner that is mutually beneficial to public and private interests.

B. Computer Crime is Different from Conventional Crime

Computer crime is different from conventional crime. It is grossly under-detected and under-reported. It is extraordinarily difficult and expensive to investigate owing to jurisdictional complexities, among other things. The laws in the area are complicated, and are evolving at a different rate from the underlying technology. As a result, responding to computer crime can severely tax even rapidly developing law enforcement response capabilities.

1. Difficulties in Detection

Roughly speaking, less than one in ten successful computer intrusions is detected.(15) There are many reasons why this may be.(16) Detection tools are still in the early stages of development. Existing security measures are, like their physical analogues, often slow and cumbersome, and as a result are only partially implemented or are implemented in ineffective ways. This leaves the difficult and often burdensome task of monitoring networks to systems personnel, who may also be overwhelmed with providing other forms of computer support. Even with monitoring, small anomalies may not be apparent. Those who use the systems may not attribute the anomalies to intrusions or other forms of unauthorized behavior. To increase intrusion detection capabilities, technological and educational solutions are required.

2. Limited Reporting

Even if intrusions are detected, victims tend not to report intrusions, particularly to law enforcement. Frequently-quoted sources place figures for reporting intrusions to law enforcement at somewhere between eleven percent and seventeen percent.(17) There are many documented reasons behind the reluctance of private businesses to report intrusions.(18) As news of a computer vulnerability could be devastating to business, companies often demand a degree of confidentiality that law enforcement rarely can promise. Businesses may also be reluctant to relinquish control over the resources they dedicate to an investigation. Because they can be extraordinarily resource-intensive, companies often prefer to remain in control of the resulting investigation—to preserve the option of terminating it before it becomes too costly. They also prefer to remain in control of their remedies, i.e., whether to take the case to civil court, criminal court, or to resolve it internally.[19] These needs—for confidentiality and control over resources and remedies—appear to be some of the principal drivers behind the development of private response alternatives.

3. Jurisdictional Complexities

Computer crimes are geographically complex, often crossing state or international boundaries. As a result, they are jurisdictionally complex, usually necessitating involvement by more than one authority, and often hindering state authorities' ability to pursue complete investigations.[20] Even novices are usually clever enough to disguise their actual location by looping through several systems before reaching their final destination. In fact, the desire to use free long-distance service for hacking activities may necessitate a certain amount of this evasiveness. The chances of an intruder remaining within one state's jurisdiction become more remote with every additional system he or she enters. And in an increasingly networked world, it is increasingly likely that an intruder would enter at least one foreign system, perhaps even without knowledge of having done so. This situation usually demands some level of involvement by federal law enforcement in the investigation.

Legal measures to ease these jurisdictional impediments are not unattainable, but are likely to place an even greater burden on a centralized federal response. For example, the U.S. government is currently working in international fora to enhance the ability of cooperating law enforcement officials to investigate computer crimes that cross international boundaries. The measures being contemplated include creating networks of law enforcement and communications carriers who can work together on investigations, and improving the legal agreements by which cooperation can be extended in time-sensitive situations.[21]

4. Resource Constraints

The burden imposed by jurisdictional complexities is aggravated by the highly resource-intensive nature of computer crime investigations. Staffing a response capability involves the cost of procuring and frequently updating hardware and software, and training and retaining qualified personnel. Perhaps most significantly, these investigations are extraordinarily time-intensive. Whereas a typical (non-"high-tech") state or local law enforcement officer may carry between forty and fifty cases at a time, a high-tech investigator has a full-time job handling three or four cases a month.[22] Considering that approximately only one tenth of all intrusions are detected and roughly one tenth of those are currently reported, the implications for building an effective response proportional to the problem—given its potential rate of growth—should be apparent.[23]

C. State of the Law

Through the 1980s, lawmakers and enforcement officials were learning how to address computer-related crime. It was seen largely as a shared responsibility between local, state, and federal government. More recently, the explosive growth of international data networks and the Internet has shifted a greater degree of enforcement responsibility and authority to the federal government.[24]

The Computer Fraud and Abuse Act ("CFAA") prohibits a range of activities involving unauthorized access to protected computers.[25] Insofar as private security experts may lack authorization to enter third-party systems, even for investigative purposes, some of the law's prohibitions may impact attempts by private parties to trace and identify unauthorized intruders.[26] Prohibitions of the Electronic Communications Privacy Act ("ECPA") may similarly restrict private intrusion response while placing carefully circumscribed conditions on law enforcement access to protected forms of communication.[27] Prohibitions in the federal wiretap statute make it unlawful to intercept real-time computer-based communications just as it is unlawful to intercept voice communications.[28] It bears noting, however, that the wiretap statute was originally drawn to pertain to telephonic voice communications and was subsequently extended to pertain to electronic (and hence computer-based) communications. The wiretap statute can, as a result, be interpreted to apply to networked computer environments in broad, unpredictable, and occasionally even counterintuitive ways.

The civil law has lagged considerably behind the criminal law in this area, leaving victims an insufficient number of middle-ground options between pursuing criminal remedies and essentially doing nothing. But effective civil remedies are beginning to appear in state codes,[29] and in 1994 the CFAA was amended to include a federal civil remedy.[30]

D. Law Enforcement Capabilities

Law enforcement techniques and capabilities appear to be improving at the state, local, and federal level. Investigators are receiving more and better training, and are hiring professional staff with relevant skills. Over time, investigators have become more acutely aware of the sensitivities of private sector victims, and are learning to conduct investigations in ways more respectful of their need for confidentiality and control over resources and outcomes. As indicated above, however, these capabilities come at considerable costs[(31)] and are inherently limited in their ability to provide confidentiality.[(32)]

At some point, it may be incumbent on law enforcement to plot the relevant growth trends, and to arrive at a realistic estimate of the resources required to continue to address these incidents. It may be, for example, that there are practical limits imposed for budgetary or policy reasons that would be placed on its ability to expand proportionately.[(33)]

E. Private Sector Capabilities

At the same time as law enforcement capabilities are increasing, the private sector response is growing very rapidly.[(34)] A series of recent mergers and consolidations in the computer security industry signals its coming of age and tremendous potential for profitability.[(35)] Practitioners estimate that there are over 600 firms currently offering some form of computer security services, including firewall installation, intrusion response, incident recovery, and backup restoration. Some but not all of these businesses offer actual incident investigation.[(36)] Though dictated, of course, by the client, the focus of these companies' efforts is not always consistent with the focus of law enforcement. The companies offer services to protect against, respond to, and mitigate the effects of harmful intrusions. Finding their source is a secondary concern.

These businesses range in size from the largest nationwide security firms to individual practitioners who, with technique, tools, and talent, are striking out on their own with increasing frequency.[(37)] Qualifications and standards of practice appear to vary accordingly. Some businesses maintain procedures to insure trustworthiness and accountability. Others primarily emphasize results.[(38)] Some businesses appear to be acutely aware of the limitations placed upon their activities by current law—civil and criminal—and conduct their businesses accordingly. Others are likely unaware of the potential implications of certain legal provisions. Still others, we fear, may even use their willingness to disregard current law to their competitive advantage.[(39)]

III. A Call for a Balanced Public/Private Approach

It may indeed be desirable to improve oversight and increase the professionalism of investigative segments of the computer security profession. But there are also certain features demanded by clients that should remain undisturbed—key features that originally fueled the rise of the profession and that remain integral to its continued growth and success. To get a sense of some of these factors and the equities underlying them, we have spoken with a number of private security practitioners, some of whose activities would likely fall within the purview of the measures contemplated here.

A. What the Industry Would Need from an Oversight Mechanism

We learned, not surprisingly, that confidentiality is the cornerstone of the computer security business. Clients insist on confidentiality above all because, for many, public confidence is their most valued asset. This includes, for example, public confidence in the ability of businesses to deliver service reliably, and to maintain with confidentiality and integrity the information they receive from clients and customers. In businesses such as these, public knowledge of an actual or apparent vulnerability, or of an event that appeared to exploit or actually exploited a vulnerability, could cause significantly more harm than the vulnerability or the event itself. Accordingly, the experts we spoke to made clear that for their services to remain valued and effective, any oversight mechanism would have to allow for services to be rendered confidentially.

Currently, confidentiality is dictated largely by the client, and is controlled through individual non-disclosure agreements often drawn by the client. A more formal type of confidentiality is not without precedent within professional licensing schemes. The doctor/patient, lawyer/client, and priest/penitent privileges are all well known and accepted. A similar type of privilege is beginning to be recognized in an area more clearly analogous to the security-service-provider/client relationship; at least one state currently recognizes a privilege for the private investigator-client relationship(40) and many other states protect client information from disclosure by law.(41)

Confidentiality would assist in the fulfillment of another condition demanded by clients—control over the outcome of the incident. It is not uncommon for clients to conduct a preliminary, internal assessment of a problem and weigh its likely causes and effects before considering additional action. When additional action is desired, customers are then able to choose among a range of available remedies, including informal resolution through private channels.(42) In such instances, the option to seek criminal investigation and prosecution still exists and in many cases may offer the only real satisfaction for the victim. There are, however, other options a client may consider. Civil remedies are available through federal and state law, for example.(43) A client may even decide that it is not worth the expenditure of resources required to identify the source of the intrusion, and may simply ask the security expert to "plug the hole."

B. Oversight Options

We noted above that certain trends and factors may make it desirable to improve oversight and increase professionalism of the computer security profession, or at least certain segments of it. But whose responsibility should it be? Is it a governmental responsibility, or should the profession police itself? How strictly should adherence to rules and norms be enforced? These are some of the fundamental policy questions that differentiate licensing schemes from less imposing, but also less effective, means of oversight.

1. Licensing

Although licensing may mean different things depending on the profession, it does have certain fairly uniform characteristics. Licensing schemes, such as those that apply to lawyers, doctors, and even most state-licensed private investigators, involve a degree of governmental involvement. The government may issue licenses directly, or establish boards of practitioners to oversee the licensing function.(44) Licensing authorities may set minimum educational and training requirements, impose professional conduct standards, and provide a mechanism for continued oversight to review the status and performance of licensees.

Licensing frameworks carry advantages. They offer robust and identifiable mechanisms to provide services related to the license. Often, licensing bodies set requirements governing receipt of the license; administer the necessary tests, background investigations, continuing education requirements, and professional conduct standards; and develop a disciplinary framework. These organizations are largely overseen by peers of the licensees—those who best understand the nature of the profession. Licensing bodies can be responsive to an ever-changing environment, as they often have considerable direct or indirect influence over licensing requirements. Licensing bodies are effective enforcers. It is generally quite clear to those obligated to keep a current license that they remain accountable to the issuing authority. A breach can substantially harm a practitioner's professional reputation and lead to monetary fines, suspension, or even revocation of the license. Whatever advantages may accrue from its ability to achieve close oversight and compliance, a licensing body has the corresponding disadvantage of requiring a fairly elaborate bureaucracy. While many bureaucratic costs can be recovered through licensing fees,(45) the mere existence of a bureaucratic framework may discourage candidates from obtaining a license or from participating in the profession.

State licensing of traditional investigators may provide both a model for and some potential obstacles to the growth of the investigative aspects of the private profession. More than forty states currently have mandatory licensing schemes for private investigators.(46) The schemes not only set forth licensing prerequisites, but make it unlawful to engage in certain specified activities without a license. Many of the licensing schemes are quite robust, and include rigorous qualifications to obtain a license, continuing education, stringent professional conduct requirements, and appropriate oversight to enforce the licensing standards.(47) Despite all of these requirements, which would seemingly ensure a high caliber of professional conduct among private investigators, abuses are not uncommon and some who are harmed by a private investigator's conduct decline to report violations to the licensing board for fear of having sensitive information publicized.[48]

State licensing schemes actually may create additional challenges, owing to an overlap between traditional private investigative services and emerging computer investigative services. Computer security experts could find themselves subject to state licensing schemes for private investigators if they are not cognizant of the law and careful to limit the services they provide. This could be problematic for a computer security expert not only because the skills and education required to be a private investigator may not be similar to his computer security skills,[49] but also because he may have to satisfy the requirements of each jurisdiction in which he conducts investigations.[50]

2. Certification

Other professions are not subject to licensing requirements, but rather require a certification or other official "seal of approval." Much like licensing, certification schemes can ensure that those practicing are properly trained and educated. To receive certification, an applicant may be required to take a prescribed set of courses, or even to pass an exam. Once the initial qualifications are met, however, few certification programs provide additional or continuing oversight. They may provide nominal professional conduct standards, but generally lack effective discipline mechanisms. They generally do not require certified professionals to stay current with new developments in order to keep their certification active.

There is a number of emerging private certification authorities in information technology. Some even address computer security services.[51] In our view, they will only accomplish part of the job. Private certification authorities may raise educational requirements, and perhaps even impose minimal liability insurance requirements and procedural guidelines to aid the private sector in obtaining trustworthy services. But without substantial incentives and/or disincentives, the government and the public are less likely to benefit. Without enforceable standards of conduct and a strong enforcement mechanism, there is little deterrence of overly intrusive investigation practices, particularly in areas where procedural shortcuts can save considerable time and resources. And without a standing body to oversee and facilitate limited information sharing, the government would likely be left in no better position to assess threats and vulnerabilities from a national perspective. It bears noting, however, that licensing and certification schemes need not be mutually exclusive. They can even coexist insofar as certain certifications may serve as necessary prerequisites to obtaining a license.

IV. Fertile Ground for Compromise

Achieving the delicate balance between public and private investigative authority involves the weighing of the needs for accountability and deterrence. We have noted the limits of a traditional law enforcement response to maximize deterrence, the inherent limits on victim confidentiality, and potential for growth of the problem. But allowing unfettered expansion of a powerful private sector response, under conditions that could jeopardize accountability, is not a satisfactory solution. Fertile ground for compromise lies somewhere in between.

A. What the Industry Could Get from an Oversight Mechanism

An oversight mechanism, such as a licensing scheme, could carry substantial benefits for those currently offering computer investigative services. Benefits include competitive advantages, a more predictable legal and liability climate, more well-defined standards of practice, and enhanced trustworthiness of those engaged in the profession.

A licensed investigator may benefit from operating in a more clearly defined legal environment. The terms of the licensing scheme might, for example, exempt a computer investigator from state licensing requirements that might otherwise apply, reducing the potential for incurring penalties for operating without a license in certain jurisdictions and reducing duplicative licensing requirements. The computer investigator may benefit not only from clarification of the administration requirements governing his activities, but also from a clarification of the substantive laws, such as the CFAA and the ECPA, that govern activities relating to computer networks. The formal recognition of the profession and its function—investigation of computer intrusions and tracking of intruders—may facilitate needed reexamination and clarification of many of the laws implicated by activities such as system monitoring, tracking of the source of an intrusion, and other attempts to identify intruders.(52) Even if clarification or modification of controlling law is slow to occur, the licensing oversight body would be available to provide guidance to investigators on the application of these laws to their practice. This guidance likely would remove much of the prevailing doubt and fear about exposure to criminal and civil penalties for violations of relevant statutes.

Consider, in addition, the potential liability that private security practitioners can currently face owing to the sensitive nature of their duties and the value of the systems and information placed in their care. The nature and extent of their duties, the conditions that might constitute a breach, and their resulting liability exposure are currently undefined in the industry. Practitioners and clients address these issues in contracts, but these contracts cannot fully address third-party liability. With increasing interconnectedness and interoperability of systems, third-party vulnerabilities, liabilities, damage, and related issues will inevitably be addressed in law—but only over time and probably only after much conflicting precedent.

Enhanced oversight may provide the quickest means of achieving a more adequately defined liability climate and to accelerating the availability of needed insurance products. An oversight mechanism could, for example, require practitioners to carry a certain level of liability coverage adequate to meet the needs of potential plaintiffs. This requirement could be scaled or structured to reflect the character of the services that were made available. This insures that those whose systems might be damaged or whose data might be compromised could recover. As importantly, and in the interest of fostering growth of the profession, an oversight scheme could bound the level of liability for the practitioner and make insurance a more estimable and predictable cost of doing business (one that could be distributed uniformly across the practitioner's client base).

Though oversight may help define the liability climate in direct ways, requiring computer security experts who track the source of an intrusion to meet certain educational criteria and to follow professional standards may ultimately, albeit indirectly, reduce the need for or reduce the cost of liability coverage. Practitioners who are educated not only about the legal limits of their ability to trace intruders, but who are also technically aware of the potential dangers and pitfalls, are more apt to stay within appropriate ranges of activity. In addition, knowing that a serious mistake, whether or not technically illegal, could have serious consequences for one's ability to practice will ensure adherence to a more rigorous standard of care.(53)

One of the more tangible benefits for all concerned that would emerge from an oversight mechanism would be the development of standard practices within the industry. Standards, defined liability, and education and training criteria would all contribute to the trust placed in a computer security expert. Thus, the license itself would likely accrue value as a marketing feature. Customers would likely opt to use a licensed investigator over an unlicenced investigator, secure in the belief that services provided will meet certain quality standards. Investigators would likely pay reasonable licensing fees if confident of the liability and marketing advantages, and the licensing fees could be used to make the governing body self-sufficient.

B. What Government Could Get from an Oversight Mechanism

Improving intrusion response should be a primary concern of both the private and public sectors. Without an adequate and effective response and the attendant benefits of stout deterrence, unlawful events may proliferate and interfere with reliable and safe operation of systems and networks. Likewise, the proliferation of even relatively "minor" events threatens to "raise the noise level"—as minor intrusion attempts may make gravely harmful events all the more difficult to detect.(54) Law enforcement is likely to express concern that a private response will not complement, but rather thwart its current law enforcement efforts. Such an argument is misdirected because it fails to recognize that a potent private sector response exists today, and that it appears to be growing at a rate at least commensurate with that of law enforcement. The question is not whether a private response should be permitted to exist—it does. In recognizing the inevitability of these private response mechanisms, law enforcement and government should instead consider the degree to which they should nurture the growth of the profession to serve the best interests of society.

That said, we do not agree that recognition or even fortification of a private investigative response will invariably interfere with law enforcement efforts. First, the mere availability of private professionals will not interfere with law enforcement's ability to investigate any incidents that continue to come to or are brought to its attention. Rather, with private professionals available to provide a supplemental response, law enforcement could be freed from the weight of preliminary work associated with distinguishing between nuisance intrusions and more serious threats to our national or economic security, such as dangerous intrusions into federal government systems and incidents involving economic espionage. Law enforcement could focus resources on investigations that, for the time being, appear to require governmental intervention, such as intrusions originating from foreign countries. Second, it is possible that law enforcement reporting might actually increase, as practitioners who are reluctant to exceed their legal authority and risk sanction or loss of license would be inclined to recommend that their clients make referrals to law enforcement when available private response options have been exhausted.

As private response capabilities grow, it is likely that an increasing number of incidents will be handled without the government's knowledge or involvement. This cannot bode well for the government's ability to obtain an accurate appreciation for the scope and nature of the threat, extant vulnerabilities, or for the development of an effective threat warning capability. The computer security industry would continue to conduct its business outside of the view of law enforcement, and the government would derive little benefit from the information, insights, or collective expertise and experience of those working within the industry. Instead, we merely suggest that the government first recognize the services being performed by computer security experts and private investigators, and second, consider aiding in the "professionalization" of the investigative portion of those services. These steps would create opportunities for increased cooperation between the industry and law enforcement.

This cooperation may involve little more than opening lines of communication between private computer specialists and the government. Increased communication, even if in the form of sanitized and generic reporting on incidents and vulnerabilities, could contribute significantly to the effort of government and law enforcement to estimate the size of the problem, to develop responsive policy, and to allocate sufficient resources to the problem area.

But cooperation likely would achieve additional benefits. Evidence collected by private computer security experts operating in an investigative capacity would be available for use in civil and criminal proceedings. Security experts may be important sources of expertise and evidence, even in cases initiated by law enforcement or cases later referred to law enforcement by the victimized system owner. They can (and do currently) loan valuable technical expertise to law enforcement in complex criminal cases. Depending on the resources available to law enforcement investigators in the future, having additional support to gather preliminary information, or to turn over to law enforcement "ready-made" cases, could be an important component of effective criminal deterrence of computer crime.(55)

C. What the Public Could Get from an Oversight Mechanism

Investigations into computer intrusion incidents are delicate undertakings. Their effectiveness is dependent on the practitioner's having a working knowledge of the latest technology, and their legality is dependent on the practitioner's having a detailed knowledge of the current state of the law. The position of investigator is not unlike that of a doctor or lawyer, who is expected to be maximally effective through their knowledge of the most recent developments in her field, but who is held strictly to the bounds of safe, ethical, and professional conduct.

Potential breaches of safe and ethical conduct can create similarly serious consequences. In the same way that the doctor protects a patient's life, and the attorney a client's liberty and property, the security specialist deals in an environment that places at risk a business' most valued commodities—its communications and information. They do their work in an environment that can be easily abused, in which abuses are extremely difficult to detect. Unlike doctors and lawyers, but more like police officers, they are in a position to place third parties in jeopardy by infringing on third-party systems and communications. Investigations should be undertaken in ways that minimize these dangers. Mitigating risks to third-party systems comes through care, equipment, and experience. It requires in-depth knowledge of the systems that are placed at risk, access to the latest technology or to a proper testing environment, and experience to guide selection of tools and techniques. Mitigating risks to personal privacy requires knowledge and appreciation of the prevailing laws, current company policies and procedures, and adherence to principled investigative practices.

Although it is certainly possible to find practitioners who meet these lofty standards, it can be made more likely through adherence to an oversight structure. Clients and customers would likely be more confident in their expectations about the safety and quality of the services provided. Third parties would likely be more confident that such services would be rendered in ways that did not impinge on their interests. In either case, the public and the profession would benefit.

D. Long-Term Benefits of a Cooperative Environment

There are still other longer-term benefits that may be realized. Facilitating the growth of a responsible investigative profession undoubtedly would expand the market for better tools to detect and identify the source of unauthorized intrusions. Increased private-sector market demand can, in turn, be expected to stimulate research and development in a way that the government market alone cannot. It is precisely because the private sector is so adept at developing and using current technology that law enforcement currently uses private computer security experts as advisors, technical consultants, and even contract support on investigations.(56) A licensed group of computer security experts trained and experienced in conducting investigations would provide even more outsourcing opportunities for government. In addition, this cadre of trained experts could be mobilized in the event of national emergencies(57) under specific arrangements defined by the oversight body.

The greatest benefit of such a supplement to current law enforcement capabilities may be the rich source of information that ultimately could help the government perform its responsibilities more efficiently and the private sector better manage its risk. Government investigators are learning more about intrusion incidents, particularly those that succeed in compromising government systems. Private computer security experts have a more intimate knowledge about the numbers and severity of intrusions into corporate computers—the vulnerabilities exploited, the tools and techniques of intruders, and the damage caused by intrusion incidents. This information is currently unaggregated. Nowhere is there a complete picture of the true size, scope, and severity of a problem that could significantly impact national and economic security. With a limited system of reporting, one that respected complete client confidentiality, a more accurate picture of this problem could be developed. This information could be used operationally, by government or the private sector, as the basis for vulnerability advisories. It could provide the insurance industry with the actuarial data it needs to develop and refine currently scarce insurance protection. At the policy level, a more complete picture of threats and vulnerabilities would allow the government properly to manage its response, to consider appropriate changes to support criminal deterrence, and to facilitate investigations and prosecutions. It would create an avenue for policy development that truly is tied to the size and nature of the problem.

V. Unanswered Questions

There are additional determinations likely to bear on the desirability or effectiveness of an oversight mechanism or authority. They involve, among other things, clarifying the scope and nature of the oversight mechanism and how participation or compliance might be enforced.

A. Who Should Be the Oversight Authority?

The principal unanswered question appears to be "who should do the licensing?" Is it a private sector or a governmental responsibility? If governmental, is it more appropriately performed at the state or federal level? Each of these options has advantages and disadvantages that can be examined effectively against the objectives of the oversight mechanism.

The private sector may offer advantages as a home for an oversight body, particularly for a profession so closely linked with sensitive private sector concerns. The ability of the private sector to influence and manage the oversight of investigative professionals may increase confidence, at least within the private sector, in those professionals' competence, qualifications, and sensitivity to concerns such as confidentiality and control of investigations.

In addition, a private board would avoid many of the bureaucratic pitfalls that may be encountered with government oversight at the state or federal level. However, a private oversight mechanism will have inherent limitations. Private bodies are not as well positioned as government to impose a mandatory licensing scheme. Most purely private professional groups offer only certifications or similarly limited forms of approval of qualifications for a profession. It is also not clear that a wholly private entity would be able to support the necessary functions attendant with licensing. Additionally, a private oversight mechanism may be reluctant (for liability-related or other reasons) to participate in many of the collaborative activities which would benefit law enforcement and government responses to the computer intrusion problem.

This is not to say that the private sector should not have a role in an oversight mechanism. In many traditionally licensed professions, oversight functions are conducted under the auspices of a state licensing board created and funded by the state government, but the board's participants are often members of the profession who also sit on a review panel. Clearly, peer review is an important aspect of any oversight scheme and could be incorporated at the state or federal level.

State governments have a long and distinguished history of licensing professionals to practice within their jurisdictions. States license doctors, lawyers, accountants, private investigators, and other professionals. While there is history of professional state licensing, problems are beginning to emerge. Many time-tested state licensing schemes are now strained, because many state licenses are effective only within the state in which they are granted. For example, to practice medicine or law within a state, a professional must either be licensed in the state, subject to a reciprocal agreement between their state of licensure and the state where he wishes to practice, or fit within narrowly circumscribed exceptions.(58) The limitations of these state-by-state licensing approaches are already being shown in professions not as intimately connected with the Internet (and other networked environments) as the computer security profession. State licensing schemes for medical doctors, for example, appear to be limiting the growth of "telemedicine," because a doctor often is not permitted to render a diagnosis on a patient located in a jurisdiction where the doctor is not licensed to practice. While technology may allow doctors, lawyers, or other professionals to practice in nearly every jurisdiction in the world, often at a reduced expense to clients, licensing requirements predicated on physical boundaries may prevent or severely restrict multi-jurisdictional practice.

Jurisdictional limitations also hamper state law enforcement officers in the investigation of computer crime. Intrusions are almost always multi-jurisdictional, even if the intruder and the victim are located next door to one another. Networks used to obtain access to a system usually cross physical and thus legal jurisdictions.(59) Because these jurisdictional complications apply with equal force to state-licensed professionals, they militate strongly against a state-level oversight mechanism for private computer security experts. Thus, in this respect, a centralized oversight mechanism is desirable.

While the federal government does not have as long and well-developed a history of licensing professions, federal licensing is not without precedent.(60) The Nuclear Regulatory Commission licenses operators at nuclear power plants. The Food and Drug Administration licenses food inspectors. There are federal licenses for everything from nuclear-material transporters and security officers to pilots.(61)

B. Who Should Be Covered by the Oversight Mechanism?

The appropriate scope of an oversight mechanism will likely be dependent on its structure. The higher the bar to obtain a licence to perform intrusion investigations, the more appropriate a limited scope may be. There are many ways to define coverage. The oversight mechanism could apply to all computer security specialists, only to computer security specialists who respond to intrusion incidents by employing defensive mechanisms, or only to those who respond "offensively" to intrusion incidents by tracing and identifying intruders. It would be neither prudent nor necessary to create an oversight mechanism so broad as to cover large numbers of employees engaged in routine system security, system design, or system maintenance. Given that security specialists often work in teams, a determination must be made as to whether the license would attach to an individual, a company, or some subset of qualified employees. A determination must also be made as to whether the oversight mechanism applies only to those who perform such services for hire or also to those who perform such services only for their employer. The oversight mechanism's scope may also depend on its requirements, such as the amount of liability coverage required.

C. Should Oversight Be Mandatory or Permissive?

Another important determination that must be made with respect to an oversight mechanism is whether it should be permissive, allowing those who wish to take advantage of its features to opt in, or mandatory, applying to all those who fall within its determined scope. Most current licensing schemes are mandatory. They set a very high level of qualifications to meet the requirements for obtaining and keeping the professional license and do not allow those without the license to practice. Some even provide criminal or civil penalties for those who practice without a license or whose activities exceed its scope. This is true for doctors, lawyers and private investigators.

By contrast, some professions make use of certification schemes that are permissive in nature. While employers may generally prefer to hire an employee with a certification, and while doing so may provide incidental liability or insurance incentives, they may accept a certain amount of education or experience as a suitable substitute. Whether a permissive or mandatory scheme is preferable may ultimately depend on the types of incentives offered licensed practitioners. Guarantees of confidentiality through privilege and limitations on liability would surely attract interested practitioners even if licensing were not made mandatory. But strict criminal prohibitions on practicing without a license could also achieve a similar result.

Although licensing offers a way to achieve a higher level of compliance with set standards of behavior, it also has the potential to drive what is already something of an "underground" practice even further below the surface. And given the difficulties already inherent in investigating criminal violations of computer crime laws, it may prove unwise to attempt to graft on top of the existing enforcement regime additional requirements to investigate and enforce mandatory licensing conditions.(62) The benefits that will flow to all parties—service providers, their customers, and government—will require participation and thus openness. A robust, open, and voluntary federal licensing scheme may be the best way to assure the availability of qualified and responsible professionals, while simultaneously encouraging them to abide by established procedures and standards of conduct.

D. Required Changes in the Law

Permitting private computer security experts to investigate intrusion incidents raises red flags for some people. Allowing the government to sanction such activity raises them for others. Is this legal? Our short answer is "yes." Many aspects of the concept detailed in this paper are based on long-standing practical and legal precedent. We may be accused of applying these precedents in new ways, but we do not believe they stretch past our understanding of current law.

Private investigations have been licensed activities for approximately fifty years and practiced for well longer. In the past, licensing was done at the state level. This paper raises as a policy option the propriety of a federal licensing scheme—one fully justified by the interstate nature of computer intrusions. These types of federal licensing schemes are not without precedent. Government agencies have adopted federal licensing techniques in several settings, including security and law enforcement, by agencies such as the Nuclear Regulatory Commission and by the Department of Agriculture.

Certain legal issues would have to be resolved in designing an oversight mechanism. The oversight function, if performed by the government, may need to be limited to ensure that the licensees will not be construed to be "agents" of the government and thus subject to the Fourth Amendment.(63)

Without the ability to go to a court and obtain a search warrant or intercept order, private response is more limited in what it is able to legally accomplish. In other respects, available civil mechanisms may provide latitude not available to the government. The question is then whether private computer security experts can do enough within the bounds of the law to make their services of use to their clients. The first and most obvious answer might come from observing the rapid growth of the profession today. Desirable services are being offered at levels that are sufficiently affordable such that demand for these services is still on the rise. Is it all being done illegally? We think not.(64) Finally, additional prosecutorial options may contribute to additional deterrence of computer intrusions. This may require expanding the availability of civil remedies to those who suffer intrusions. A robust private investigative response undoubtedly will contribute to the effectiveness of civil adjudicative functions, and vice versa. Both will serve the interest of enhancing deterrence.

E. International Implications

We have not studied in detail the comparative legal barriers that face public and private investigators operating through networks linked to foreign countries. But given that private parties, including private investigators, currently enjoy unfettered access to international networks and systems—access which, if performed by law enforcement for investigative purposes, may raise issues of international law—this area should be considered more thoroughly. We certainly would not want to see the activities of law enforcement or private parties interpreted by foreign powers as hostile acts.(65) These concerns—the need to exercise particular caution in investigations implicating computers in foreign nations—may provide yet another compelling reason to impose additional oversight and standards on the investigative portions of the computer security profession.

VI. Conclusion

As FBI Director Louis Freeh recently noted, the future of effective computer crime enforcement lies in the creation of workable partnerships.(66) We agree that a coordinated effort to clarify the roles of the public and private sectors with respect to investigation and responsive legal action is warranted. We believe such a clarification could be done in a manner that is mutually beneficial to public and private interests. Professional licensing offers a novel venue for cooperation and compromise. And the debate that would accompany such a proposal will, regardless of the result, undoubtedly raise the level of awareness and depth of understanding of the gravity and complexity of the computer intrusion problem we face.

Footnotes

1. Not used.

2. This paper explores ways to supplement investigative capabilities. Private investigators may be both an alternative and a supplement to traditional law enforcement responses to computer crime. In some cases, companies may opt to use a private method to resolve an intrusion problem instead of pursuing criminal remedies. In other instances, a criminal remedy may be unavailable for practical reasons, requiring a private response capability to supplement law enforcement. In still others, private services may be used at the preliminary stages of an investigation to gather evidence for subsequent legal action. We recognize, of course, that enhanced investigative capabilities ideally should be coupled with enhanced opportunities for prosecution, either through the civil or criminal law.

See Security in Cyberspace: Hearings Before the Senate Comm. on Governmental Affairs, Permanent Subcomm. on Investigations, 104th Cong. 73-74 (1996) [hereinafter *Security in Cyberspace Hearings*] (statement of Minority Staff discussing the reluctance of local law enforcement to develop computer-related expertise, due at least in part to the required technical expertise and need for special training and equipment, and noting the dependence on federal law enforcement created by the lack of a local response capability).

4. Reliable statistics remain elusive owing to definitional ambiguities, methodological inconsistencies, and limited reporting. Investigators have conveyed anecdotally their sense that the volume of potentially criminal incidents is increasing. See, e.g., Sharon Walsh & Robert O'Harrow Jr., Trying to Keep a Lock on Company Secrets, Wash. Post, Feb. 17, 1998, at D1 (comments of FBI Section Chief William Perez). Attorney General Janet Reno recently requested $64 million in increased funding to "expand efforts to protect the nation's critical infrastructures from cyber-attacks and to combat cybercrime." *Hearings Before the Subcomm. on Commerce, Justice, and State of the House Comm. on Appropriations*, 105th Cong. 16 (1998) (statement of Attorney General Janet Reno). Most of the funding is intended to support 75 new FBI agents and 24 new federal prosecutors to track down and prosecute computer criminals. *See id.*

5. For example, some violations of the Computer Fraud and Abuse Act are misdemeanors, and law enforcement officers overwhelmingly prefer to dedicate resources to the investigation of more serious felonies. *See* 18 U.S.C.A. 1030(c)(2)(A) (West Supp. 1998) (enumerating punishment under 1030(a)(2), (a)(3), (a)(5)(C), and (a)(6) as a fine or imprisonment for not more than one year, or both).

6. On November 21, 1997, Mathew Bevan, a.k.a. "Kuji," one of two then teen-aged hackers responsible for the celebrated Rome Labs intrusion incident, was freed after London prosecutors declined to go forward with the prosecution. The decision not to go forward appeared to be based on the cost of trying the case and the uncertainty of prevailing due to evidentiary problems. See Duncan Campbell, More Naked Gun Than Top Gun, Guardian, Nov. 27, 1997; Stephen Farrell, Hacker Who Broke into NASA Walks Free, Times (London), Nov. 22, 1997.

7. Predominant reasons given by security experts and survey respondents for non-reporting consistently include fear of negative publicity, fear of competitors using information to their advantage, and loss of productivity. *See, e.g.*, Computer Security Institute, 1997 CSI/FBI Computer Crime and Security Survey (1997) (unpublished report on file with *Harvard Journal of Law & Technology*); *see also* Computer Security Institute, Annual Cost of Computer Crime Rises Alarmingly: Organizations Report $136 Million in Losses (Mar. 4, 1998) (press release on file with *Harvard Journal of Law & Technology)*;WarRoom Research, LLC, 1996 Information Systems Security Survey (1996), available at ttp://www.warroomresearch.com/wrr/SurveysStudies/ 1996ISS_Survey_SummaryResults.htm. Such fears may not be unjustified. After Citibank received publicity for the 1995 intrusion into

its system, six of its competitors targeted the bank's top 20 customers, claiming their systems were more secure than Citibank's. *See Security in Cyberspace Hearings* (statement of Minority Staff), *supra* note 2, at 34-35.

8. *See infra* notes 28-29 and accompanying text.

9. *See, e.g.*, Anne W. Branscomb, *Rogue Computer Programs and Computer Rogues: Tailoring the Punishment to Fit the Crime*, 16 Rutgers Computer & Tech. L.J. 1, 57 (1990) (considering as alternatives strict liability for service and software providers, compulsory insurance coverage, and establishment of higher ethical values); Michael P. Dierks, Computer Network Abuse, 6 Harv. J.L. & Tech. 307, 337-39 (1993) (proposing greater emphasis on prevention through regulation or tax incentives for greater computer security); Robert L. Dunne, Deterring Unauthorized Access to Computers: Controlling Behavior in Cyberspace Through a Contract Law Paradigm, 35 Jurimetrics J. 1 (1994) (examining the advantages of using contract law to address unauthorized access to computers); David L. Gripman, Comment, The Doors Are Locked but the Thieves and Vandals Are Still Getting in: A Proposal in Tort to Alleviate Corporate America's Cyber-Crime Problem, 16 J. Marshall J. Computer & Info. L. 167, 172 (1997) (proposing that courts should impose on corporations a duty to have adequate computer network security to prevent intrusions that can damage the corporation or third parties).

10. *See generally, Security in Cyberspace Hearings* (statement of Minority Staff), *supra* note 2.

11. Several sources offer statistics on the current scope and projections of the future growth of the computer misconduct problem. However, the divergence of the results of these surveys and the methodological pitfalls associated with them limit their utility. *See*, M.E. Kabay, ISCA White Paper on Computer Crime Statistics (visited Apr. 8, 1998)<http://www.ncsa.com/knowledge/research/comp_crime.htm ("Given these problems of ascertainment, computer crime statistics should generally be treated with scepticism."). One obvious problem with statistics that attempt to measure the frequency and costs of computer intrusions is that many—though no one can really know how many—go undetected. We have used, for purposes of discussion, the assumption that roughly 1 in 10 successful intrusions is detected, and of those, roughly 1 in 10 is reported to law enforcement. At least one published source agrees with this estimate, although some security professionals have referred to it as optimistic. *See id.*

Losses from individual intrusion incidents also vary considerably depending on the nature of the intrusion and the intention of the intruder. Most figures place the average loss in the neighborhood of $40,000 per successful external intrusion, but the consequential damages associated with such an attack have gone as high as $10 million. *See* WarRoom Research, *supra* note

6; David Bernstein, *Industry Survey,* Infosecurity News, May 1997, at 20; Reuters, *Sabotage Suspect Charged,* CNET News.Com (Feb. 18, 1998) http://www.news.com/News/Item/0,4,19245,00.html; *see generally Security in Cyberspace Hearings, supra* note 2.

As long as firm data on the numbers of intrusion incidents that occur are unavailable, it will be almost impossible to determine accurately the scope and magnitude of the computer misconduct problem. This will hamper not only law enforcement efforts to build an adequate response capability, but also the development of national policies to protect our information infrastructure. *See Security in Cyberspace Hearings* (statement of Minority Staff), *supra* note 2, at 37.

12. The term "computer-related misconduct" is used in a broad sense to refer not only to unauthorized intrusions, but also to unauthorized interceptions of communications, routine instances of trespass that may result from exceeding authorized access, or incursions into privacy caused by, for example, e-mail "snooping."

13. *See, e.g., Hearings Before the Subcomm. on Terrorism, Tech. and Governmental Info. of the Senate Comm. on the Judiciary,* 105th Cong. (Mar. 17, 1998) (statement of Sen. Kyl) (on file with the *Harvard Journal of Law & Technology); id.* (statement of Jamie S. Gorelick & Sam Nunn) (on file with the *Harvard Journal of Law & Technology*).

14. The analogy is imperfect, as it fails to account for issues relating to enforcement of computer-related misconduct across international boundaries.

15. *See* KaBay, *supra* note 10. It is difficult to estimate the percentage of intrusion incidents that are actually detected. For example, in a 1996 General Accounting Office report, the Defense Information Systems Agency (DISA) estimated that during vulnerability assessments of Department of Defense systems only 4% of attacks were detected. *See* General Accounting Office, Information Security: Computer Attacks at Department of Defense Pose Increasing Risks, GAO/AIMD-96-84 (1996).

16. *See* Dierks, *supra* note 8, at 332-33, for discussion of difficulties of detecting computer crimes.

17. The Computer Security Institute reports 17% of detected intrusions are reported. Previous FBI estimates have been at 11%. Estimates of the rates of reporting for the Department of Defense are even more disparate. The range is from between 1 in 8 (12%) and 1 in 140 (0.7%). *See generally,* the discussion of DISA and Air Force Information Warfare Center statistics in John D. Howard, An Analysis of Security Incidents on the Internet:1989-1995, 174-77 (1997) (unpublished Ph.D. dissertation, Carnegie Mellon University, on file with *Harvard Journal of Law & Technology*).

18. *See, e.g.*, Branscomb, *supra* note 8, at 55-56; Scott Charney & Kent Alexander, *Computer Crime*, 45 Emory L.J. 931, 938 (1996); Dierks, *supra* note 8, at 335 (financial disincentives to reporting); *see also* James A. Fagin, Computer Crime: A Technology Gap, 15 Int'l J. Comp. & Applied Crim. Justice 285 (1991); B.J. George, Contemporary Legislation Governing Computer Crimes, 21 Crim. L. Bull. 389 (1985).

19. We understand it is not uncommon for companies to hire private investigative specialists to track and identify sources of intrusions or other anomalies. Often, as a cost-effective alternative to law enforcement, investigators will issue intruders a warning, which often provides satisfactory resolution of the problem for that victim. Similar methods were noted during discussion with computer security experts in conjunction with the Security in Cyberspace Hearings. *See Security in Cyberspace Hearings* (statement of Minority Staff), *supra* note 2, at 48-49.

20. *See generally*, Joel R. Reidenberg, *Governing Networks and Rule-Making in Cyberspace*, 45 Emory L.J. 911 (1996) (discussing the inadequacy of traditional, territory-based regulatory regimes for governing activities in cyberspace).

21. *See, e.g.*, Communique of the Meeting of Justice and Interior Ministers of The Eight (Dec. 10, 1997), *available at* http://www.qlinks.net/comdocs/washcomm.htm (communication from meeting to discuss enhancing the abilities of the participant nations to investigate and prosecute high-tech crimes).

22. *See* Ingrid Becker, *Cybercrime: Cops Can't Keep Up with Technobandits*, Cal. Law., June 1995, at 47, 91 (quoting Bill Spernow of the System for Electronic Analysis and Retrieval of Criminal Histories ("SEARCH") Group).

23. Assuming that existing law enforcement capabilities were sufficient to be perfectly responsive to all incidents currently reported, these figures suggest that a 10% increase in either the number of incidents detected or the number reported would effectively double the resources required to investigate. A 10% increase in detection and reporting would require a quadrupling of existing resources. Considering that computer crime specialists only appear to be able to manage roughly one tenth of the number of cases handled by non-high-tech investigators, the number of investigators required to fully staff a response capability—one that would keep pace with anticipated improvements in intrusion detection technologies—would be staggering.

24. By the late 1980s, 49 states had computer crime laws on the books, and Congress had passed (in 1984) and amended (in 1986) the Computer Fraud and Abuse Act ("CFAA"). 18 U.S.C.A. 1030 (West Supp. 1998). Early iterations of the CFAA recognized the shared responsibilities of local, state, and federal law enforcement. Congress, reluctant to preempt state computer

crime enforcement, drew the law to protect only a relatively narrow class of "Federal interest computers." The term applied roughly to government computers, banking computers, and computers involved in offenses that crossed state lines. *See* 18 U.S.C.A. 1030(e)(2) (West Supp. 1998). Successive amendments to the CFAA in 1994 and 1996 have now expanded federal jurisdiction over government computers, banking computers, and computers "used in interstate or foreign commerce or communication." 18 U.S.C.A. 1030(e)(2)(B) (West Supp. 1998).

25. *See* 18 U.S.C.A. 1030 (West Supp. 1998).

26. The CFAA, in its most general sense, prohibits *unauthorized* intrusions into computers protected by the statute. But if an intruder, through her own act of unauthorized intrusion, were to implicitly consent to reciprocal actions by the victim, then the intruder would be hard-pressed to argue that these reciprocal actions were not authorized. "Banner" warnings posted on computer systems currently serve a similar purpose with respect to consensual monitoring of unauthorized intruder activity. Note, however, that such consent, even if upheld as valid, would not necessarily constitute consent to pass through third-party systems to identify the source of an intrusion.

The fact remains that the full implications of the CFAA on the activities of computer security professionals have not been adequately spelled out. Legal experts can reasonably disagree on the interpretation of key provisions, and few technology professionals are fully aware of the scope and implications of alternative interpretations. For example, a major telecommunications company recently announced its intent to release free software that can track down "hackers" by following paths back through several servers to locate the source of the attack. *See* MCI, Information on DoSTracker, (Oct. 9, 1997) (press release), *available* at http://www.security.mci.net/dostracker/prelease.html. Some applications of this software might be construed by the Department of Justice to constitute a criminal violation of applicable computer crime laws.

27. *See* 18 U.S.C.A. 2701-2711 (West 1970 & Supp. 1998).

28. *See* 18 U.S.C.A. 2510-2522 (West 1970 & Supp. 1998). Thus, some applications of commonly-used "packet sniffer" devices, without proper consent, may be construed as violations of the federal wiretap statute.

29. Representative state civil remedies include Cal. Penal Code 502(e) (West 1988) (permitting recovery of compensatory damages); Conn. Gen. Stat. 52-570b (1997) (allowing civil recovery independent of criminal actions for acts done recklessly); Ga. Code Ann. 16-9-93(g) (1996) (allowing victim to recover damages, including lost profits).

30. *See* 18 U.S.C.A. 1030(g) (West Supp. 1998). To date, we have located one attempt to invoke the federal civil remedy, albeit unsuccessfully and in an unreported case. *See* Letscher v. Swiss Bank Corp., No. 94 CIV. 8277 LBS, 1997 WL 304895 (S.D.N.Y. June 5, 1997).

31. *See supra* Part II.B.

32. Law enforcement may be willing to modify its current investigative procedures to allow for greater confidentiality and control by victimized companies. There are, however, legal and, more importantly, constitutional limitations on the extent to which such procedures can be modified. For example, while the private sector may crave confidentiality, the Constitution and laws of Congress quite properly require a large degree of transparency and openness. *See, e.g.*, Darryl C. Wilson, *Viewing Computer Crime: Where Does the Systems Error Really Exist?*, 11 Computer/L.J. 265, 284 (1991).

33. *See* discussion of growth of incidents *supra* notes 14 & 22.

34. As early as 1994, estimates of the annual growth rate for the information security industry were as high as 70% to 100%. *See Infosec Growth to Continue—Internet Security Hot,* Security Tech. News, Oct. 21, 1994, available in 1994 WL 8715532.

35. *See, e.g.*, Malcolm Maclachlan, *Security Market is Maturing, but Needs Standards,* TechWeb News (Mar. 9, 1998) http://www.techweb.com/wire/story/0398iwld/TWB19980309S0015 ; Todd Spangler, *Rapid Consolidation in Security Market,* Webweek (Dec. 8, 1997) http://www.internet-world.com/print/1997/12/08/ news/19971208-rapid.html; Wylie Wong, Security Software Companies Continue Consolidation, TechWeb News (Feb. 24, 1998) http://www.techweb.com/ wire/story/TWB19980224S0011.

36. It is unclear, however, what percentage of computer security firms might make available response capabilities, or even how a useful conceptual distinction might be expressed to separate those who provide more conventional computer security services from thos who offer investigative service. Some firms attempt to maintain a hard distinction between "intrusion response" and "incident investigation." "Intrusion response" may refer to responsive and restorative actions taken by security specialists within systems that belong to their clients. "Incident investigation" may refer to actions taken outside a client's system, in an attempt to track and identify sources of disruption. Given the close interdependence of many systems, potential shared vulnerabilities, and muddled conceptions of system ownership, we question whether this can always be a serviceable distinction.

37. A search of the Internet reveals that an increasing number of members of the private investigations field are branching out into computer-related areas. *See, e.g.*, In fact, the National Association of Investigative Specialists

"Investigator of the Year" for 1996 specializes in computer-related cases. *See* Ralph D. Thomas, *The Nation's Cutting Edge Cyber Detective-A New Kind of Private Eye* (visited Apr. 9, 1998) <http://www.pimall.com/nais/n.seanor.html>.

38. This tension is exemplified by the debate within the computer security field over whether or not to hire hackers as consultants. Some find the practice irresponsible and fear liability; others stand behind the practice as an important way to secure needed expertise. For further discussion, *see* Richard Power, Current and Future Danger: A CSI Primer on Computer Crime & Information Warfare 12-13 (2d ed. 1996).

39. In instances where private investigative practices may run afoul of current criminal law, law enforcement resources would be doubly taxed by having to investigate the conduct of the intruder and the investigator.

40. *See* Mich. Comp. Laws 338.840 (1997) (stipulating investigator/client privilege).

41. *See* Ariz. Rev. Stat. Ann. 32-2455 (West 1996); Conn. Gen. Stat. 29-156q (1996); Haw. Rev. Stat. 463-15 (1996); 225 Ill. Comp. Stat. 446/195 (West 1996).

42. *See supra* note 18.

43. *See supra notes* 28-29.

44. For example, the Nuclear Regulatory Commission ("NRC") sets out the terms and conditions for licensing and issues licenses directly to operators of nuclear facilities. *See* 10 C.F.R. 55 (1997). In the professions typically licensed at the state level there is often an additional layer between the licensee and the state. This is often a board of professionals that assists the state in setting educational and other qualification standards and reviewing character and other disciplinary matters. *See, e.g.,* N.Y. Educ. Law 6521-6529 (McKinney 1997) (licensing for medical doctors).

45. California's private investigator licensing framework is at least partially funded through licensing fees. *See* Cal. Bus. & Prof. Code 7520 (West 1997).

46. *See, e.g.,* Mass. Gen. Laws ch. 147 23 (1994); N.J. Stat. Ann. § 45. 19-10 (West 1995).

47. *See, e.g.,* 225 Ill. Comp. Stat. 446/1-299 (1997).

48. *See* Michael A. Braun & David J. Lee, *Private Police Forces: Legal Powers and Limitations,* 38 U. Chi. L. Rev. 555, 559-60 (1971).

49. The flip side of this is that some who are currently offering computer intrusion investigative services are in fact state-licensed private investigators. It is unlikely that the skill sets of computer security experts and private investigators are similar. In fact, the danger posed by those who may be "licensed" computer investigators already, but who are not experts

in computer technology, may be one of the most powerful arguments in favor of requiring a unique license to conduct this specialized type of investigation. *See supra* Part II.B.

50. States, in general, do not recognize any kind of reciprocity for private investigator licensing. In fact, some states even allow local jurisdictions to add requirements on top of the state requirements. *See, e.g.*, Ohio Rev. Code Ann. 4749.09 (Anderson 1997). This jurisdictional complexity is one of the more vexing issues complicating state investigations of computer misconduct that crosses state lines.

51. *See, e.g.*, International Information Systems Security Certification Consortium (visited Mar. 14, 1998) <http://www.isc2.org. The International Information Systems Security Certification Consortium ("ISC^2") certification framework is based on an exam, and requires either continuing education over a three-year period, or passing the exam again, to maintain certification. The ISC^2 also has a code of ethics.

52. *See supra* note 25 and accompanying text.

53. We can imagine that if there were substantial and enforceable consequences for breach of or damage to third-party systems in the course of an investigation, it would provide an additional incentive for private investigators to contract with or obtain consent from those parties. Obtaining consent in this way could also serve the interest of comporting with the dictates of the criminal law. *See* 18 U.S.C.A. 2511(2)(d) (West 1970 & Supp. 1998) (consent exception to wiretap statute); 18 U.S.C.A. 2702(b) (West 1970 & Supp. 1998) (consent exceptions to ECPA).

54. *See* Trent D. McNeeley, *Hackers, Crackers & Trackers*, The American Legion Magazine, Feb. 1997, at 34 (quoting FBI agent that those who pose real threats "tend to hide among the noise—the everyday, 'OK' hackers who just enjoy penetrating and exploring systems"), *available at* <http://www.legion.org/pubs/1997/hackers.htm .

55. In a growing number of areas, particularly those involving economic crime, private investigators—many of them former law enforcement officers—are employed by companies and trade associations to continue to pursue their chosen profession, albeit from the private side. This is common practice, for example, in the insurance and intellectual-property-based industries. In such instances, industry investigators conduct preliminary investigations and compile evidence of incidents that might otherwise escape the attention of law enforcement. They may present to law enforcement evidence so gathered, and often continue to provide assistance to law enforcement after cases are referred.

56. George Vinson, Supervisory Special Agent of the FBI's computer crime squad, was quoted in an Internet newsletter as saying that his unit outsour-

ces most of the technical work on their investigations to computer security experts. *See* Joel Deane, Digital Dragnet: The Hacking Crackdown, at The Hacking Crackdown (visited Apr. 26, 1998) http://www.zdnet.com/zdtv/thesite/0597w3/life/life550_051297.

57. Although the Report of the President's Commission on Critical Infrastructure Protection found no current, imminent threat of a successful nation-state or terrorist attack sufficient to affect large portions of the United States or U.S. infrastructure, it nonetheless recommended initiation of a series of measures to promote the development of a protective environment. *See* The President's Commission on Critical Infrastructure, Critical Foundations: Protecting America's Infrastructure (1997). In response, President Clinton issued Presidential Decision Directive 63, on May 22, 1998. *See* The White House, Office of the Press Secretary, *Fact Sheet: Summary of Presidential Decision Directives 62 and 63* (May 22, 1998)http://http://www.pub.whitehouse.gov/uri-res/I2R?urn:pdi://oma.eop.gov.us/1998/5/22/ 6.text.1

58. Such as when lawyers try a case *pro hac vice.*

59. These jurisdictional complications have placed much of the responsibility for investigating incidents on the federal government, which is currently struggling with jurisdictional issues relating to national boundaries. *See* discussion of international implications *infra* Part V.E.

60. In fact, new technologies may require some new thinking. A recent bill on the use of digital signatures proposed a certification scheme very similar to state licensing schemes, but to be implemented at a national level. *See* H.R. 2937, 105th Cong. (1997).

61. *See, e.g.,* 7 C.F.R. 868 (1997) (food inspectors); 10 C.F.R. 55 (1997) (NRC operator licenses); 10 C.F.R. 71 (1997) (packaging and transportation of radioactive material); 10 C.F.R. 95 (1997) (nuclear-facility security officer licenses); 14 C.F.R. 61 (1997) (pilot licenses); 46 C.F.R. 10 (1997) (licensing of maritime personnel).

62. *See supra* note 38 and accompanying text.

63. Though private individuals are not subject to the Fourth Amendment and other laws restricting law enforcement activities, this does not mean that they are free to break into and search homes, offices, or computer systems. Rather than being subject to the procedural restrictions on law enforcement, private individuals are subjected to criminal and/or civil liability for breaking and entering, theft, assault, kidnaping or false imprisonment, battery, and other actions that are outside legal boundaries. Under some state licensing schemes for private investigators, committing such acts may be grounds for losing one's license or other disciplinary action. *See, e.g.,* Ind. Code 25-30-1-18 (1993); Nev. Rev. Stat. 648.150 (1997); N.C.

Gen. Stat. 74c-10 to -12 (1989); Okla. Stat. tit. 59, 1750.7 (1989); S.C. Code Ann. 40-17-140 (Law. Co-op. Supp. 1997).

64. First, it bears noting that successful investigation of even high-tech crimes is often dependent on sound conventional investigative practices. Second, under existing law, there is still room for a private investigator to operate, although she is required to proceed cautiously, and frequently only with the consent of the parties involved.

We do not intend to suggest that there is no room for improvement in the current legal structure that governs networked environments. The laws that currently apply to computer networks are laws that were originally intended to apply only to telephone conversations. Over time they have been expanded in the breadth of their coverage to include computer networks, but the statutes themselves have not been substantially revised to account for differences in public expectations when communicating in networked environments. The evolution of this legal framework has not yet taken a turn to address more adequately and contend with the unique aspects of the Internet and other computer networks. This does lend a degree of ambiguity to the determination of what is legal in such an environment and what is not. This should not be seen as a barrier to erecting an oversight structure for computer security experts, but rather, such an oversight structure should be seen as an opportunity to create a context and an impetus for the reexamination of the laws in light of their new applications and possible outcomes.

65. For example, in the often-cited "Rome Labs" case, intruders into Air Force computers at Rome Labs in New York used the Air Force system as a vehicle by which to access a system called the "Korean Atomic Research Institute." The intruders copied material back to the Rome Labs system. U.S. officials had considerable cause for concern; they were not sure whether the Atomic Research Institute belonged to North or South Korea and did want the apparent intrusion by a U.S. Air Force computer to be considered an attack on the North Koreans. *See Air Force Investigative Office Deemed Incompetent During Rome Labs "Info-War" Break In*, Crypt Newsl., Jan. 1998 http://sun.soci.niu.edu/~crypt/other/crpt46.htm>.

66. "Clearly these problems and issues cannot be solved unilaterally by law enforcement, no more than they could be solved unilaterally by the private sector. If we are to identify and respond to these various problems, we've got to unite the efforts of industry and law enforcement on an international scale." Louis J. Freeh, Director of the FBI, Speech at the 1997 International Computer Crime Conference (Mar. 4, 1997), *available at* http://www.fbi.gov/dirspch/compcrim.htm>.

CHAPTER 4

THE FEDERAL LEGAL LANDSCAPE

A "LEGAL FOUNDATIONS" STUDY

Report to the
President's Commission
on Critical Infrastructure Protection
1997

This report was submitted to the President's Commission on Critical Infrastructure Protection, and informed its deliberations and recommendations. This report represents the opinions and conclusions solely of its developers.

CONTENTS

INTRODUCTION

This report represents the first effort of its kind to characterize the legal and regulatory structures of the critical infrastructures as they pertain to infrastructure assurance. As a first step, surveys were sent to the ten agencies represented on the Commission: the Departments of Commerce, Defense, Energy, Justice, Transportation, and the Treasury, The Central Intelligence Agency, the Federal Bureau of Investigation, the Federal Emergency Management Agency, and the National Security Agency.[1] The surveys asked agency counsel to identify the legal authorities and mechanisms associated with their agency that related to infrastructure assurance. The survey also asked that existing or planned programs related to infrastructure assurance and other agencies with a stake in the activities be identified. After the initial agency submissions were received, it was apparent that the federal government's authorities with respect to infrastructure assurance were spread across a vast number of agencies and departments. Assistance was then requested from another twenty-five federal organizations.

The first part of this report provides a high-level overview of the legal landscape of each of the critical infrastructures. Though this effort does incorporate some additional sources,[2] it relies primarily on the submissions received thus focusing on the federal government's legal and regulatory authorities. The report then presents the results of the two-stage agency survey process. It proceeds agency-by-agency and, in some cases, office-by-office and describes each of the relevant body's authority and jurisdiction and its current mechanisms and programs relating to infrastructure assurance. It also attempts to identify authorities or mechanisms and, in some instances, to merely spot infrastructure assurance-related issues. The efforts to characterize the legal landscape of infrastructure assurance has lead to the identification of many legal and policy issues and will help drive the implementation process for the Commission's recommendations.

---

1 Exec. Order 13010 (July 15, 1996); 61 Fed. Reg. 37347, 37348 (July 17, 1996).

2 The PCCIP Legal Authorities Database includes a broader range of materials including relevant authorities at federal, state, and international levels as well as materials from the private sector.

PART ONE
THE CRITICAL INFRASTRUCTURES

Overview

A number of difficulties arise in any effort to characterize the legal landscape of infrastructure assurance. The first and most fundamental challenge arises with regard to identifying and collecting infrastructure assurance-related authorities. Typically an attorney confronted with a research project will be able to recognize the issue and identify relevant bodies of law. Research then may consist merely of typing a word or phrase into the search engine of a legal database, or even the Internet. However, for infrastructure assurance there is no existing body of law-no keyword under which to search. The task thus requires an upfront recognition of the types of issues which may impact, positively or negatively, the fulfillment of certain infrastructure assurance objectives.

With eight infrastructures named in the Executive Order and the disparate subparts contained within each of those critical infrastructures, the next challenge is in many ways one of sheer volume. Massive amounts of law have been promulgated and implemented under the auspices of the critical infrastructures. None of these laws were specifically directed at "infrastructure assurance," but many impact directly on infrastructure assurance concerns. Physical security, information security, reliability and safety of services, emergency authorities, law enforcement, intelligence and defense activities are just a few of the legal considerations which should be taken into account in compiling the legal landscape.

In other areas, the law and thus the structure and assurance level of the infrastructures are in flux. Recent deregulation and restructuring efforts in such industries as telecommunications and electric power may impact these infrastructures' abilities to address infrastructure assurance concerns. It is difficult, though, to analyze the impacts of such changes on infrastructure assurance and the legal landscape until the authorities and mechanisms have been more fully implemented.

This preliminary effort should make clearer the areas where federal involvement is historically and legally precedented and those areas which are better left to the states, associations, and individual owners and operators. It may be necessary to examine these roles, particularly as they evolve in this area of increased competition, deregulation and restructuring, in the future. The current legal landscape of infrastructure assurance appears to provide a rich body of resources tapped and untapped for meeting infrastructure assurance objectives.

Information and Communications Infrastructure

To the extent any of the critical infrastructures are owned by the federal government, full authority exists to protect them from potential threats and vulnerabilities. If the infrastructure is privately owned, a federal role is generally premised on the infrastructures involvement in interstate commerce.

This is true in the area of telecommunications. For example on the government side, the GSA administers the federal telephone system under the FTS 2000 program.[3] NCS ensures the availability of telecommunications service during emergencies through GETS. NSA is charged with the security of the government telecommunications services.[4] Programs are in place for other means of communications as well as, such as NTIA's role in spectrum allocation and use of satellites for the government[5] and BLM's emergency radio capabilities.[6] Other federal government agencies also play more narrow and defined roles, such as the U.S. Secret Service's role in investigating access device fraud.[7]

The Federal Legal Landscape: *Information & Communications*

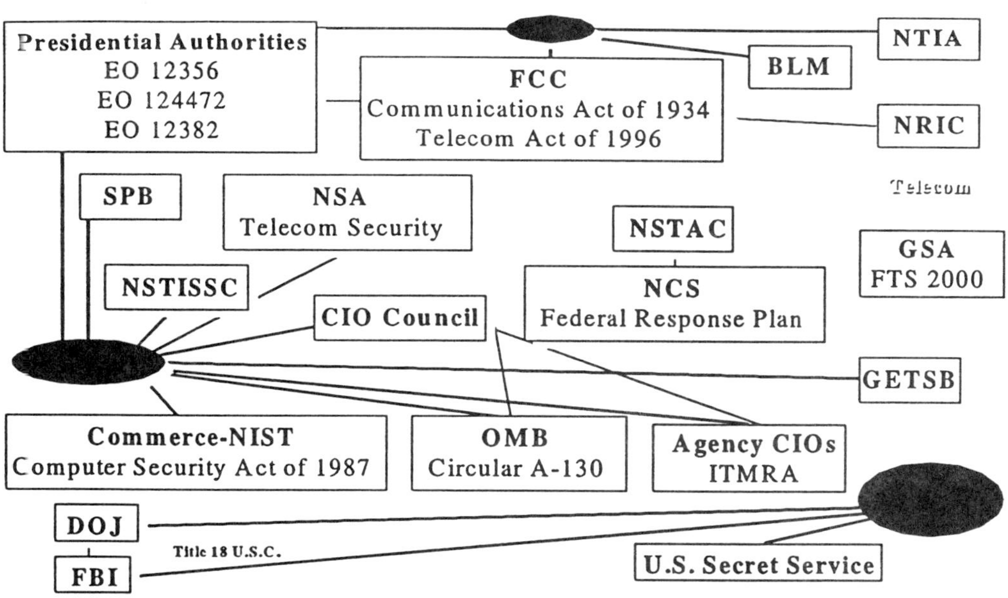

---

3 *See infra* p. 37.

4 *See* discussion of NSA authorities *infra* pp. 31-32.

5 *See infra* p. 25.

6 *See infra* p. 45.

7 *See infra* p. 35.

Jurisdiction over privately-owned communications media is given to the FCC based on the interstate commerce implications of such communications. The exact nature of the FCC's jurisdiction does vary depending on the communications media in question.[8] However, the FCC's general mandate is to ensure that safe and reliable communications services are widely available. To accomplish this mission, the FCC has regulatory authority to issue licenses, make appropriate regulations, issue fines and other penalties. The FCC uses many of these authorities to ensure the safety and reliability of communications, however there may be room for a greater focus on infrastructure assurance objectives. In addition, state Public Utility Commissions (PUCs) generally regulate facility and equipment siting and rate setting. While the focus of both the FCC, at the federal level, and the PUCs, at the state level, have traditionally been on providing universal service, these bodies are now moving their focus toward access and continuity of service. For example, the FCC chartered Network Reliability and Interoperability Council currently has reporting requirements in place for outages, but is considering whether the thresholds for reporting should be changed in light of industry restructuring and security and reliability concerns. Also, the impact that some of the provisions of the Telecommunications Act of 1996, such as new obligations for interconnection, unbundling of services, and resale of infrastructure access on existing providers of local exchange services, will have on infrastructure objectives is not yet clear and may depend on the manner of FCC implementation, a process that is still underway. In the meantime, new technologies are being introduced in the communications arena everyday.

For protection of information systems specifically, much the same pattern is found. Government systems have a wide and diverse group of federal entities charged with some aspect of their protection and proper functioning. NIST has the primary responsibility for setting standards for information security,[9] but other entities such as NSA, SPB, CSSPAB, CIO Council, and GITSB, provide inputs and play a role in federal information security endeavors. Individual agencies, through their CIOs, also bear considerable responsibility for their own systems. Some agencies have risen to the challenge more readily than others. FCC, for example, has its own computer emergency response capability and conducts "red teaming" activities.[10]

In contrast, there is no federal government focal point for privately-owned critical infrastructure information security. NIST and NSA may provide technical expertise, but any regulatory requirements are derived from the overall regulatory scheme of the oversight agency. FERC, for example,

---

8 *See* discussion of FCC authorities *infra* pp. 38-43.

9 *See infra* p. 27.

10 *See* discussion of FCC Computer Security Program *infra* pp. 42-43.

requires interstate transmitters of energy to link to the OASIS system and has put in place requirements for the security of that system.[11] Few agencies have mandated such requirements, though they most likely have the statutory authority to do so in order to protect the safe and proper operation of the critical infrastructures they regulate.

Whether government or private communications systems, two needs are critical to infrastructure assurance objectives—law enforcement and education and awareness. The Secret Service,[12] to a limited extent, and the Department of Justice[13] both have jurisdiction over crimes in the communications infrastructure. Education efforts for proper use of computers and computer ethics may be funded for the private sector through Departlment of Education programs.[14] OPM and individual agencies are generally responsible for training government employees on such topics.

Physical Distribution

The transportation infrastructure's legal landscape is particularly difficult to characterize because of the unique nature of each of the various modes of transportation involved. Rather than a centralized and cohesive structure, the legal landscape is fragmented. The Department of Transportation is itself broken into a number of individual agencies to address each of the modes of transport.[15] These include the Federal Railroad Administration, the Federal Aviation Administration, the Federal Highway Administration, the Federal Transit Administration, the Surface Transportation Board, the U.S. Coast Guard, and the Maritime Administration. More specialized federal bodies also exist to address specific elements of the transportation infrastructure, such as the St. Lawerence Seaway. This mode-specific approach to governance is echoed in the state structures which regulate transportation.

---

11 For further discussion of OASIS system and FERC regulations *see infra* p. 55.

12 *See infra* p. 34.

13 *See* FBI summary *infra* pp. 15-16.

14 *See infra* pp. 35-36

15 *See infra* pp. 19-22 for further discussion of DOT authorities.

The Federal Legal Landscape: *Physical Distribution*

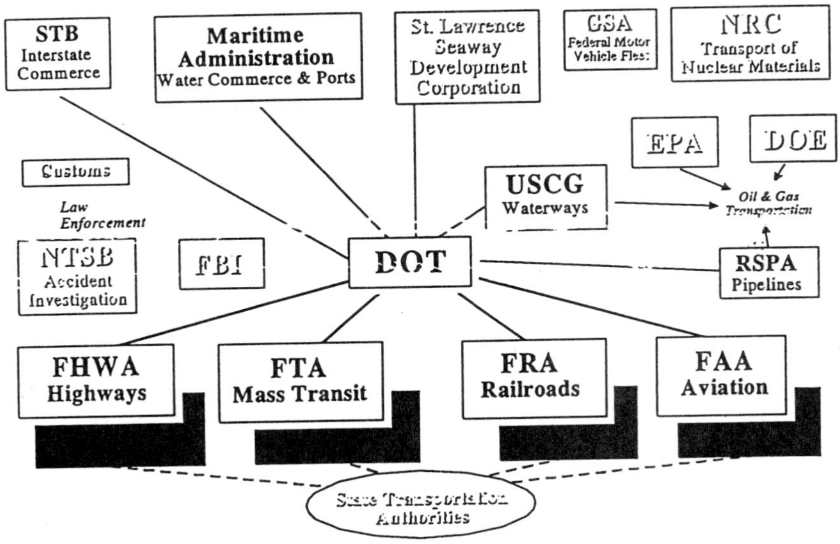

The nature of these agencies' authorities over the modes of transportation they concern also differ tremendously. While air transportation is more heavily regulated by the FAA, the Federal Highway Administration and Federal Transit Authority play a more influential role through grant programs and studies aimed at safety. This may be due to the *intra*state nature of many of these modes of transportation, necessitating a more state-centric approach to oversight. States and local governments bear tremendous responsibility for the transportation infrastructure, much of it being owned by those government entities. Highways and roads, bridges, even airports are usually owned and maintained by municipal or state authorities. However, to the extent transportation involves a specific federally-regulated material, such as nuclear materials, a strong federal role will be defined for the agency primarily responsible for such items—in this case the Nuclear Regulatory Commission.[16]

An effective legal strategy for infrastructure assurance may need to be tailored to each individual mode of transport. Authorities to conduct inspections, require training, and security measures may exist at the federal or state level for any of the elements of the transportation infrastructure. There may also be room for improvement in their current implementation. For example, there are no security regulations for rail passenger terminals similar to those for airports.[17] In addition, the impact of information tech-

---

16 *See infra* pp. 55-52.

17 *See* Argonne National Laboratory, *Overview of the Impacts of Regulatory Agency Practices on Critical Infrastructure Protection*, Prepared for PCCIP § 7 (July 15, 1997).

nology and automation will have to be carefully studied to determine the proper use of legal authority for infrastructure assurance objectives. Electronic systems already control air traffic and rail traffic, and are quickly moving into urban traffic control centers.[18]

Banking and Finance

Despite the complex web of financial institutions and other entities involved in the banking and finance infrastructure, federal regulation and control of the infrastructure is fairly tightly centered and maintained. The Federal Reserve[19] and FDIC[20] regulate many aspects of the operations of the banks they oversee. These organizations have already had to respond to important infrastructure-related concerns through their regulatory structure with the increasing popularity of electronic commerce. The Federal Financial Institutions Examinations Council is working to standardize the practices among the federal regulators regarding electronic fund transfers.[21]

The Federal Legal Landscape: *Banking and Finance*

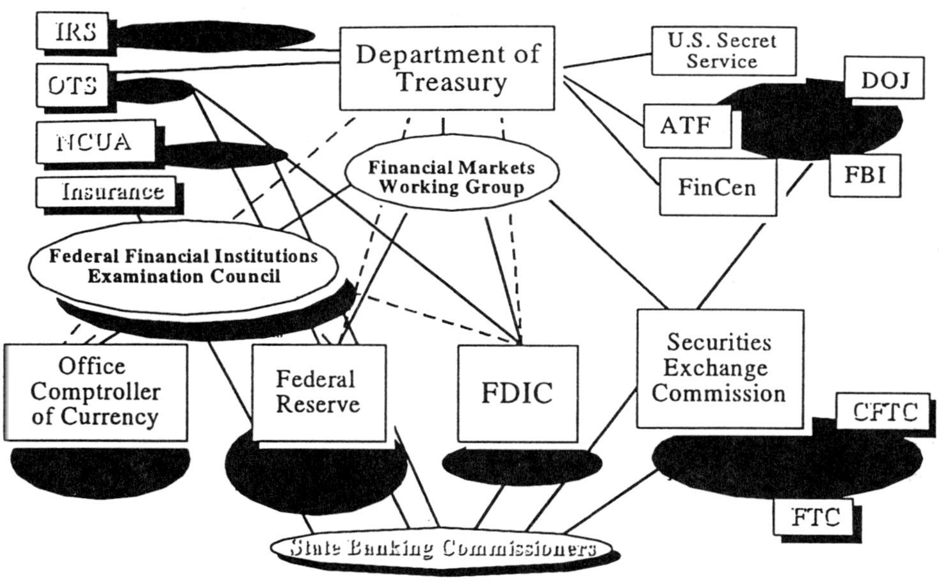

18 *Id.*

19 *See infra* pp. 57-58.

20 *See infra* pp. 58-59.

21 *See infra* p. 59.

The SEC also has the necessary authority to respond to incidents disrupting the trading of securities, though it has not yet been forced to invoke such authority.[22] The department of Treasury has attempted to play an integrating role for its many constituents—OCC, FDIC, IRS—through such efforts as the Financial Markets Working Group which is designed to share information and to find solutions to various problems in the banking and finance area.[23] This group has lately begun to focus on infrastructure assurance-related topics such as disruptions. The Secret Service has jurisdiction over crimes involving currency and other financial instruments.[24] The FBI and Department of Justice also have jurisdiction over federal crimes in the banking and finance infrastructure.[25]

Energy

As was demonstrated in the telecommunications arena, federal authority differs considerably from government to privately owned systems. DOE has the statutory authority to require government energy facilities to meet any requirements necessary to meet infrastructure assurance objectives.[26] While these currently involve some physical security requirements, it is not otherwise clear to what extent such objectives are currently taken into account.

Both DOE and FERC have significant authority over interstate entrgy facilities even if privately owned.[27] State PUCs have jurisdiction over retail sale of electricity. This includes the authority to issue permits and set standards for construction. This authority differs depending on the energy source. In fact, nuclear energy is regulated at the federal level by the NRC almost exclusively.[28]

---

22 *See infra* pp. 56-57 for discussion of SEC authorities.

23 *See infra*p. 28.

24 *See infra*pp. 34-35 for discussion of Secret Service jurisdiction.

25 *See infra* pp. 15-16.

26 Government-owned energy facilities are discussed *infra* pp. 16-17.

27 *See infra* pp. 16-19 (discussion of DOE) authorities) and pp. 53-56 (discussion of FERC authorities.

28 *See* discussion of NRC authorities infra pp. 51-52.

The Federal Legal Landscape: *Energy*

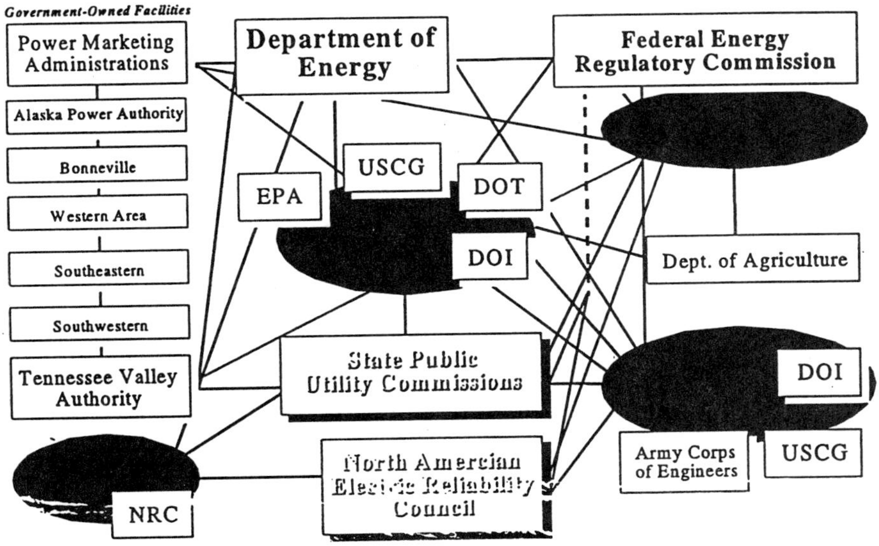

Generally, federal energy regulators have licensing authority, ability to impose reporting requirements and to conduct investigations, and to make regulations for operation of energy facilities. The DPA provides powers to assist in reconstitution following an emergency.[29]And specific powers such as ordering interconnection or expanded service and tariff review could be used for infrastructure assurance objectives.[30] While sufficient authority is in place, the federal government has not yet required substantial physical, or for that matter cyber, security measures for non-nuclear energy facilities.

Pipelines pose more complex legal questions than do other types of energy-related infrastructures. FERC has authority over the interstate transportation and sale of gas and oil.[31] The Department of Transportation is responsible for the safety and construction standards for pipelines, including those which transport oil and gas.[32] The EPA also has a role in the transport and storage of oil and gas. Before infrastructure assurance objectives related to pipelines are implemented, the roles of these agencies should be clarified.

---

29 *See, e.g.*, discussion *infra* p. 17.

30 The potential uses for such regulatory authorities are discussed within the FERC summary. *See infra* pp. 53-56.

31 *See infra* p. 55.

32 *See infra* p. 19.

There is sufficient authority at the federal level to enhance infrastructure assurance within the energy infrastructure. However, as the industry continues its move toward competition in electric generation, careful consideration must be given to using those authorities to ensure that infrastructure assurance objectives are met. In addition to DOE, FERC and NRC, unique roles are also played by the Department of Agriculture in development of universal service in rural areas,[33] BLM in the protection of utilities on public lands and leasing of lands for production of oil and gas,[34] state public utility commissions, and associations such as the North American Electric Reliability Council (NERC). NERC, a voluntary association of utilities formed in 1968, is becoming an increasingly important presence in the energy industry.[35] NERC just recently voted for its reliability standards to be obligatory for its members. The roles of the state PUCs will also be important to follow as the infrastructure undergoes this change. Due to this fragmentation of the jurisdiction over the energy infrastructure; all of these legal and regulatory entities should be taken into account when evaluating the legal landscape.

Vital Human Services (Including Emergency Services and Continuity of Government Services)

Vital Human Services refers to a wide range of activities at federal, state and local levels that are both crucial for everyday life and needed in the event of an emergency. The responsibilities for provision of services related to police, fire, and medical care belong principally to local and state governments. Under unique circumstances, such as a major disaster that strains state response capabilities, the Federal government has the authority to respond under the Stafford Act.[36] Emergency functions are administered by FEMA, but involve many of the federal agencies under a management structure set out in the Federal Response Plan. Other avenues, such as the Nunn-Lugar-Domenici Weapons of Mass Destruction legislation of 1996, provide training or access to equipment for first responders.

---

33 *See* discussion of Department of Agriculture programs *infra* p.56.

34 *See infra* p. 47.

35 NERC is discussed extensively in Argonne National Labs' Regulatory Landscape Report to the President's Commission on Critical Infrastructure protection. *See supra* note 16 § 4.

36 *See infra* pp. 23-24.

The Federal Legal Landscape: *Vital Human Services*

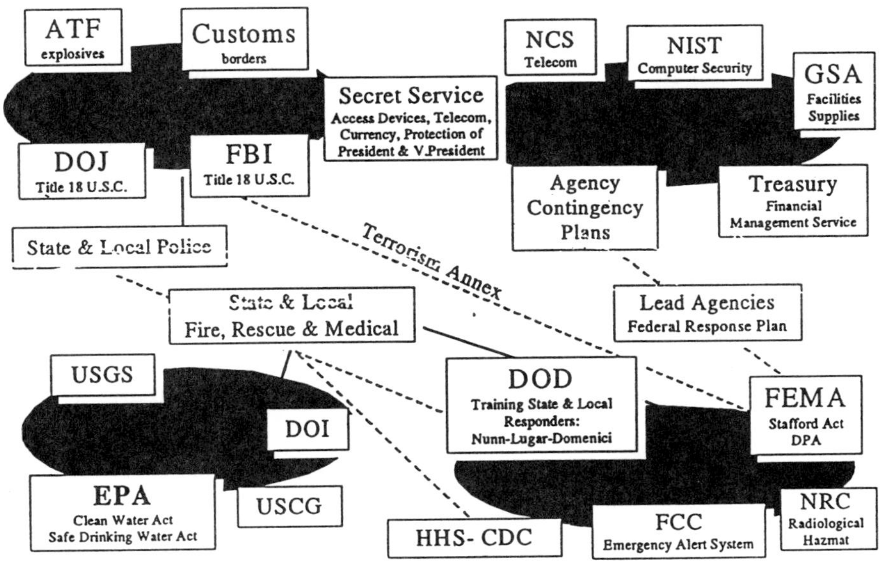

The government has planned for continuity of its own communications capabilities through GSA, NCS and other avenues for the provision of telecommunications service during an emergency.[37] The Federal Response Plan and other documents divide responsibilities during emergencies among the federal agencies. It is not currently clear whether the federal government has adequately taken into account interdependencies among the critical infrastructures in its own planning for emergencies.

Individual agencies, in addition to GSA, have taken measures in the wake of the Oklahoma City bombing to increase physical security of their facilities and to prepare for emergencies. Some agencies have prepared evacuation plans and plans for reconstitution after an incident.[38] Disparities between agencies are similar to those in the area of information security, where some are well prepared to handle a contingency and others have not taken the need for security precautions and advanced planning as seriously. As provision of government services moves increasingly "on-line," agencies that provide such services will undoubtedly feel pressure to ensure the integrity and reliability of those services from potential cyber-related disruptions. However, no government entity has clear and

37 *See infra* p. 37.

38 *See, e.g.*, summary of U.S. Postal Service authorities and mechanisms *infra* p. 44.

authoritative enforcement authority over the implementation of such requirements.

If a critical infrastructure incident reaches a level where it impacts national interests, several federal mechanisms and authorities are applicable. Cross-infrastructure services such as law enforcement and intelligence are important parts of averting national security level incidents pertaining to infrastructures. The FBI[39] and CIA[40] have adequate authority to include the critical infrastructures within the ambit of their normal investigatory and intelligence related functions. In fact, the FBI has taken several steps to improve its awareness and ability to respond to critical infrastructure incidents such as the Key Asset Program and the CITAC which houses the Infrastructure Protection Task Force.[41] The Department of State, Customs. Federal Emergency Management Agency, Secret Service, the prosecutors of the Department of Justice, and others may also be involved in an infrastructure-related incident.

The federal government plays only a limited role in protection of the drinking water supply. Local Governments and state PUCs are the primary authority for protection of the drinking water supply. However, the EPA has a strong role related to keeping drinking water safe and clean.[42] The EPA sets standards adopted and followed by most local water infrastructure operators. These standards generally address issues such as vulnerability to industrial or natural conditions, but do not address intentional sabotage of drinking water quality. Monitoring of water quality will likely continue to expand under the existing legal regime. The U.S. Geological Survey also runs a National Water Quality Assessment program.[43] The Bureau of Reclamation also is involved in water operations though usually only at the onset of development of a processing operation.[44]

Cross-Infrastructure Issues

Many of the legal authorities collected during the survey process do not fit neatly into one of the critical infrastructures, rather they cut across all of the infrastructures in special ways. Law enforcement and intelligence functions are two such examples which are well documented within the agency submissions.[45] However, there are other examples that pervade the way all

---

39 *See* discussion of FBI authorities *infra* pp. 15-16.

40 *See* discussion of CIA authorities *infra* pp. 14-15.

41 *See infra* pp. 15-16 for further discussion of these FBI programs.

42 *See supra* note 16 at 97 (discussion of EPA role in drinking water infrastructure).

43 *See infra* p. 46.

44 *See infra* pp. 48-49.

45 *See, e.g., infra* pp. 14-15 (CIA); pp. 15-16 (FBI); pp. 33-34 (U.S. Secret Service); pp. 31-32 (Customs Service); pp. 32-33 (ATF).

of the federal agencies surveyed are carrying out their missions, but were not included within the agency submissions. Recent "Reinventing Government" initiatives, such as the Government Performance and Results Act,[46] the Information Technology Management Reform Act,[47] and other authorities that seek to include risk management techniques in the regulatory structure, set out a new way of conducting government business.[48] These acts set the tone for future legal and regulatory efforts related to infrastructure assurance.

PART TWO

SUMMARY OF AGENCY AUTHORITIES

Central Intelligence Agency

The CIA has the authority to undertake intelligence and counterintelligence gathering and dissemination activities involving any of the critical infrastructures to the extent that there is foreign intelligence involving those infrastructures.[49] If the investigation or gathering of such intelligence and counterintelligence is to take place within the United States, the CIA efforts must be coordinated with the Federal Bureau of Investigation.[50] The CIA does not have police, subpoena, law enforcement or internal security powers.[51] However, a recent amendment to the National Security Act allows the Intelligence Community to collect information abroad about non-U.S. per-

---

46 Pub. Law No. 103-62 (1993).

47 Pub. Law No. 104-106 (1996).

48 *See e.g.*, Pub. Law No. 104-304 (1996) (Accountable Pipelines Safety and Partnership Act of 1996); *see also supra* note 16 at 54-55.

49 The National Security Act of 1947, 50 U.S.C. §§ 401 *et seq.*, and the Central Intelligence Agency Act of 1949, 50 U.S.C. §§ 403a *et seq.*, define the CIA's authority with regard to collection of intelligence data. Executive Order 12333 (1981), however, contains the specific scope of the intelligence activities in which the CIA may engage and limits those activities to the gathering of "foreign intelligence" and "counterintelligence." "Foreign intelligence" is specifically defined in the Executive Order as "information relating to the capabilities, intentions and activities of foreign powers, organizations or persons, but not including counterintelligence except for information on international terrorist activities." Exec. Order 12333 § 3.4(d). The Executive Order also defines counterintelligence as "information gathered and activities conducted to protect against espionage, other intelligence activities, sabotage, or assassinations conducted for or on behalf of foreign powers, organizations or person, or international terrorist activities, but not including personnel, physical, document or communications security programs." Exec. Order 12333 § 3.4(a).

50 Exec. Order 12333 § 1.8(a) & (c). *See also infra* pp. 15-16 (Federal Bureau of Investigation Summary). Similarly, the FBI, which is tasked with collecting counterintelligence primarily within the United States, must coordinate counterintelligence gathering outside the U.S. with the CIA. Exec. Order 12333 § 1.14(a) & (b).

51 50 U.S.C. § 403-3(d)(1).

sons in order to support a U.S. law enforcement or counterintelligence investigation upon request of a U.S. law enforcement agency.[52]

Actual mechanisms in place relating to the various infrastructures are limited to the protection and reconstitution of CIA's own telecommunications, power, emergency services, and critical government operations. The CIA may engage in covert actions to "influence political, economic, or military conditions abroad" with the approval of the President.[53] The Director of the Central Intelligence (DCI) plays a pivotal role in intelligence community, sitting on the Committee on Transnational Threats,[54] approving assignment of funds under the National Foreign Intelligence Program,[55] and coordinating relationships with foreign governments.[56] In addition, the DCI reports to Congress on potential threats to the national information infrastructure from foreign powers and, semi-annually reports to Congress on foreign nations' acquisition of technology relating to weapons of mass destruction and advanced conventional munitions.[57]

Federal Bureau of Investigation

The Federal Bureau of Investigation (FBI) has been delegated the authority to investigate all crimes against the United States.[58] In fulfilling its statutory mandate, the Bureau is tasked with acting as the lead agency in all crimes for which it has concurrent or primary jurisdiction and which involve terrorist activities.[59] In addition, FBI is to take charge of investigations involving espionage, sabotage, subversive activities and related matters. The only substantial limits on the scope of the FBI's investigatory authority over violations of federal law come from specific grants of investigatory power to other investigative agencies.[60] Among such provisions is Executive Order 12333 which requires that the FBI conduct counterintelligence outside the United States in coordination with the CIA.[61] The FBI suggests that 28 U.S.C. § 533 is sufficiently broad that it may be extended beyond federal crimes in that it contains a provision to deal with "unanticipated exigent circumstances."[62] In combination, the various federal statutes addressing

---

52 50 U.S.C. § 403 (1996).

53 50 U.S.C. § 413b(a) & (e).

54 50 U.S.C. § 402(I)(1) (1996).

55 50 U.S.C. § 403-3(c).

56 50 U.S.C. § 403-4(e).

57 H.R. 3259 §§ 310 & 711.

58 28 U.S.C. § 533 vests the Attorney General with the power to appoint officials to "detect and prosecute crimes against the United States." The Attorney General has delegated the investigatory authority to the FBI. 28 C.F.R. Part 0.85(a).

59 28 C.F.R. Part 0.85 defines "terrorist activities" as "unlawful use of force and violence against persons or property to intimidate or coerce a government, the civilian population, or any segment thereof, in furtherance of political or social objectives."

60 *See* 28 C.F.R. part 0.85(a).

61 Exec. Order 12333 § 1.14(b).

protection of infrastructures[63] and the authorization to conduct counterintelligence activities, which specifically include protection of certain infrastructures,[64] provide a broad basis for FBI involvement in infrastructure assurance.

The broad nature of the FBI's authority is evident from the variety of mechanisms currently in place within the Bureau aimed at infrastructure assurance. The Strategic Intelligence Operations Center allows the FBI to monitor incidents and coordinate responses on a national and potentially worldwide level. The FBI has developed plans to deal with unconventional crisis situations including a Nuclear Terrorism Response Plan, a Chemical/Biological Incident Contingency Plan, the Critical Incident Response Group, and the Hostage Rescue Team. The Computer Intrusion and Infrastructure Threat Analysis Center was established to handle computer/communications related infrastructure assurance issues, but its role has been expanded to develop the expertise necessary to deal with any infrastructure threat. In addition to these programs, the FBI has developed an Infrastructure Vulnerability/Key Asset Program as a counter terrorism mechanism. Unfortunately, this program has not reached its full potential and currently has an out-of-date database on the key assets and an insufficient warning system to alert the general public.[65]

Department of Energy

Electrical Power Systems

The Department of Energy (DOE) has statutory authority to implement measures for protection of transmission and related facilities, the minimization of consequences of attack, and reconstitution following attack for DOE-owned facilities.[66] This authority is vested in five different Power Marketing Administrations (PWAs)—Alaska Power Authority, Bonneville Power Administration Southeastern Power Administration and Western

---

62 *See* FBI Authorities for Critical Infrastructure Incident Response 1, fn. 1 (October 11, 1996).

63 Examples of such statutes include: 18 U.S.C. § 1030(a)(5)(A) (computer fraud and related activities); 18 U.S.C. § 2511 (interception of communications); 18 U.S.C. §§ 793-94 (espionage); 18 U.S.C. § 641 (theft of government property); 18 U.S.C. § 2153 (sabotage).

64 The National Security Threat List (1995) includes foreign power sponsored or coordinated intelligence activities directed at denial of communications services, unauthorized monitoring of communications systems, unauthorized disclosure of information on a communications system, etc.

65 *See* FBI Authorities for Critical Infrastructure Incident Response at 6-7.

66 This authority is held by various Power Marketing Administrations (PWAs) and is contained in their respective enabling acts. *See* 16 U.S.C. § 832 *et seq*; 16 U.S.C. § 838; 16 U.S.C. § 837-837h; 16 U.S.C. § 839; 43 U.S.C. § 620 *et seq.*; 16 U.S.C. § 837g-1; 16 U.S.C. § 825s.

Area Power Administration, Southwestern Power Administration which developed the mechanisms for using their infrastructure assurance authority on an individual basis. In addition, the Defense Production Act of 1950[67] may be useful in the case of an energy emergency. It allows the President to require performance of contracts on a priority basis and to allocate materials and resources as "necessary or appropriate to promote the national defense."[68] 50 U.S.C. app. § 2071(c) authorizes the President to make the prioritization of contracts or allocation of resources in order to maximize domestic energy supplies. Features of the DPA, such as Voluntary Agreement Authority and the DPA loan and grant fund, could also be used for infrastructure assurance.

While DOE has no specific authority for infrastructure assurance over non-Federal facilities, there are two statutes which potentially could be used for such purposes. The Public Utilities Regulatory Policies Act (PURPA) gives DOE authority to gather information and make recommendations to industry regarding electric energy transmission system reliability.[69] Executive Order 10485 allows conditions to be attached, "as the public interest may require," to permits for electrical power transmission across the U.S. border.[70] Statutory vehicles available for the reconstitution of electrical power systems include the Defense Production Act of 1950 and the Federal Power Act.[71] The latter authorizes DOE to order temporary interconnections, generation and transmission of electric energy in emergency situations.

DOE also has authority with respect to nuclear energy facilities. The Atomic Energy Act of 1954 (AEA)[72] provides the general authority for maintenance and protection of nuclear facilities owned by the United States and under DOE control. The AEA allows the DOE to promulgate regulations and orders to: protect real and personal property, restrict data, guard against loss or diversion of special nuclear material, and govern design, location, and operations related to such facilities. Included among these powers is a limited arrest authority and authorization to allow certain employees to carry firearms. The AEA also authorizes the President to use the powers of any government agency as necessary to protect against unlawful dissemination of restricted data, and to safeguard DOE facilities, equipment, materials and other property. DOE nuclear facilities are also protected under the AEA from unauthorized photography, mapping or drawings. The Presi-

---

67 50 U.S.C. app. § 2061 *et seq.*

68 50 U.S.C. app. § 2071(a).

69 16 U.S.C. § 824-2(b).

70 Exec. Order 10485 § 1(a)(3).

71 16 U.S.C. § 791S *et seq.*

72 42 U.S.C. § 2011 *et seq.*

dent's powers in emergency situations under the Defense Production Act extend to nuclear facilities, both under DOE control and private, and may be used to reestablish vital capabilities. DOE has no further authority over non-federal nuclear facilities to require infrastructure assurance measures, but the Nuclear Regulatory Commission does have such authority.

Other government entities with a stake in electrical power systems protection include: The Federal Energy Regulatory Commission, the Tennessee Valley Authority, the Department of the Interior, and the U.S. Army Corps of Engineers.

Gas and Oil Storage and Transportation

With respect to DOE-owned facilities, the Energy Policy and Conservation Act[73] and the Department of Energy Organization Act[74] authorize the Secretary of Energy to establish, operate, and protect the physical security of the Strategic Petroleum Reserve (SPR) oil storage and transportation systems in Louisiana and Texas. The authority to protect the SPR includes the power to take action to deter trespass at those facilities.[75] The Secretary of Energy also has authority to operate and protect the Naval Petroleum and Oil Shale Reserve's oil and gas production systems in California, Utah and Wyoming.[76] With respect to reestablishing vital capabilities related to gas and oil storage and transportation, the Defense Production Act of 1950 may be used in much the same way as with electrical power systems.

DOE has no specific authority to require infrastructure assurance measures of non-federal facilities. However, the Natural Gas Act, under which DOE has the power to grant permits for import and export of natural gas, may have potential uses for infrastructure assurance measures in that it allows the Federal Energy Regulatory Commission (FERC)[77] to grant the permits subject to conditions and modifications that it deems necessary or appropriate.[78] Significant statutory authority is in place for the reestablishing of vital

---

73 42 U.S.C. §§ 6231-6241.

74 42 U.S.C. §§ 7270a-7270b.

75 The statute conferring authority to make regulations prohibiting trespass, 50 U.S.C. § 7270b(a), and the regulations that actually prohibit trespass, 10 C.F.R. § 1048.5, define trespass slightly differently. The Secretary of Energy is given the broad power to issue regulations relating to "the entry upon or carrying, transporting, or otherwise introducing or causing to be introduced any dangerous weapon, explosive, or other dangerous instrument or material likely to produce substantial injury or damage to persons or property." The actual regulation prohibiting trespass addresses "willful unauthorized entry" and "willful unauthorized introduction of weapons or dangerous materials."

76 41 U.S.C. § 251 *et seq.*

77 *See infra* p. 53-56 discussing FERC's authorities.

78 15 U.S.C. § 717b(a).

capabilities of the gas and oil energy infrastructure. The Defense Production Act of 1950, as with electrical systems, gives the President authority to create priorities for contract performance and allocation of resources under appropriate conditions. The Act also facilitates cooperation between gas and oil service companies in the case of emergency by providing a limited antitrust defense for industry participants voluntarily helping the U.S. with emergency preparedness.[79] The same Act also provides the President with the authority to train and employ persons from the private sector to facilitate planning for and responding to emergencies.[80] In the case of an existing or imminent "severe natural gas shortage, endangering the supply of natural gas," the President is authorized to approve purchases of natural gas and to allocate supplies of natural gas in interstate commerce.[81] Under identical circumstances, the President is also authorized to prohibit the burning of natural gas by electrical power plants or major fuel-burning installations.[82] The Powerplant and Industrial Fuel Use Act[83] allows the President control over the allocation and burning of coal during "a severe energy supply interruption."

Department of Transportation

The Office of Intelligence and Security (OIS) operates under the Secretary of Transportation and is charged with assessing intelligence information relating to long-term transportation security, developing policies and strategies for dealing with threats to transportation security, coordinating countermeasures with appropriate Federal departments, and serving as the primary liaison with the intelligence and law enforcement community. The primary law enforcement agency under the Department of Transportation is the Coast Guard, which has jurisdiction over enforcement of criminal laws over the high seas and waters of the United States.

Telecommunications

The Coast Guard and the Federal Highway Administration have internal and/or department-wide computer security measures in place. The Saint Lawrence Seaway Development Corporation (SLSDC) has established an Operations and Contingency Plan concerning security and disaster/backup recovery. The FAA also has an internal security program for its computer systems as required the Computer Security Act of 1987.[84]

---

79 50 U.S.C. app. § 2158.

80 50 U.S.C. app. § 2160.

81 15 U.S.C. §§ 3362-3363.

82 15 U.S.C. § 717z & 16 U.S.C. § 2601 *et seq.*

83 42 U.S.C. § 8301 *et seq.*

84 Pub. Law No. 100-235.

Gas and Oil Storage and Transportation

The Secretary of Transportation has the statutory authority to prescribe minimum safety standards for transportation of gas and hazardous liquids and for operation of pipeline facilities generally.[85] The Research and Special Programs Administration (RSPA) is charged with regulating pipelines and has implemented numerous regulations to that end. RSPA has in place requirements for design and installation of pipe components, operation and maintenance procedures addressing continuing surveillance and emergency response, line patrolling, training, firefighting equipment and security, and damage prevention programs.[86] Similar statutory authority and regulatory requirements exist for liquefied natural gas pipeline facilities as well.[87] The Office of Pipeline Safety (OPS) is charged with enforcing pipeline safety measures.[88] Among the enforcement measures available to OPS are civil penalties, compliance orders, and hazardous facility orders. In addition, 49 U.S.C. § 60123 makes it a crime to knowingly and willfully damage or destroy or attempt to damage or destroy, an interstate transmission or hazardous liquid pipeline facility. OPS coordinates its transportation safety responsibilities with the Department of Energy. Currently, OPS is working to develop a contingency plan to address security and terrorism issues within the DOT, as well as participating in interagency security information meetings.

The Coast Guard also has authority over the transportation of oil and gas (as well as other hazardous liquids) to the extent they are transported over the high seas.[89] Safety regulations include additional requirements for vessels that transport oil or hazardous material in bulk as cargo or cargo residue, including double hull requirements. The Coast Guard has also implemented regulations and requirements regarding the transportation of oil as part of an environmental initiative and oil pollution response.

SLSDC also has an emergency response plan in place for events such as oil spills that affect the St. Lawrence Seaway.

---

85 49 U.S.C. § 60102.

86 *See, e.g.*, 49 C.F.R. Part 192, subpart D; 49 C.F.R. §§ 195.116, 195.254, & 195.258 (design and installation of components); 49 C.F.R. § 192.613; 49 C.F.R. § 192.615; 49 C.F.R. § 195.402 (c)-(d) (surveillance and emergency response); 49 C.F.R. §§ 192.705 & 192.721 (line patrol); 49 C.F.R. § 195.403 (training); 49 C.F.R. §§ 195.430 & 195.436 (firefighting and security); 49 C.F.R. §§ 192.614 & 195.442 (damage prevention).

87 *See* 49 U.S.C. § 60103; 49 C.F.R. Part 193.

88 49 C.F.R. Part 190.

89 *See, e.g.*, 33 U.S.C. § 1225, 1231; 46 U.S.C. § 3306; 46 U.S.C. § 3703; 49 U.S.C. app. §§ 1801-19.

Transportation

The Coast Guard's role in the protection of the transportation infrastructure relates primarily to boating. The Coast Guard Commandant has authority over boating safety and may set and require compliance with manufacturing requirements and standards. Both civil penalties and criminal sanctions are available as enforcement mechanisms. The Coast Guard may engage in surveillance and boarding activities to monitor compliance.

The Federal Highway Administration (FHWA) has a three-part role in transportation infrastructure assurance. FHWA provides federal financial assistance to States for construction and improvement of highways; it administers the highway bridge inspection and rehabilitation program; and the emergency relief program to assist in the repair or reconstruction of Federal-aid highways and certain federal roads. As part of the Emergency Relief Program, FHWA may provide funding for emergency relief after serious damage is caused to highways, roads, and trails that has occurred as the result of a natural disaster or catastrophic failure from any external cause.[90]

The National Highway Traffic Safety Administration (NHTSA) is charged with setting and enforcing safety standards for motor vehicles and making grants to states to help organize highway safety programs in order to reduce deaths, injuries and economic losses resulting from motor vehicle crashes.

While the Federal Transit Administration (FTA) does not currently have any measures in place specifically addressing infrastructure assurance, the statutory grants of authority to the Secretary of Transportation relating to the FTA mission[91] encompass infrastructure assurance measures. 49 U.S.C. § 5321 provides the general authority for the Secretary of Transportation to make grants for crime prevention and security. Grants are also authorized for mass transportation security projects such as increased lighting at transit stations, camera surveillance, and emergency phones.[92] The Violent Crime Control and Law Enforcement Act of 1994 makes available up to ten million dollars in grants for these same purposes.[93]

SLSDC has an emergency response plan in place to deal with damage to tunnels under the canals or vessel accidents in the Seaway.

---

90 23 U.S.C § 125; 23 C.F.R. Part 668

91 The FTA provides financial and planning assistance for the design, building, and operation of rail, bus and paratransit systems.

92 *See* 49 U.S.C. § 5307(d)(1)(J)(i).

93 49 U.S.C. § 13931.

The Maritime Administration (MARAD) conducts a preparedness program for industrial mobilization of all materials related to shipping via voluntary agreements with vessel and equipment owners. During national security emergencies, MARAD is charged under the Defense Production Act with controlling ports. In addition, MARAD maintains the National Defense Reserve Fleet which, when called to duty, operates under the Navy and is an essential part of the national defense waterborne supply network. MARAD is also responsible for development and maintenance of the Merchant Marine at levels sufficient to carry the nation's waterborne commerce as well as foreign commerce, and to act as a military auxiliary during a national emergency. MARAD chairs the Secretary of Transportation's *Ad Hoc* Working Group on Maritime Security Awareness and produces the Maritime Security Report on a quarterly basis. Together with Secretary of Defense, MARAD works to promote and develop ports and transportation facilities in connection with water commerce and to alleviate port congestion.

The Federal Railroad Administration (FRA) has the responsibility for the establishment of safety standards for railroads as well as investigatory authority over rail accidents. The FRA has a two-part role in transportation infrastructure assurance. First, during national emergencies the FRA is responsible for the acquisition of transportation requirements. In doing so, the FRA may draw on the President's prioritization authority under the Defense Production Act. In addition, the FRA has been active in a counter-terrorism effort designed at protecting the rail system from attack. Criminal statutes are already in place which prohibit targeting of organizations engaged in interstate commerce or property of the United States. Through the Railroad Police Rule, implemented under the Crime Control Act of 1990, railroad police officers are authorized to enforce the laws of any state where their employer owns property so long as their enforcement action is for the limited purpose of protecting railroad property, personnel, passengers, or cargo. The FRA has also drafted an anti-terrorism bill to afford the rail system greater protection from terrorist attack.

The Federal Aviation Agency (FAA) is authorized to adopt and enforce reasonable rules, regulations, orders and minimum safety standards for civil aviation. The FAA has also been granted authority over regulation and enforcement of screening procedures of air passengers and cargo for weapons, overall federal responsibility for air piracy, airport security including foreign airports, transportation of hazardous materials, and drug trafficking.

Water Supply Systems

Under the Federal Water Pollution Control Act and the Coast Guard's federal law enforcement activities, DOT may have some responsibility for assuring the safety of the water supply.

Federal Emergency Management Agency (FEMA)

FEMA, unlike many of the other agencies discussed, is not a regulatory agency and relies primarily on grants to elicit compliance with its programs. FEMA's programs are not limited to disaster response, and indeed encompass preparation, mitigation and coordination activities. FEMA's preparedness and mitigation authority and activities may be more beneficial to infrastructure assurance objectives than its disaster relief function.

FEMA is authorized to study and develop emergency preparedness measures to protect life and property, including research and study of the "best methods of treating the effects of hazards," "developing shelter designs," and "developing equipment or facilities and effecting the standardization thereof to meet emergency preparedness requirements."[94] This grant of power is particularly broad when the statutory definitions of "hazard" and "emergency preparedness" are considered. A "hazard" is defined as an emergency or disaster resulting from either a natural disaster or an accidental or human-caused event."[95] "Emergency Preparedness," defined in the same section, includes all activities "to prepare for or minimize the effects of a hazard upon the civilian population, to deal with the immediate emergency conditions which would be created by the hazard, and to effectuate emergency repairs to, or the emergency restoration of, vital utilities and facilities destroyed or damaged by the hazard."[96] Specific measures mentioned in the statute include: development of organizations, operational plans and training and recruitment of personnel; control of traffic; fire, rescue and medical services; monitoring for danger of special weapons; unexploded bomb reconnaissance, and essential repair or restoration of vital facilities.[97] In addition, the powers granted to the Director of FEMA for emergency preparedness measures include coordination among federal agencies, promoting development of interstate compacts for disaster response and development of emergency communications and warning systems.[98]

When an area is declared a disaster by the President, FEMA is authorized by the Stafford Act to make grants to the states of up to 75 percent of the cost of measures designed to substantially reduce the risk of future loss in that area.[99] Additionally, FEMA may take appropriate action to mitigate future hazards when it makes emergency repairs to, or rebuilds, buildings

---

94 42 U.S.C. § 5196(e).

95 42 U.S.C. 5195a(a)(1).

96 *Id.* § 5195a(a)(3).

97 *Id.*

98 *See* 42 U.S.C. § 5196(c), (g) & (h).

99 42 U.S.C. § 5170.

damaged by disaster.[100] FEMA has statutory authority for mitigation of flood[101] *et seq.* and earthquake hazards,[102] and for dam safety.[103]

FEMA is authorized by the President to coordinate, review, and evaluate plans for emergency preparedness by executive agencies.[104] This charge includes such facets of emergency preparedness as warning systems, state emergency preparedness plans, and planning to reduce the consequences of terrorist incidents. With respect to national security preparedness, FEMA is to work with the National Security Council and coordinate activities with federal agencies and state and local governments.[105]

FEMA is also charged with responding to disasters by rendering assistance to state governments and other federal agencies in the following circumstances. FEMA may assist a federal agency during a disaster by making repairs to Federal facilities.[106] In addition, FEMA may render disaster assistance when the President determines that there is an emergency in an area where the United States exercises exclusive or preeminent responsibility.[107]

Department of Commerce

The Department of Commerce has limited, but important, legal authority for protecting critical national infrastructures. Its role in infrastructure assurance, in many instances, is related to the export of products. The Department plays a crucial role in continuity of government services through internal and external regulations regarding information and technology security.

Telecommunications

The National Telecommunications and Information Administration (NTIA) is authorized to assign radio frequencies to federally-owned systems under 47 U.S.C. § 902(b)(2)(A). This power is administered through the Office of Spectrum Management. All government frequencies are collected in a database called the Government Master File (GMF). This database will be duplicated and stored in an additional location in order to

---

100 *Id.* § 5171(c).

101 42 U.S.C. § 4001

102 42 U.S.C. § 7701 *et seq.*

103 33 U.S.C. § 467h.

104 *See* Exec. Order 12148 (1979).

105 *See* Exec. Order 12656 (1988).

106 42 U.S.C. § 5171.

107 42 U.S.C. § 5191(b).

create redundancy. NTIA does not allow remote access to its mainframe in order to protect this database.

Executive Order 12472 governs management and allocation of telecommunications resources during both non-wartime and wartime crises and emergencies. The Joint Telecommunications Resources Board (JTRB), as created by the Office of Science and Technology Policy (OSTP), within the Executive Office of the President, is given the primary authority for managing telecommunications resources during non-wartime emergencies. NTIA is a member of the JTRB. The telecommunications function for the National System for Emergency Coordination (NSEC) is performed by JTRB.

During wartime, NTIA may work with the National Security Council (NSC) and OSTP in advising the President with regard to radio spectrum priorities.[108] The NTIA Emergency Readiness Plan for Use of the Radio Spectrum is currently under review. The plan addresses actions that may be taken to prepare for, respond to, or recover from all emergency situations. Part III of the plan will contain telecommunications service priorities for Radiocommunications for use in a Presidentially-declared emergency under the Communications Act of 1934, as amended.[109]

Under the Communications Satellite Act of 1962,[110] NTIA may make recommendations to the President to ensure the availability and appropriate utilization of the communications satellite system for general government purposes. No mechanisms are currently in place for making or implementing recommendations.

The Bureau of Export Administration (BXA), through the Defense Priorities and Allocations System (DPAS), administers the Defense Production Act (DPA), as amended by the Stafford Act. Under the authority of these statutes, DPAS may set priorities in contracts and orders for industrial resources to meet national defense requirements. "National defense," as defined by the Stafford Act, includes civil "emergency preparedness." Thus, priority for government contracts and orders may be used to prepare for, respond to, or recover from both natural and human made disasters.

Electrical Power Systems

The BXA may use its power under the DPA for creating priority for government contracts and orders for industrial resources for national defense requirements to reconstitute critical infrastructures after attack. The Secretary of Commerce has additional authority under the International Emergency Economic Powers Act[111] to impose export controls on dual use

108 Exec. Order 12472 § 2(c).

109 47 U.S.C. § 606.

110 47 U.S.C. § 721(a)(6).

items.[112] A dual use item is one that has both potential commercial and military applications. The export controls are administered by the BXA and may be used to protect national security, foreign policy, and domestic shortages. Current regulations in place include export controls for most types of weaponry, crude oil, and exports to countries supporting terrorist organizations.[113]

Gas and Oil Storage and Transportation

See discussion of priorities in contracts under Electrical Power Systems *supra* p. 118.

Transportation

See discussion of priorities in contracts under Electrical Power Systems *supra* p. 118.

Water Supply Systems

See discussion of priorities in contracts under Electrical Power Systems *supra* p. 118.

Continuity of Government Operations

The Department of Commerce has measures in place for the continuity of its specific government services. The Bureau of Census and the National Technical Information Service have mechanisms in place for the continuity of their own functions and for the prosecution of the information they store. Census has three basic mechanisms for protecting information and ensuring continued operation. 13 U.S.C. § 9 ensures the confidentiality of data submitted to Census from public disclosure. In addition, the Bureau has a plan for the protection of vital records. These records include plans and procedures for accomplishment of mission, standby emergency documents and delegations of authority, lines of succession, program documentation and budget estimates, vital ADP records, programs, software, specifications, customer directories and mailing lists, survey designs, demographic and economic data, Current Population Survey, Census of Manufacturing, and geographic and cartographic reference files. Census also has a disaster recovery/contingency plan. It is designed to ensure

---

111 50 U.S.C. § 1701 *et seq.*

112 The structure of these export controls is based on a prior law, since lapsed, the Export Administration Act of 1979 (EAA), 50 U.S.C. app. § 2401 *et seq.*The International Emergency Economic Powers Act (IEEPA), 50 U.S.C. § 1701 *et seq.*, has replaced the EAA as the relevant legal authority, but the implementing regulations have remained unchanged.

113 *See* 15 C.F.R. Part 730.

continued normal operation during emergencies. The plan includes a number of elements such as priorities, emergency chain of command, computer system information, a plan for dealing with physical disruptions including electric power, descriptions of alarm systems, and the functions of the various post-disaster teams that are needed for recovery.

The National Institute of Standards and Technology (NIST) plays a key role in setting government-wide standards and guidelines to assist Federal agencies in protecting their information technology resources under authority granted by the Information Technology Management Reform Act of 1996,[114] and the Computer Security Act of 1987.[115] The latter charges NIST with creating computer security standards for the federal government. The Office of Management and Budget Circular No. A-130 mandates that government agencies follow the NIST standards. NIST is given additional authority to conduct research and development, to coordinate efforts with other agencies, and to assist the private sector as necessary to effectuate the goals of the Computer Security Act. Most recently, NIST has entered into a partnership arrangement with the National Security Agency to establish testing methods and objective measures for computer security products and to accredit private labs to certify such products.[116]

The NTIA also has a Continuity of Operations Plan which is currently in draft form. It was created pursuant to Department Administrative Order 210-10, February 23, 1996. Additional requirements for NTIA's continuity plan come from Federal Preparedness Circular 60, which requires that NTIA be able to function out of three locations in case of emergencies.

Department of the Treasury

The Department of the Treasury's legal authority relevant to infrastructure assurance is concentrated in the area of banking and finance. The Treasury's authority and responsibility for the federal government's financial transactions, both payments and collections, also impacts the continuity of government operations.

Banking and Finance

The Department of the Treasury can influence or control the efficiency and security of the financial marketplace through different mechanisms. The President's Working Group on Financial Markets (FMWG), created by Executive Order 12631, is chaired by the Secretary of the Treasury. The

---

114 Pub. Law No. 104-106.

115 Pub. Law No. 100-235.

116 *See also* discussion of NSA authorities and the National Information Assurance Partnership *infra* p. 112.

FMWG facilitates information exchange among the government agencies with responsibility for the financial markets and regulation of securities and commodity firms and banks. FMWG has already begun to focus on disruptions of financial markets or major market participants from both physical and cyber causes. The Group is in a position both to mitigate the effects of disruption and to work to limit systemic risk and lower the damage that accompanies disruption. While FMWG has no statutory authority of its own, it is nonetheless in the position to influence policy with regard to banking and finance infrastructure assurance.[117]

Through its statutory authority to regulate government securities brokers and dealers, the Treasury may be able to enact measures for greater infrastructure assurance. Currently, Treasury has recordkeeping and reporting requirements in place, in addition to rules regarding financial responsibility and related practices. The Treasury submission suggests that the recordkeeping and reporting requirements, and its statutory authority, may be broad enough to impose standards for protection of information and telecommunications systems.

Continuity of Government Operations

Treasury has two functions that relate to the continuity of government operations in the face of infrastructure disruption. First, the Department of Treasury is charged with the issuance of Treasury debt which is administered through the Bureau of Public Debt. The Treasury's authority in this area is derived from a variety of statutory sources including 31 U.S.C. § 3121 *et seq.* and 12 U.S.C. § 391. Second, the Financial Management Service (FMS), located within Treasury, is responsible for making payments for government programs and providing the system for collecting government receipts, including federal taxes. Both of these functions rely extensively on the Federal Reserve Banks. The banks used by the Financial Management Service are required by agreement to comply with the Computer Security Act of 1987 and related standards and guidelines. FMS conducts audits to ensure compliance with these requirements.

Department of Justice

The Department of Justice (DOJ) plays a unique role in infrastructure assurance. Unlike many of the other agencies and departments surveyed

---

117 The Treasury submission notes that FMWG can wield influence both by raising visibility of an issue and through the rule making roles of its member agencies. The members of FMWG include: the Board of Governors of the Federal Reserve System, the Securities and Exchange Commission, and the Commodity Futures Trading Commission. Other agencies that participate in FMWG meetings include: the Comptroller of the Currency, the Federal Deposit Insurance Corporation, Federal Reserve Bank of New York, the National Economic Council, and the Council of Economic Advisors.

which have roles relating to particular infrastructures, the Department of Justice and the Attorney General are expected to play a primary role in responding to terrorist incidents under authorities such as Presidential Decision Directive 39 (PDD-39) and the Terrorism Incident Annex to the Federal Response Plan. For terrorist matters that occur within the United States and have no significant foreign involvement, the Department of Justice, through the FBI, handles coordination of federal agency efforts.

Perhaps the most visible role for the Department of Justice with respect to infrastructure assurance is in enforcing federal laws through the investigatory authorities of the FBI and prosecutorial resources of DOJ and the United States Attorneys' Offices. The Department's 94 United States Attorneys' offices are acquiring their own expertise in these important areas. Within the past several years, These offices have appointed, and the Department has specially trained, attorneys to handle crisis response issues (Attorney Critical Response Group Attorneys—ACIRGs) and computer and telecommunications-related crimes (Computer and Telecommunications Coordinators—CTCs).

Other Department components have responsibilities with important ties to infrastructure assurance policies. Within DOJ's Criminal Division, the Computer Crime and Intellectual Property Section and the Terrorism and Violent Crime Section play a pivotal role not only in prosecuting offenses against the critical infrastructures, but in developing policy initiatives and legislative proposals to improve laws and enforcement in these areas.

Increasingly, the Department of Justice, in coordination with the Department of State, is involved in international agreements to facilitate U.S. enforcement efforts concerning cyber-crime and physical threats to critical infrastructures. Bilateral and multilateral agreements address issues such as extradition, mutual legal assistance, and evidence sharing. Increasingly, such agreements address substantive and investigative issues relating to computer crime. The Department of Justice also administers a rewards program designed to encourage cooperation with counterterrorism enforcement efforts.

The Office of Intelligence Policy and Review works closely with law enforcement and the intelligence community to facilitate sharing of information between the communities, and also assists in the investigation of terrorism-related cases under the Foreign Intelligence Surveillance Act. The elements of the Office of Justice Programs—the National Institute of Justice, the Bureau of Justice Statistics, and others may be involved in future studies of infrastructure assurance issues and efforts to develop a better understanding of the scope of the problem facing the nation and the proper role of law enforcement. The Office of Justice Programs also provides funds to state and local law enforcement agencies to bolster preparedness in the event of a terrorist attack.

With respect to information sharing, the Civil Division of DOJ should play a role in predicting the liability climate surrounding effective implementation and operation of an information sharing and threat warning capability. The Antitrust Division can serve as a focal point for issues relating to private-sector to private-sector sharing of threat and vulnerability information.

National Security Agency

The National Security Agency (NSA) identified its role in infrastructure assurance as deriving from three legal sources—Executive Order 12333, "United States Intelligence Activities," National Security Directive 42, "National Policy for the Security of National Security Telecommunications and Information Systems," and the Computer Security Act of 1987. These three documents give NSA substantial authority for the communications security of the United States.

Executive Order 12333 designated the Secretary of Defense as the executive agent for the United States for communications security activities, except as limited by the National Security Council.[118] NSA is authorized to carry out the Secretary of Defense's responsibilities for communications security under the executive order.

Consistent with Executive Order 12333, National Security Directive 42 assigns the Secretary of Defense responsibility for the government's National Security Telecommunications and Information Systems Security. As part of this Directive, the Director, NSA, is named the National Manager for this function. Among the duties this role entails are: (1) examining national security systems and evaluating their vulnerability to foreign interception and exploitation; (2) acting as the U.S. government focal point for telecommunications and information systems security for national security systems; (3) assessing the overall security posture of and disseminating information on threats to and vulnerabilities of national security systems; (4) reviewing and approving all standards, techniques, systems, and equipment related to the security of national security systems; and (5) coordinating with the National Institute of Standards and Technology (NIST) in accordance with the provisions of the Computer Security Act of 1987.

Executive Order 12333 and NSD 42 limit NSA's role to national security systems, which NSA recognizes as a potential limit on its role in assuring the security of the nation's critical information infrastructure. However, NSA has assumed a role in assisting NIST with its responsibilities under the Computer Security Act of 1987. NSA and NIST have agreed on a division of labor, memorialized in "Memorandum of Understanding Between the Director of the National Institute of Standards and Technology and the Director

---

118 Exec. Order 12333 § 1.11(e) (1981).

of the National Security Agency Concerning the Implementation of the Public Law 100-235," March 24, 1989. Under this MOU, NSA provides technical guidelines to NIST for sensitive information in federal computer systems, conducts assessments of hostile intelligence and foreign exploitation threats to information systems upon request of federal agencies, provides technical guidance and endorses products to secure information systems, and supports NIST in its role in assisting the private sector with information security issues.

The NSA-NIST partnership has been formalized recently as the National Information Assurance Partnership. This Partnership is designed to combine the expertise of the two entities to assist both government and the private sector in improving the quality of their computer security. To accomplish this goal, the Partnership is working to establish testing methods and objective measures for computer security products and to accredit private sector labs to conduct such testing and, ultimately, certify products.

United States Customs Service

The primary missions of the Customs Service of the Department of Treasury are to ensure that all merchandise coming into the United States is lawfully imported and to protect the borders of the United States. The Customs Service enforces laws relating to the importation and exportation of merchandise including laws requiring: the collection of duties and tax on imported goods; monitoring the clearing of arriving and departing vessels, vehicles and aircraft; detecting and preventing smuggling operations and fraudulent importations, and seizing merchandise and conveyances involved in such activities; excluding prohibited merchandise (*e.g.*, unlicensed munitions) from entry; controlling imports subject to quotas (*e.g.*, textiles), other international agreements, and intellectual property protection; and enforcing export laws pertaining to critical technology, munitions and embargoes.[119]

In addition to the authority to enforce customs laws, the Customs Service is authorized to enforce all federal laws at the borders of the United States. The Customs Service's unique position at the nation's borders allows it to exercise broad search and seizure and arrest functions. Customs officers may search persons, conveyances, baggage, cargo, and merchandise entering the United States without a search warrant and without suspicion of criminality.[120] Customs officers may stop vehicles and board vessels and aircraft without warrants to perform inquiries and searches.[121] Customs

---

119 *See* Title 19, United States Code.

120 *See* 19 U.S.C. §§ 482, 1499, 1581 & 1582.

121 *See* 19 U.S.C. §§ 482, 1467 & 1581; 19 C.F.R. Part 122.

officers carry firearms and are authorized to make warrantless arrests for any federal violations occurring in their presence.[122] In addition, a customs officer may conduct undercover investigations and seek and obtain search warrants, court orders authorizing interception of communications, and administrative summonses. Any merchandise or conveyance involved in a Customs violation is generally subject to civil forfeiture and may be seized by Customs officers without a warrant.[123] These authorities also apply to outbound goods, conveyances and people.[124]

The Customs Service also has emergency powers. Customs may take the following actions to protect against attacks and to minimize the damage from such attacks: close ports; remove a customhouse to a secure location; seize and forfeit vessels seeking to enter a closed port; and refuse to clear for export vehicles with suspect cargoes.[125] The President may authorize the Secretary of Treasury, who has delegated the power to the Commissioner of Customs, to extend the time provided under the Tariff Act of 1930 for the performance of any act and also to allow for the importation free of duty of certain supplies used in emergency relief work. Under the International Emergency Economic Powers Act,[126] the Secretary of Treasury, or his designee, may restrict entry, prevent movement, and take other authorized action during time of national emergency. In addition, during a Presidentially declared emergency, restrictions on imports and exports may be imposed when there is "any unusual and extraordinary threat, which has its source in whole or in substantial part outside the U.S. to the national security, foreign policy, or economy of the United States." During emergencies such as war, insurrection, invasion, or disturbance to international relations, the Secretary of Treasury may allow the Customs Service to take possession and control of vessels away from officers and crews.[127] The Secretary may also impose severe controls during times of insurrection—a power which has not been used since the Civil War.[128]

Bureau of Alcohol, Tobacco and Firearms

The Bureau of Alcohol, Tobacco and Firearms (ATF) operates under the umbrella of the Department of Treasury. The ATF's primary authorities

---

122 19 U.S.C. § 1589a.

123 *See* 19 U.S.C. §§ 482, 1581 & 1595a.

124 *See, e.g.*, 31 U.S.C. § 5317(b); 22 U.S.C. § 401; United States v. Ajlouny, 629 F.2d 830 (2d Cir. 1980).

125 50 U.S.C. §§ 219-226.

126 50 U.S.C. § 1701 *et seq.*

127 50 U.S.C. § 191. During World War I, Customs officials physically seized German vessels in U.S. ports under this law.

128 50 U.S.C. §§ 205-226.

relevant to infrastructure assurance pertain to emergency services and continuity of government operations.

ATF's mission, as a law enforcement agency, includes reducing violence, collecting revenue, and protecting consumers. Its jurisdiction extends to alcohol, tobacco, firearms and explosive matters. The most important aspects of ATF's role in infrastructure assurance pertain to its enforcement authority for provisions of title 18, United States Code, related to importation, manufacture, distribution and possession of firearms and ammunition,[129] explosives,[130] and arms and implements of war.[131]

United States Secret Service

The Secret Service is a law enforcement agency under the Department of Treasury and is charged with the protection of the President and Vice President, their families, foreign heads of state and the White House. The Secret Service is also charged with the detection and arrest of individuals who violate criminal laws relating to coins, obligations, and other securities of the United States and foreign governments, electronic fund transfers, credit or debit card fraud, false identification documents, and fraud committed against certain financial institutions, such as the Federal Deposit Insurance Corporation.

Banking and Finance

The Secret Service has exclusive jurisdiction for investigations involving the counterfeiting of coins, obligations and other securities of the United States and foreign governments.[132] Among the obligations and securities investigated by the Secret Service are U.S. coins and currency, Treasury checks, Department of Agriculture food stamps coupons, and postage stamps. The counterfeiting of these obligations and securities is criminalized by the provisions of Chapter 25 of Title 18, United States Code. In order to reduce opportunities for counterfeiting, the Secret Service works extensively with other federal agencies, including the Bureau of Engraving and the Federal Reserve Bank, and banking institutions to raise awareness, provide counterfeit detection seminars, improve security of currency design, and maintain the effectiveness of currency sorting equipment.

Under the authority granted by 18 U.S.C. § 3056 and Public Law 101-509, the Secret Service has jurisdiction over the investigation of other crimes within the banking and finance industry, including financial institution

---

129 18 U.S.C. Chapter 44.

130 18 U.S.C. Chapter 40.

131 22 U.S.C. § 2778.

132 18 U.S.C. § 3056.

fraud, access device fraud, and computer crime. The Secret Service uses investigations into financial crimes as an important source of data regarding systemic vulnerabilities. This information is shared with industry in an effort to raise security and awareness.

Telecommunications

The Secret Service's investigative authority under 18 U.S.C. § 1029, Access Device Fraud, and 18 U.S.C. § 1030, Computer Fraud, allows it to investigate criminal activity affecting components of the Telecommunications Infrastructure.

Continuity of Government Operations

The Secret Service is responsible for the safety of the President, Vice President, their immediate families, major Presidential and Vice Presidential candidates, former Presidents, and visiting foreign heads of state.[133] The Secret Service is also charged with protecting the White House Complex and the Vice President's residence. As part of its charge to protect the President, the Secret Service investigates threats against the President, successors to the Presidency, former presidents and certain other persons protected by the Secret Service.[134]

Department of Education

The Department of Education may provide funding for the education of school children on computer ethics and use of technology. The funding authorities that may be used for these purposes include the Fund for the Improvement of Education,[135] the Partnerships in Character Education Pilot Project,[136] and the Technology for Education Act of 1994.[137] In addition, state educational agencies may seek funding to educate school children in computer ethics through Goals 2000: Educate America Act.[138]

---

133 *Id.*; 3 U.S.C. § 202.

134 18 U.S.C. § 871 & 879.

135 *See* 20 U.S.C. § 8001 (authority to carry out programs directly or to use grants or contracts with state and local educational agencies, institutions of higher learning, and other public and private agencies and organizations on programs authorized under 20 U.S.C. § 8003).

136 *See* 20 U.S.C. § 8003 (authority to provide grants to state educational agencies on character education programs incorporating civil virtue).

137 *See* 20 U.S.C. § 6801 *et seq.* (designed to promote use of technology in education directly or through contracts or grants with state or local educational agencies, institutions of higher education, or other public or private organizations).

138 20 U.S.C. § 5801 *et seq.*

In addition to the funding available for educating school children on computer ethics and technology, the Department of Education also plays a role in continuity of government operations. The Department has issued internal policies relating to cyber disruptions.

General Services Administration

The General Services Administration (GSA) plays a critical role in the continuity of government operations in maintaining federal buildings and government telecommunications. These activities, as well as others carried out by GSA, also have an impact on the assurance of other critical infrastructures.

Continuity of Government Operations

GSA, under authority granted by the Federal Property and Administrative Services Act of 1949 and the Public Buildings Act of 1959, acquires and manages public buildings and space.[139] Under GSA's current appropriations act,[140] GSA is required to use its funding for constructing, altering, managing and protecting public buildings to meet minimum security standards provided by current law. GSA has indicated that its public buildings-related responsibilities can be leveraged to create structures better able to avoid or survive infrastructure attack.

In addition to the management of federal buildings, GSA is also responsible for procuring and supplying personal property and nonpersonal services for the federal government. Under 41 U.S.C. §§ 405 & 421, GSA, DOD and NASA have joint responsibility for issuing the Federal Acquisition Regulation (FAR), under the overall direction of the Office of Federal Procurement Policy. The FAR[141] is the basic regulation for acquiring services and supplies.[142] GSA has indicated that government purchase of supplies can be used to encourage private industry to develop products and services better able to achieve the infrastructure goals of the PCCIP.

Under Executive Order 12656 of November 1988, GSA is given lead responsibility for developing National Security/Emergency Preparedness (NS/EP) plans and procedures related to federal buildings (including shelter management programs), and supplies and public utility services needed by federal agencies during national emergencies.

---

139 *See* 40 U.S.C. § 471 *et seq.* & 40 U.S.C. §§ 601-619, 490. Implementing regulations are found in the Federal Property Management Regulations (FPMR) system, 41 C.F.R. Chapter 101.

140 Pub. Law No. 104-208.

141 48 C.F.R. Chapter 1.

142 GSA's Federal Supply Schedules are located at FAR Subpart 8.4 and Part 38.

Continuity of Government Operations and Telecommunications

GSA has the current responsibility for managing the FTS 2000 program and coordinating the follow-on program.[143] The current program provides government-wide telecommunications services for federal users, principally under two long-distance service contracts expiring in 1998. In addition to management under normal circumstances, GSA is also directed to ensure that federally owned or managed domestic communications facilities and services meet the National Security and Emergency Preparedness requirements of Federal civilian departments, agencies and entities.[144] GSA's current appropriations law makes funding available for this purpose, specifically for "the interagency funding of [NS/EP] telecommunications initiatives which benefit multiple Federal departments, agencies, or entities, as provided by [Executive Order 12472]."[145]

In the implementation of FTS2000 and the planned follow-on contracts NS/EP is a priority. In order to ensure that the network is maintained in a state of readiness for national emergencies, contractors must meet requirements related to grade of service, service reliability, interoperability, and restoration capabilities (including government-controlled priorities). Certain critical users may also use National Communications System (NCS) Government Emergency Telecommunications Service (GETS).[146] The implementation of the FTS program has been designed to meet the goals of the National Information Infrastructure; to support implementation of IT recommendations of the National Performance Review (NPR); and to provide the bulk of the telecommunications services for the emerging Government Services Information Infrastructure (GSII).

Telecommunications

By President's Memorandum, "Facilitating Access to Federal Property for the Siting of Mobile Services Antennas," August 10, 1995, the President directed GSA, and other federal agencies, to assist in "infrastructure buildout" by developing "procedures necessary to facilitate the appropriate access to Federal property for the siting of mobile services antenna." This Memorandum directs the implementation of the authority to "outlease" federal rooftops under the Public Buildings Cooperative Act of 1976[147] to provide federal property, rights-of-way, and easements for siting of new

---

143 *See* Pub. Law No. 104-106, section 5124(b) (The Information Technology Management Reform Act, or the Clinger-Cohen Act).

144 *See* Exec. Order 12472, "Assignment of National Security and Emergency Preparedness Telecommunications Functions."

145 Pub. Law No. 104-208, Title VI at § 619.

146 *See* Exec. Order 12472 § 1.

147 40 U.S.C. § 490(a)(17)-(19).

telecommunications services dependent on Federal spectrum rights for transmission contained in the Telecommunications Act of 1996.[148] A Federal Property Management Regulation Bulletin was scheduled for issue in May of 1997.

Transportation and Gas and Oil

Pursuant to the Federal Motor Vehicle Fleet, GSA acquires and manages alternative fueled vehicles.[149] GSA has indicated that this program plays a modest role in infrastructure assurance by reducing the nation's dependence on oil while expanding the alternative fuel infrastructure. But, it is a valuable example of how the government's purchasing power can be used to diversify critical components of the economy, thereby lessening the harm that would befall one component's incapacitation.

Federal Communications Carriers

The Federal Communications Commission (FCC) is an independent federal agency that reports directly to Congress. The FCC has general jurisdiction over interstate and international communications by radio, television, wire, satellite and cable.[150] The extent of the FCC's regulatory jurisdiction within each of these areas does vary. The passage of the Telecommunications Act of 1996 and the adoption of the February 1997 World Trade Organization Treaty on Basic Telecommunications may both effect infra structure assurance objectives, although the results are, at this time, unknown.

Telecommunications Carriers

FCC jurisdiction over common carriers is set out in Title II of the Communications Act of 1934.[151] This jurisdiction is for the purpose of regulating the provision of interstate and foreign (international) telecommunications service by carriers. "Telecommunications" is defined by the Act as "the transmission, between points specified by the user, of information of the user's choosing, without change in the form or content of the information as sent and received."[152] This definition of "telecommunications" creates

---

148 Pub. Law No. 104-104 § 704(c).

149 *See* 40 U.S.C. §§ 481 & 491 and 41 C.F.R. Part 101-38 (Federal Property and Administrative Services Act of 1949); 42 U.S.C. § 6201 et seq. (Energy Policy and Conservation Act); Pub. Law No. 102-486 (Energy Policy Act of 1992); Exec. Order 13031, December 13, 1996 (Federal Alternative Fueled Vehicle Leadership); and Exec. Order 12844, April 21, 1993 (Federal Use of Alternative Fueled Vehicles).

150 *See* 47 U.S.C. § 1 *et seq*. (The Communications Act of 1934).

151 47 U.S.C. §§ 201-276.

152 47 U.S.C. § 2(43).

an obligation on the part of carriers to protect the integrity of the information being carried. This obligation implicitly requires that the carrier protect its infrastructure in order to safeguard the information transmitted.[153] In addition, the FCC under its general statement of purpose from Congress may be able to take further action as necessary[154] to fulfill its mandate to "make available, so far as possible, to all the people of the United States . . . a rapid and efficient . . . Communications service with adequate facilities . . . for the purpose of the national defense, and for the purpose of promoting safety of life and property through the use of wire and radio communication."[155] The Communications Act of 1934 also grants power to the President in times of emergency for the setting of preferences or priorities for communications and the suspension of rules and regulations applicable to stations capable of emitting electromagnetic radiations.[156]

Wireline Telecommunications Carriers

The FCC requires wireline telecommunications carriers to report major service disruptions (generally disruptions that cause loss of service to more than 30,000 customers for more than 30 minutes).[157] The FCC's Federal Advisory Committee, the Network Reliability and Interoperability Council (NRIC), is responsible for analyzing this data. Of the 700 disruptions reported in 1996, all were traceable to physical causes. The most substantial physical "threat" to wireline carriers are dig-ins usually attributable to the failure of the excavator to contact the facility operator before beginning work. While no reported disruptions have to date been attributed to cyber threats, fraud against telecommunications carriers and their customers is believed to cost between 1 and 5 billion dollars annually. The Department of Justice and the Secret Service have primary jurisdiction over these fraud cases. Among the causes of these losses are unauthorized access to customers' PBX network for making long-distance calls, calling card theft and various kinds of pay phone fraud. NRIC is currently studying whether the Telecommunications Act of 1996 has increased the vulnerability of wireline networks to cyber attack, by providing greater access for competitors to telecommunications carrier transmission elements, switching elements and operating systems. Its final report was due Summer of 1997.

---

153 *See* 47 U.S.C. §§ 3(44) & 3(46).

154 47 U.S.C. § 4(i) provides that the Commission may perform "any and all acts, make such rules and regulations, and issues such orders not inconsistent with this Act as may be necessary in the execution of its functions."

155 47 U.S.C. § 1.

156 *See* 47 U.S.C. § 706.

157 *See* 47 C.F.R. § 63.100.

The FCC has also attempted to provide protection from harm caused by connection of terminal equipment and associated wiring through uniform standards.[158] Using terminal equipment to defeat telephone billing systems is a harm to the network, and to prevent these harms the Commission has, for example, required PBX machines to return answer supervision signals to the central office. Under the authority of section 208 of the Act, the FCC may hear complaints against carriers. This authority is used to address billing complaints involving fraud caused by third parties, most often through PBX systems.[159]

The FCC's regulations include a couple of seldom used provisions which could be used for infrastructure assurance objectives. 47 U.S.C. § 605 prohibits any person assisting in receiving or transmitting interstate or foreign communication from divulging the communications except through authorized channels. Also, under the authority of 47 U.S.C. § 214(d), the FCC may require, after an opportunity for a full hearing, that a carrier provide itself with adequate facilities for the expeditious and efficient performance of its service as a common carrier. This section has a few exceptions which include those carriers who provide interstate service only through interconnection with facilities of unrelated carriers.[160]

Wireless Telecommunications Carriers

The primary focus of the FCC's regulation of wireless communications is the avoidance of radio frequency interference. The Commission has broad authority over radio services licensees.[161] In addition to the general authority over licensees, the FCC is specifically granted the authority to "suspend the license of any operator upon proof sufficient to satisfy the Commission that the licensee—(C) has willfully damaged or permitted radio apparatus or installations to be damaged."[162] This authority could support additional infrastructure assurance-related regulations.

NRIC has recommended that wireless carriers establish processes for reporting service disruptions, but has not recommended, nor has the FCC required, wireless carriers to follow the reporting requirements placed on wireline carriers. Terrestrial wireless carriers do report disruptions of service of 90 continuous days or discontinued service.[163] Satellite wireless

---

158 *See* 47 C.F.R. Part 68.

159 Generally, the FCC has found the customer, rather than the telephone company, liable for fraudulent charges where the customer had the capability to restrict access to and egress from its PBX at all times, particularly where the customer took no steps available to it to detect or prevent unauthorized calls.

160 *See* 47 U.S.C. § 2(b)(2).

161 *See* 47 U.S.C. § 303(r).

162 47 U.S.C. § 303(m)(1)(C).

carriers file semi-annual reports which include discussions of the causes of any unscheduled transponder outages lasting more than 30 minutes.[164]

Terrestrial wireless carriers are protected from fraud by criminal provisions such as 18 U.S.C. § 1343. Law enforcement agencies have jurisdiction over the criminal prosecutions of those who steal phones or alter electronic serial numbers or mobile identification numbers. Under 47 U.S.C. § 302(d), the FCC has prohibited the intercepting of cellular radio transmissions using certain "scanning devices." Satellite wireless carriers are protected from unauthorized access by 18 U.S.C. § 1367 which criminalizes the intentional or malicious interference with the operation of a satellite.[165] Other provisions prevent unauthorized use, interception and disclosure of communications such as 47 U.S.C. § 605 (prohibits use of radio communications without being "entitled"); 18 U.S.C. §§ 2511-21; and 47 U.S.C. § 333 (prohibits willful or malicious interference with licensed radio communications).

Non-Telecommunications Communications Facilities

The FCC also regulates broadcast television, broadcast radio (AM and FM), shipboard radio, other safety and special radio services, and cable television. The Emergency Alert System (EAS)[166] requires cast networks; cable networks and program suppliers; AM, FM, and TV broadcast stations, Low Power TV stations, cable systems and other entities to operate on an organized basis during emergencies. This system allows the President, and local officials during local emergencies, to address the general public. The Plan for the Security Control of Air Traffic and Air Navigation Aids[167] defines the responsibilities fo the Commission for the security control of non-federal air navigation aids. The Public Safety National Plan[168] specifies special

---

163 *See* 47 C.F.R. § 22.317. NRIC has studied terrestrial wireless service disruptions in cellular radio service based on voluntarily provided data and has concluded that the primary causes of service outages are software design, hardware failure, and procedural error. Network Reliability—The Path Forward, http://www.fcc.gov/oet/nric, Network Reliability Performance at 6.3. In addition, NRIC has reviewed the performance of terrestrial wireless networks after natural disasters. The Council found that the networks were fairly robust because of the presence of multiple service providers and the mobility of the infrastructure. Portable infrastructure can be imported into an impacted area and service can be quickly restored.

164 *See* 47 C.F.R. § 25.210(j)(2). New technologies in satellites—onboard management of individual switched subscriber connections, moving tracking antennas—will create more complex reliability issues and may necessitate further study.

165 18 U.S.C. § 1367 was in response to an incident of unauthorized access to a satellite in 1986—the so-called "Captain Midnight" event. There was a second incident in 1987.

166 47 C.F.R. Part 11.

167 47 C.F.R. Part N, §§ 87.393-397.

168 47 C.F.R. § 90.16.

policies and procedures governing the Public Safety Radio Services and the Special Emergency Radio Service.

Continuity if Government Operations

The FCC has a well-developed internal program for protecting the security fo its own computer networks and its continued operation. FCC Instruction 1479.1, the FCC Computer Security Program Directive for employees and contractors, sets the policy for ensuring the availability of data and computer systems, the integrity of data and system, and the confidentiality of Commission data.[169] The FCC Computer Security Officer maintains a Two-Year Strategic Plan outlining major Information Technology Center (ITC) tasks and establishes appropriate milestones and anticipated expenditures. The Computer Security Officer also conducts training in compliance with the Computer Security Act of 1987. The FCC requires that contractors work with the Commission to abide by the Computer Security Program Directiave and has incorporate this requirement into their contracts.

On a quarterly basis, the Computer Security Officer conducts automated analysis of key systems, in fulfilment of the requirements of the Computer Security Act of 1987. The results are used to eliminate potential system threats and to brief management. Periodic account audits are also conducted on critical systems as another method for collecting threat information. In 1995, the FCC Inspector General and the Computer Security Program began using "Tiger Teams" to identify system security deficiencies. The Computer Security Officer, FCC Telecommunication System Manager and Inspector General are working on addressing similar security concerns for the FCC telecommunications system.

FCC computer hardware is protected through the FCC Physical Security and Guard Force program. Through a series of sensors, monitor stations, closed circuit cameras, and alarms, FCC Physical Security Guard forces are notified when key electronic assets are placed near strategically located sensors. Computer networks managed by ITC also have protective barriers. According to preset conditions, intruders are locked out automatically when the conditions or thresholds are met. Systems either automatically reset after a period of time or require system management intervention for resetting. The FCC also has a five layer program for controlling computer

169 The FCC Computer Security Program is designed to meet all requirements set forth in Office of Management and Budget (OMB) Circular A-130, Appendix III, Security of Federal Information Resources, Public Law 100-235, The Computer Security Act of 1987, and other federal mandates for providing computer security support to Federal agency systems. The document details proper use and penalties for misuse of FCC systems, use of identification and authentication for controlling access, as well as guidance on use of electronic mail, intranet and Internet services.

viruses. The ITC has prepared for emergencies by establishing a Computer Emergency Response Team (CERT). The ITC CERT responds to incidents ranging from inappropriate e-mail to computer intrusions. The FCC also participates in the NIST, Federal Computer Incident Response Capability (FedCIRC) and the NIST Federal Information System Security Manager Forum.

United States Postal Service

Continuity of Government Operations

The U.S. Postal Service is an important element in the continuity of government operations, as well as the continued communication of private businesses and individuals. The Postal Service, an independent establishment of the Executive Branch, has broad authority to accomplish its mission—the delivery of parcels throughout the world.[170] In addition to the authority to prescribe regulations necessary and appropriate for carrying out its function, the Postal Service also has authority to determine its expenditures.[171] Postal revenues are permanently appropriated to the Postal Service Fund and the Postal Service may also borrow funds within certain prescribed limits.[172]

The Postal Service also has the authority for the investigation and enforcement of all civil and criminal statutes of the United States pertaining to the Postal Service and the mails.[173] This authority is delegated to the Postal Inspection Service,[174] which investigates and enforces both generic statutes governing federal property, employees, and processes as they pertain to the Postal Service, as well as provisions that deal specifically with the Postal Service and the mails.[175] The Postal Service is currently negotiating a memorandum of understanding with the Department of Justice and the Federal Bureau of Investigation which will give the Postal Service jurisdiction over crimes involving electronic commerce services offered by the Postal Service. The general powers of the Postal Service also include security guards who may exercise the authority of special police.

---

170 *See* 39 U.S.C. §§ 201, 403(a); *see also* 39 U.S.C. § 407 (international postal arrangements).

171 39 U.S.C. § 401(2), (10); 39 U.S.C. § 401(3).

172 39 U.S.C. §§ 2003(b), 2401(a); 39 U.S.C. § 2005(a).

173 39 U.S.C. § 404(a)(7); 18 U.S.C. § 3061.

174 39 C.F.R. § 224.233.

175 Specific criminal statutes relating to the mails include: 18 U.S.C. §§ 876-77 (mailing threatening communications); 18 U.S.C. §§ 1701-03 (obstruction of the mails); 18 U.S.C. § 1705 (destruction of a letter box or mail); 18 U.S.C. § 2115 (breaking into post office).

Under the Part 26 of Executive Order 12656, Assignment of Emergency Preparedness Responsibilities, the Postal Service has lead responsibility for plans and programs to provide essential postal services during a national security emergency. The Postal Service would also assist other agencies in locating and leasing privately owned property for federal use in such an emergency. Federal Preparedness Circular 60, Continuity of the Executive Branch of the Federal Government at the Headquarters Level During National Security Emergencies, designates the Postal Service as a Category I organization, required to provide a capability for uninterrupted emergency operations during a national security emergency. Procedures for individuals post offices and Headquarters for dealing with disruptions of normal operations are set out in the Postal Service's Headquarters Emergency Plan (February 1991) and the Emergency Planning Manual (December 1981). Both of these plans are currently being considered for update.

Department of the Interior

The bureaus and organizations that comprise the Department of the Interior are the principal federal authority for the proper use and protection of federal lands. The Department's jurisdiction ranges from national parks to Indian reservations. While the Department of the Interior has little direct authority related to infrastructure assurance, the authority that it exercises over federal lands affects many of the critical infrastructures identified in Executive Order 13010. The Department of the Interior also has a limited role in law enforcement for the protection of public safety and public lands and resources. This role provides for cooperation with other law enforcement bodies at the federal, state and local levels.[176] Authorities related to Indian lands are administered by the Bureau of Indian Affairs (BIA). These authorities may have an impact on the infrastructure physically located on Indian lands or infrastructure services provided to the inhabitants of Indian Reservations.[177] Similarly, the National Park Service (NPS) and U.S. Fish and Wildlife Service (FWS) have authority to regulate the conduct of certain infrastructures physically located on National Park and FWS lands.[178]

---

176 *See* 446 DM Chapter 1; 43 U.S.C. § 1733, and BLM Law Enforcement General Order 23.

177 *See, e.g.*, 25 U.S.C. §§ 311-328 and 25 C.F.R. Part 169 (right-of-way through Indian lands for telecommunications purposes); Indian Mineral Development Act of 1982, 25 U.S.C. §§ 2101-09 and 25 C.F.R. Parts 225-27 (oil and gas agreements); American Indian Trust Fund Management Reform Act, 25 U.S.C. §§ 4001-4061; Indian Dams Safety Act, 25 U.S.C. §§ 3801-3804 (maintenance and repair of dams located on Indian lands); Indian Law Enforcement Reform Act, 25 U.S.C. §§ 2801-2809.

178 16 U.S.C. § 1 *et seq.; See, e.g.*, 16 U.S.C. § 5 & § 79 and 36 C.F.R. § 14 (telecommunications facilities and electric power).

Emergency Services and Telecommunications

The Bureau of Land Management (BLM) owns a radio, satellite, and microwave communications system, the National Interagency Fire Center (NIFC), designed to coordinate information and communications relevant to wildland fire-fighting operations. The BLM also responds to natural disasters using the National Incident Radio Support Cache, a system of equipment and human resources trained and maintained to travel worldwide. BLM's personnel and telecommunications network cannot assist any state or local jurisdiction absent an interagency agreement or request by the jurisdiction.

BLM also has authority for the physical protection of infrastructures under 18 U.S.C. § 1362, which prohibits the willful or malicious injury or destruction, or willful or malicious attempt to injure or destroy, any of the works, property, or material of any radio, telegraph, telephone or cable, line, station, or system or other means of communication operated or controlled by the United States, whether such system is constructed or under construction. It is also a crime to hinder, delay or obstruct the transmission or communication over such a line or system. BLM has authority to enforce criminal laws protecting satellites and satellite transmissions.[179] Its law enforcement officers also have authority over specific acts on federal land.[180]

Emergency Services

BLM has trained personnel and equipment ready to respond to emergencies involving BLM-owned facilities. These resources include law enforcement, firefighting and environmental response personnel as well as police cars and fire engines. Among the more important authorities granted to the Secretary of Interior for emergencies is the power to enter into cooperative agreements for the protection of federal lands, which includes management as well as law enforcement functions.[181] Under certain circumstances, the Department is authorized to respond to emergencies absent a cooperative arrangement. 42 U.S.C. § 1856 allows mutual aid in fire protection and emergency assistance in extinguishing fires in the vicinity of agency facilities.

Other important cooperative arrangements exist between Interior and other federal agencies. Via an interagency agreement, NRC is able to use BLM's telecommunications system and personnel for the protection of nuclear facilities under its jurisdiction. BLM has also made an interagency

---

179 *See* 18 U.S.C. § 1367.

180 *See, e.g.*, 18 U.S.C. § 1864 (prohibits use of a hazardous or injurious device on federal land).

181 *See* 43 U.S.C. 1737; Chapter XI-Law Enforcement Cooperation., H-9260-1 BLM Law Enforcement Management.

agreement with FEMA to assist state and local governments during emergencies or disasters upon the request of such governments. This interagency agreement is also triggered by Presidentially declared disasters. BLM and NPS are also signatories to the Inter-Departmental MOU allowing Cross-Deputation of Law Enforcement power within the federal government in appropriate cases, including requests for assistance made through the Incident Command System. This System is a federal government wide method that is used in appropriate cases for any federal officer or group of federal officers requesting assistance in an emergency.

At the state level, BLM has MOUs in place with local and state law enforcement and fire prevention offices. Such MOUs and ongoing relationships could be relevant to the provision of emergency services in the event of an attack or natural disaster. One example is the 1993 Cooperative Fire Protection Agreement in the State of Alaska. This agreement designates pre-determined fire protection responsibility areas in Alaska regardless of whether the lands are administered by the federal or state governments, native corporations, boroughs, or the military. The National Park Service has entered into similar agreements for cooperation with law enforcement and fire prevention agencies at the state and local levels. 4 D.C. Code § 201 *et seq.* provides special authority for the Park Police in the District of Columbia and its environs. The NPS may also close park areas for public safety purposes.

Under the National Wildlife Refuge System Administration Act of 1966 certain FWS personnel have law enforcement authority on Service lands.[182] A bill has been pending in Congress since 1995 to give Bureau of Reclamation personnel law enforcement authority to better administer Reclamation lands.

The United States Geological Survey (USGS) research and data collection functions may assist the federal, state and local governments during emergencies. Among the more relevant functions to infrastructure assurance are the Global Seismographic Network (a network of high-quality digital instruments that measure record ground shaking, which can serve as the principal source of public information on earthquakes and explosions occurring internationally),[183] National Water Quality Assessment,[184] and Committee on Civil Applications of Classified Overhead Photography (CAC).[185]

---

182 16 U.S.C. § 668dd(f).

183 *See* 15 U.S.C. §§ 5652 (The Land Remote Sensing Policy Act of 1992); 42 U.S.C. §§ 7701 *et seq.* (The Earthquake Hazards Reduction Act of 1977).

184 *See* Pub. Law No. 102-580 (Water Resources Act of 1992).

185 Authority for this program is set out is a classified Presidential Directive from 1975.

Electric Power

BLM law enforcement officers are authorized to enforce 18 U.S.C. § 1366 which makes it a crime to knowingly or willfully damage or attempt to damage an energy facility. "Energy facility" includes a company involved in the production, storage, transmission, or distribution of electricity, fuel, or another form or source of energy, or research, development, or demonstration facilities related thereto.

The Secretary of the Interior is responsible for preventing "unnecessary or undue degradation of the [public] lands."[186] The Secretary also has the power to grant rights-of-way over, public lands for systems for the generation, transmission, and distribution of electrical energy.[187]

Oil and Gas Storage and Transportation

Under the Mineral Leasing Act[188] similar to that in the electric power arena has been granted to the Secretary of the Interior for oil and gas pipelines. BLM enforces 18 U.S.C. § 1366 for oil and gas facilities as well as electric power facilities. In this context, the power to grant rights-of-way for pipelines also includes the authority to impose stipulations addressing damage to public or private property and hazards to public health and safety.[189] If necessary to protect public health, safety, or the environment, the Secretary may order the immediate temporary suspension of activities within the right-of-way or permit area.[190]

The Secretary also has the authority to regulate drilling and surface-disturbing activities on public lands. The Secretary is required to set standards for reclamation and restoration after the cessation of oil and gas activities. This standard-setting authority could be used to regulate the rebuilding and the cleanup process on public lands.[191] The Minerals Management Service (MMS) is responsible for all of DOI's oil and gas leasing activities on the Outer Continental Shelf (OCS). Under the Outer Continental Shelf Lands Act,[192] the Secretary of the Interior not only may lease OCS tracts for gas and oil exploration and production, but may also regulate and approve all production plans, including safety and construction specifications; inspect and require reports sufficient to mandate compliance with regulatory standards; and impose civil penalties on lessees not in

---

186 43 U.S.C. § 1732(b).

187 FLPMA § 501(a)(4).

188 30 U.S.C. § 185.

189 30 U.S.C. § 185(h).

190 30 U.S.C. § 185(o).

191 *See* 30 U.S.C. § 226(g).

192 43 U.S.C. § 1331 *et seq.*

compliance. This regulatory authority extends to operating procedures, pipelines, and offshore storage of oil and gas. This authority is implemented by 30 C.F.R. Part 250, Oil and Gas and Sulphur Operations in the Outer Continental Shelf. Overlap in the authority of the MMS and the Coast Guard is governed by a MOU providing for coordination. Coordination of overlapping jurisdiction with the Department of Transportation is governed by an MOU dated December 10, 1997, entitled "Memorandum of Understanding Between the Department of Transportation and the Department of Interior Regarding OCS Pipelines."

Transportation

The Department of the Interior only has authority over transportation related to federal lands. However the authority over the transportation infrastructure on federal lands is quite broad, including the authority to grant rights-of-way,[193] enforce criminal laws protecting federal lands,[194] to permit or even require construction and maintenance,[195] and to order immediate temporary suspension of activities on a right-of-way area.[196]

Water

The Bureau of Reclamation (Reclamation) develops, manages, and protects water and related resources in an environmentally sound manner in the interest of the American public. The general authority for Reclamation's activities is found in the Reclamation Act of 1902[197] and the Reclamation Project Act of 1939.[198] Under these acts the Secretary of the Interior is authorized to "perform any and all acts and to make such rules and regulations as may be necessary and proper for the purpose of carrying the provisions of this Act into full force and effect." This includes developing and delivering water supplies for authorized purposes. Under 43 U.S.C. §§ 390b, 485h(c) & 521, the Secretary may enter into contracts to supply water for purposes other than irrigation.[199] Responsibility for operation and management of Reclamation projects may be transferred from the government to water user organizations. This transfer may include provisions for the protection of water supply systems via contract or regulation.[200]

---

193 43 U.S.C. § 1763.

194 18 U.S.C. § 1361 (willful injury of any property of the U.S.); 18 U.S.C. § 1864 (use of a hazardous or injurious device on federal lands).

195 43 U.S.C. § 1762(a)-(b).

196 43 C.F.R. § 2803.3.

197 43 U.S.C. § 373.

198 43 U.S.C. § 485i.

199 Authority to enter into contracts for irrigation is provided by 43 U.S.C. § 485h(d) & (e).

200 *See* 43 U.S.C. §§ 498-499.

The Reclamation Safety of Dams Act[201] authorizes the Secretary of the Interior to perform such modifications on Reclamation dams as reasonably required to preserve their "structural safety." In the case of an emergency, or unusual conditions, the Commissioner of Reclamation may use emergency funds set aside from the Reclamation fund.[202] For this purpose, "unusual or emergency conditions" means, "... generator failures, damage to transmission lines, or other physical failures or damage, or acts of God, or of the public enemy, fires, floods, drought, epidemics, strikes, or freight embargoes, or conditions, causing or threatening to cause interruption in water or power service."[203] Specific regulations govern the conduct at Reclamation dams, including prohibitions on carrying explosives or willful destruction.[204]

Continuity of Government Operations

On or about October 1, 1997, all BLM law enforcement manual and electronic files, both at the national level and in the field, will be switched on line to a system called LAWNET. The system has been built to deny access to "unauthorized personnel" and to protect the information contained therein from penetration.

There are also several initiatives being sponsored at the Secretary level which may impact infrastructure assurance. The DOI All-Hazard Coordination Group (AHCG) has representatives from all significant DOI bureaus and offices. AHCG was chartered in 1995 to facilitate interbureau and interagency standardization prior to and during the course of a major, complex incident. The objectives of the group are to provide timely information to policy officials in the Department, Administration and Congress; minimize the loss of life, property and resources; minimize costs of response; function on an interagency basis with other Federal Agencies as well as interstate, state, local and tribal officials; and respond to incidents effectively and in a coordinated manner through sharing of resources, information and personnel within the Department. In addition an interagency agreement with the USDA provides for cross-designation of law enforcement officers to allow supplementation of agency law enforcement personnel resources for law enforcement emergencies, violations in progress, or needs specifically identified. FEMA has delegated its role for all stages of Presidentially declared disasters under the Stafford Act to DOI until FEMA arrives (usually after the first 24 hours). DOI has also created an Emergency Communications Plan for Alaska Operations. DOI is cur-

---

201 43 U.S.C. § 506.

202 43 U.S.C. § 502.

203 43 U.S.C. § 503.

204 *See, e.g.*, 43 C.F.R. § 421 (Rules of Conduct at Hoover Dam).

rently in the process of assessing the maintenance and repair issues for its facilities. The study is being conducted by the DOI Planning, Design, and Construction Council and should be completed in June 1997.

Department of State

The Department of State administers itself, or in coordination with other agencies, authorities directed toward detection and control of international terrorism that could be targeted against any or all of the infrastructure areas listed in Executive Order 13010.[205]

Telecommunications

22 U.S.C. § 2707 provides for oversight of foreign policy related to international communications and information policy, and the Department operates the world-wide Diplomatic Telecommunications System serving posts overseas and linking them to Washington.

Transportation, Gas and Oil Storage, Water Supply Systems

Executive Order 11423 authorizes the Secretary of State to receive applications for permits and to grant or deny permits for facilities at borders connecting the U.S. with a foreign country, specifically pipelines for petroleum, water or sewage facilities, and facilities for the transport of person or property (*e.g.*, monorail). The Secretary of State also is authorized to administer travel controls into and out of the U.S. for aliens and citizens.[206]

Emergency Services

22 U.S.C. § 2709 authorizes special agents of the Department of State and Foreign Service to exercise specific protective and arrest functions.

Continuity of Government Operations

The Department of State has lead responsibility under Part 13 of Executive Order 12656 for continuity of government and other national emergency activities that affect foreign relations.

United States Information Agency

The U.S. Information Agency (USIA) is responsible for overseas information and education and cultural exchanges. Under the President's mandated reorganization of Foreign Affairs Agencies, USIA will be integrated

---

205 *See, e.g.*, 18 U.S.C. § 113B; Pub. Law No. 99-399; 22 U.S.C. §§ 2711-12 & 2708; 22 U.S.C. Chapter 64 (aviation security); Pub. Law No. 101-298 (Biological Weapon Control); 22 U.S.C. § 3244 & Pub. Law No. 99-399 (Nuclear Terrorism).

206 *See* 22 U.S.C. § 211a; 8 U.S.C. § 1185; 18 U.S.C. §§ 1541-1546.

with the State Department within the next two years. Much of USIA's functions are performed using communications systems, including the Internet and electronic databases and radio and television broadcasts. USIA uses electronic means to link with foreign media, government and academia. It also controls all U.S. nonmilitary international broadcasting, including the Voice of America. Many of the broadcast services are dependent on satellites. However, USIA does not currently have in place any authorities relating to infrastructure assurance though they do have many infrastructure related assets.

Nuclear Regulatory Commission

Electrical Power Systems

The Atomic Energy Act of 1954 (AEA), as amended,[207] and section 201 of the Energy Reorganization Act of 1974[208] provide the NRC with licensing and related regulatory authority over commercial nuclear power plants. Under section 103 of the AEA,[209] the NRC issues commercial licenses for production or utilization facilities including nuclear power plants. The NRC may issue additional standards or instructions to govern the possession and use of special nuclear material as necessary or desirable to promote the common defense and security or to protect health or minimize danger to life or property.[210] The NRC also has authority under the AEA for the protection of restricted data,[211] to guard against the loss or diversion of special nuclear material, and to govern any activity authorized by the AEA. The NRC may require reports, recordkeeping and inspections for activities conducted under section 103 licenses.[212] The NRC may revoke licenses,[213] or impose civil penalties for violations of rules, regulations, or licensing provisions.[214] Criminal penalties are also available for violation of AEA provisions or acts of sabotage. The Attorney General enforces such provisions.[215]

---

207 42 U.S.C. § 2011 *et seq.*

208 42 U.S.C. § 5811.

209 42 U.S.C. § 2133.

210 *See* 42 U.S.C. § 2201(b).

211 The NRC has further authority for prescribing rules and regulations necessary to prevent unauthorized disclosure of a licensee's or applicant's safeguards information. *See* 42 U.S.C. § 2167.

212 This includes a provision of the Energy Reorganization Act of 1974 that requires that directors and responsible officers of companies constructing, owning, operating, or supplying components for AEA licensed activities report to the NRC that a facility, activity, or basic component supplied to such facility fails to comply with AEA provisions or any applicable NRC rule, regulation, order or license pertaining to safety hazards. 42 U.S.C. § 5846.

213 42 U.S.C. § 2236.

These statutory authorities are implemented by chapter 1 of Title 10 of the Code of Federal Regulations. The NRC has promulgated regulations governing the criteria and procedures for determining eligibility for access to restricted data, national security information, special nuclear material and employment clearance; reporting of defects and noncompliance; domestic licensing of production and utilization facilities; physical protection of plants and materials; material control and accounting of special nuclear material; and security facility approval and safeguarding of national security information and restricted data.

Transportation

Implementation of authority to impose requirements for transportation of nuclear materials is limited to safe packaging requirements to ensure radiological health and safety, and to safeguard requirements to assure the security of designated shipments of nuclear material from sabotage. The NRC has the authority to establish by rules, regulation, or order such standards and instructions to govern the possession and use of special nuclear material as the NRC may deem necessary or desirable to promote the common defense and security or to promote health or to minimize danger to life or property.[216] The NRC may also promulgate rules or regulations to guard against the loss or diversion of special nuclear material and to prevent such use or disposition thereof as the NRC determines to be inimical to the common defense and security.[217]

Federal Energy Regulatory Commission

The Federal Energy Regulatory Commission (FERC) was established by the Department of Energy Organization Act.[218] FERC has authorities relating to many aspects of interstate energy generation, transmission and sale covering electric facilities, hydroelectric facilities, and oil and gas facilities.

Electric Facilities (Non-Hydroelectric)

Under the Federal Power Act (FPA),[219] FERC has the following authorities over interstate transmission and wholesale sale of electric power: (1) re-

---

214 42 U.S.C. § 2282. Injunctive relief for violations of the AEA, or regulations thereunder, is also available with the assistance of the Attorney General of the United States. 42 U.S.C. § 2280.

215 *See* 42 U.S.C. § 2271 *et seq.*

216 42 U.S.C. § 2201(b).

217 42 U.S.C. § 2201(i). The implementing regulations are located at 10 C.F.R. Parts 71 (packaging and transportation of radioactive material) & 73 (physical protection of plants and materials).

218 42 U.S.C. §§7101-7352.

porting requirements; (2) investigatory authority; (3) authority to require interconnection and services; (4) rate and service oversight authorities; and (5) consultation and other general authorities.

More specifically, FERC has the general authority, delegated from DOE, to require public utilities to file periodic or special reports "as necessary or appropriate to assist the Commission in the proper administration of this Act [FPA]."[220] In addition, FERC has the authority to require public utilities to make reports regarding anticipated shortages, to make and periodically revise contingency plans to deal with shortages, and to accommodate any shortages in a manner that gives due consideration to public health and safety.[221] Such statutory authority and the implementing rules were written to insure continuity of service to customers of public utilities. FERC also requires information to be transmitted annually by transmitting utilities regarding available transmission capacity and known constraints.[222]

FERC has authority to conduct investigations regarding violations of the FPA or its regulations and to acquire information to serve as a basis for recommending further legislation.[223] FERC's rules relating to investigations are contained in 18 C.F.R. Part lb.

Upon application of a State commission or any person engaged in the transmission or sale of electric energy, FERC may direct a public utility to establish physical connection of its transmission facilities with other facilities, and to sell and exchange energy. This authority is subject to a public interest standard, and may not be used to compel the enlargement of generating facilities, or if the connection would impair the ability of the utility to render adequate service to its customers.[224] Under a similar procedure, after a complaint by a State Commission, FERC may require a utility to provide an adequate level of service as determined by FERC.[225]

FERC's rate and service oversight authorities allow review of rules, regulations, contracts and practices affecting rates within it jurisdiction to determine whether they are unjust, unreasonable, or unduly discriminatory. Upon such a finding, FERC may determine just and reasonable terms to be thereafter observed.[226] FERC noted this authority as having potential appli-

---

219 16 U.S.C. §§ 791a-825r

220 16 U.S.C. § 825c (by delegation from DOE per Delegation Order No. 0204-1)

221 *See* 16 U.S.C. § 824a(g); 18 C.F.R. Part 294.

222 16 U.S.C. § 8241; 18 C.F.R. § 141.300 (requiring filing of Form No. 715).

223 16 U.S.C. § 825f; 16 U.S.C. § 825j (by delegation from DOE per Delegation Order No. 0204-1).

224 *See* 16 U.S.C. § 824a(b); *see also* 16 U.S.C. § 824i (additional interconnection authorities).

225 16 U.S.C. § 824f.

226 *See* 16 U.S.C. §§ 824d & 824e.

cations for addressing reliability concerns. FERC has used related authorities to promulgate rules regarding the requirements for rate schedules, offer of non-discriminatory open access transmission,[227] and utility participation in electronic information systems.[228]

Under PURPA, DOE is to consult with FERC with respect to requesting, from time to time, the reliability councils or other appropriate persons (including federal agencies) to examine and report concerning any electric utility reliability issue; and recommending industry standards for reliability to the electric utility industry, including standards with respect to equipment, operating procedures, and training of personnel.

In addition to these authorities, FERC also has general authority to "perform any and all acts . . . as it may find necessary or appropriate to carry out of the provisions of this Act [FPA]."[229] FERC may seek court orders to enforce provisions of the FPA or any rule or order thereunder.

FERC Order No. 889[230] requires that public utilities that own, control, or operate interstate transmission facilities to participate in an electronic information system (OASIS) in order to disseminate transmission information on a uniform basis. FERC included in the rule security measures that were proposed by industry including data encryption, firewalls, and use of passwords. The initial phase of compliance with these requirements was to be completed in January 1997. FERC is considering a second phase of implementation that would require additional functions and performance requirements. FERC does not view potential cyber threats as posing a significant danger to the OASIS system because it is not directly linked to industry operating systems.

Hydroelectric Facilities

FERC's authority to issue licenses for hydroelectric facilities includes regulations regarding the keeping of records, investigatory authorities, and requirements for construction, maintenance and operation. Many of FERC's regulations in this area require coordination or adherence to other federal entity rules or regulations. FERC works closely in this area with many other interested parties, such as the U.S. Coast Guard, U.S. Army Corps of Engineers, the Bureau of Reclamation, NRC, DOE, FEMA, and the Washington State Department of Ecology. These entities, and additional

---

227 *See* 18 C.F.R. Part 35.

228 *See* 18 C.F.R. Part 37.

229 16 U.S.C. § 825h.

230 61 Fed. Reg. 21,737 (May 10, 1996), FERC Stats. & Regs. Para. 31.035 (1996), *order on reh'g*, Order No. 889-A, 62 Fed. Reg. 12,484 (March 14, 1997), FERC Stats. & Regs. Para. 31,049 (1997), *reh'g pending*.

regional agencies, were all identified by FERC as having a stake in infrastructure assurance for hydroelectric facilities.

Gas and Oil Storage and Transportation

FERC has authority, delegated from DOE, to approve or disapprove the siting, construction and operation of new domestic facilities at the place of entry for imports or exit for exports, unless the Assistance Secretary exercises this disapproval authority. FERC also has authority over the rates and charges of natural gas companies. Rates must be filed with FERC and must be just and reasonable. It has authority to alter natural gas tariffs if found not to be just and reasonable, and to require natural gas companies to extend or improve their gas transportation facilities and to establish physical connections with other facilities. FERC must provide approval before any natural gas company can undertake or complete construction, operation, acquisition, extension, or improvement of facilities used for the transportation of gas in interstate commerce. It may impose reasonable conditions on such authorizations. All such authorities are derived from the Natural Gas Act (NGA).[231]

FERC also has authority over the rates and practices of oil pipelines pursuant to the Interstate Commerce Act.[232] FERC's authority to review oil company tariffs and to impose terms and conditions could be used to further infrastructure assurance objectives.

Department of Agriculture

The Department of Agriculture (Agriculture) has numerous grant and loan programs in place to assist rural areas in receiving important services that touch on many of the critical infrastructures. They include the Rural Utilities Service, the Rural Housing Service, and the Rural Business-Cooperative Service. These loan and grant programs are used to foster telecommunications, electric, oil and gas, water, transportation and bankirg and finance related services in rural areas. Loans and grants are also available to improve emergency services such as fire and rescue squads. These programs usually require compliance with Agriculture regulations in constructing or operating facilities with loan or grant funds. The Department of Agriculture currently notes that these regulations do not specifically take into account the need to withstand physical or cyber attack, or other infrastructure assurance related issues. Additional programs, such as the Distance Learning and Telemedicine project may also need to be reviewed to ensure that

---

231 15 U.S.C. §§ 717b *et seq.*

232 49 U.S.C. §§ 1(5), 2, 6, 15(1) & 15(7).

they adequately take into account cyber and other threats in their operations and regulations.

The Securities and Exchange Commission

The Securities and Exchange Commission (SEC) has various authorities to respond to emergencies under the Securities Exchange Act of 1934 and the Investment Company Act of 1940. The SEC's emergency authorities may generally be invoked when there is a major market disturbance characterized by the occurrence or substantial threat of either sudden or excessive price volatility or substantial disruption of the safe or efficient operation for clearance and settlement.[233] In the event of such an emergency, the SEC may: (1) suspend all trading in nonexempt securities on an exchange or off-exchange for up to 90 calendar days; (2) suspend trading of nonexempt individual securities and groups of securities for no more than 10 business days; (3) alter existing Commission and self-regulatory organization (SRO) rules or impose new restrictions and approve SRO rules on an expedited basis; and (4) take emergency action with respect to mutual fund redemptions. The Commission is also authorized to approve SRO rules which allow some exchanges to conduct trading halts in response to national security situations. Individual SRO trade halts have been used infrequently, but have been successful enough to obviate the need for the SEC to invoke its authority to halt trading.

Board of Governors of the Federal Reserve System

The Board of Governors of the Federal Reserve System (Federal Reserve) has authority related directly to only one of the critical infrastructures being studied by the Commission—banking and finance. However, the Federal Reserve also may exercise certain powers, such as making properly secured loans to individuals, partnerships or corporations, which may indirectly impact the other critical infrastructures listed in Executive Order 13010.[234]

Banking and Finance

The Federal Reserve has authority to examine certain depository institutions and other financial institutions for the purpose of addressing the safety and soundness of their business practices. The institutions include member banks of the Federal Reserve System;[235] registered bank holding

---

233 15 U.S.C. § 781(k)(6).

234 *See, e.g.*, 12 U.S.C. §§ 343 & 347c.

235 12 U.S.C. §§ 325-326.

companies;[236] Edge Act corporations and agreement corporations;[237] foreign branches of U.S. member banks;[238] U.S. branches, agencies, and affiliates of foreign banks;[239] representative offices of foreign banks;[240] and Federal Reserve banks.[241] Among the enforcement actions that the Federal Reserve may take against these financial institutions are cease and desist orders,[242] orders of removal and prohibition,[243] and orders of assessment of civil monetary penalties.[244] Bank Service Corporations are also subject to examination and regulation by appropriate federal banking agencies.[245]

The Bank Protection Act[246] permits the Federal Reserve, the Comptroller of the Currency, the Federal Deposit Insurance Corporation, and the Director of the Office of Thrift Supervision, to issue rules establishing minimum standards for banks and savings and loan associations with regard to the installation, maintenance, and operation of security devices and procedures, in order to discourage robberies, burglaries, and larcenies and to assist in the identification and apprehension of persons who commit such acts.[247]

The Federal Reserve has several unique authorities that may allow for reconstitution of the banking and finance infrastructure. Under sections 10(b) and 13(8) of the Federal Reserve Act,[248] a Federal Reserve Bank may make appropriately secured advances to member banks at appropriate rates subject to rules and regulations of the Reserve Board.[249] The Board of Governors may suspend, for a period not to exceed thirty days, any reserve requirements specified in the Federal Reserve Act.[250] The Board is also responsible for the circulation of sufficient currency to meet the needs of the economy.[251] In addition to its rulemaking authority for the auto-

---

236 12 U.S.C. § 18439c.

237 12 U.S.C. §§ 625 & 602.

238 12 U.S.C. § 602.

239 12 U.S.C. § 3105.

240 12 U.S.C. § 3107.

241 12 U.S.C. § 248(j), 485.

242 12 U.S.C. § 1818(b).

243 12 U.S.C. § 1818(e).

244 12 U.S.C. §§ 1818(i) & 1847.

245 *See* 12 U.S.C. §§ 1861-67.

246 12 U.S.C. §§ 1881-84.

247 *See* Regulation P, 12 C.F.R. Part 216.

248 12 U.S.C. §§ 347b & 347.

249 *See generally* Regulation A, 12 C.F.R. § 201.

250 12 U.S.C. § 248(c).

251 12 U.S.C. §§ 411-421 & 428(o).

mation of the check processing system,[252] the Federal Reserve may suspend certain requirements regarding the availability of funds deposited by check in the event of any interruption of communication facilities, suspension of payments by another institution, war, or other emergency beyond the control of the depository institution.[253]

Federal Deposit Insurance Corporation

The Federal Deposit Insurance Corporation (FDIC) takes a very active role in the assurance of the banking infrastructure. Through its regulatory authority to conduct examinations of insured banks and depository institutions,[254] the FDIC has put in place regulations and policies intended to respond to the risks posed to banks by the integration of computer technology into retail banking operations. In February 1997, the FDIC issued the Electronic Banking Safety and Soundness Procedures and Financial Institution Letter (FIL) 14-97 to all insured financial institutions. The issuance resulted from requests of the banking industry for guidance on safety and soundness of practices in electronic banking. The Procedures not only identify risks, but also require examiners to monitor bank practices to ensure the appropriate safeguards are in place. The safeguards include capabilities to detect and prevent unauthorized access and duplicate transactions. The Procedures address: (1) system planning and deployment; (2) system operating policies and procedures; (3) transactional audit procedures; (4) legal and regulatory matters; (5) administration and systems operation; and (6) vendors and outsourcing.

In addition, FDIC issued guidance in July 1997 emphasizing the need to plan for the restoration of service resulting from any interruption of a key information system.[255] The FDIC policy reflects the revised policy guidance of the interagency Federal Financial Institutions Examination Council (FFIEC) on Corporate Business Resumption and Contingency Planning. The FDIC policy specifically names the board of directors as being responsible for ensuring that appropriate plans have been implemented for each financial institution. Contingency plans are evaluated by examiners during regular supervisory reviews of institutions.

The FDIC submission also noted that banks are required to report known or suspected criminal violations of federal law or suspicious transactions related to money laundering activity or a violation of the Bank Secrecy Act.[256] The Suspicious Activity Reporting (SAR) is currently administered

---

252 12 U.S.C. § 4008 (Expedited Funds Availability Act).

253 12 U.S.C. § 4003(d); 12 C.F.R. § 229.13(f).

254 *See, e.g.*, 12 U.S.C. § 1819-1820 & 1868.

255 *See* FDIC, Financial Institution Letter 68-97 (July 14, 1997).

by the Financial Crimes Enforcement Network (FinCEN) on behalf of federal banking and credit union regulators (including FDIC, OCC, OTS, Federal Reserve, and the NCUA). FinCEN acts as a central collection point and includes all SARs in a database available to law enforcement and regulatory agencies.

International Trade Commission

The International Trade Commission (ITC) is an independent, quasi-judicial agency established to govern trade-related matters. ITC has various investigative and adjudicative powers related to unfair trade practices, including import practices that harm U.S. businesses, patent infringement, economic effects of trade agreements and collection of trade-related data and statistics.

The primary area of ITC operations concerned with infrastructure assurance relates to the internal security of the agency, which includes both physical and information security. Information security is governed by a series of ITC Directives. U.S. ITC Directive 1345.0 established an Information Security Committee composed of management, general counsel, and external relations representatives. The Committee is responsible for developing, defining, inspecting and advising on the agency's facilities, procedures and controls for safeguarding national security information, sensitive information, and confidential business information in the possession and control of the ITC. Included within the ITC information security directives are requirements that employees report all possible compromises to ITC data and specific penalties for violations of ITC policy.

CONCLUSION

There are some very general conclusions that can be drawn from this federal "landscaping" exercise. First, there are many robust laws and regulations in place that serve to protect and assure our nation's critical infrastructures. Many of these laws and regulations were put in place with objectives other than infrastructure assurance in mind, but they nonetheless contribute to the infrastructures' resistance to threats of many kinds. However, the emerging cyber threat has only been addressed in very discrete areas. And as agencies begin to address the cyber threat they do so in a largely uncoordinated effort across the federal government. Coordination is going to be crucial for successful infrastructure assurance efforts because of the large number of federal organizations involved, not to mention state and local, international, and private concerns. This report can act as an important roadmap in considering who should be brought to

---

256 *See* 12 U.S.C. §§ 1818-1819; 31 U.S.C. § 5318; 12 C.F.R. Part 353.

the table from the federal government to plan and coordinate use of particular sets of legal authorities.

As infrastructure assurance efforts progress into the future, additional legal work is going to be crucial. This first attempt to survey and present the legal authorities of federal agencies with missions related to infrastructure assurance has only been a small part of the much larger task. The larger task that should be accomplished is closer review of the relevant authorities, mechanisms, and programs to ensure that they are appropriately compatible, that there are not unnecessary redundancies, that jurisdictional boundaries have been drawn properly, and that each crucial component has found a home within a responsible federal organization. This report may act as a starting point for this activity by focusing attention on such areas as priorities in restoration of service, new infrastructure assurance concerns such as interdependencies, and demonstrating the link between infrastructure assurance and agency missions that may be suitable for inclusion in agency strategic planning efforts. In addition, future efforts should begin to fold state and local, international and private authorities, mechanisms and policies into the large picture of the legal landscape for infrastructure assurance.

TABLE OF AUTHORITIES

Statutes

12 U.S.C. § 1818(b)
12 U.S.C. § 1818(e)
12 U.S.C. § 18439c
12 U.S.C. § 248(c)
12 U.S.C. § 248(j), 485
12 U.S.C. § 3105
12 U.S.C. § 391
12 U.S.C. § 4003(d)
12 U.S.C. § 4008
12 U.S.C. § 602
12 U.S.C. §§ 1818(i) & 1847
12 U.S.C. §§ 1818-1819
12 U.S.C. §§ 1819-1820 & 1868
12 U.S.C. §§ 1861-67
12 U.S.C. §§ 1881-84
12 U.S.C. §§ 325-326
12 U.S.C. §§ 343 & 347c
12 U.S.C. §§ 347b & 347
12 U.S.C. §§ 411-421 & 428(o)
12 U.S.C. §§ 625 & 602
12 U.S.C. § 3107
13 U.S.C. § 9
15 U.S.C. § 717b(a)
15 U.S.C. § 717z
15 U.S.C. § 781(k)(6)
15 U.S.C. §§ 3362-3363
15 U.S.C. §§ 5652
15 U.S.C. §§ 717b *et seq.*
16 U.S.C. §1 *et seq.*
16 U.S.C. § 2601 *et seq.*
16 U.S.C. § 5 & § 79
16 U.S.C. § 668dd(f)
16 U.S.C. § 791 a *et seq.*
16 U.S.C. § 8241
16 U.S.C. § 824-2(b)
16 U.S.C. § 824a(b)
16 U.S.C. § 824a(g)
16 U.S.C. § 824f
16 U.S.C. § 824i

16 U.S.C. § 825c
16 U.S.C. § 825f
16 U.S.C. § 825h
16 U.S.C. § 825j
16 U.S.C. § 825s
16 U.S.C. § 837g-1
16 U.S.C. § 839
16 U.S.C. §§ 791a-825r
16 U.S.C. §§ 824d & 824e
18 U.S.C. § 1029
18 U.S.C. § 1030
18 U.S.C. § 113B
18 U.S.C. § 1343
18 U.S.C. § 1361
18 U.S.C. § 1362
18 U.S.C. § 1366
18 U.S.C. § 1367
18 U.S.C. § 1705
18 U.S.C. § 1864
18 U.S.C. § 2115
18 U.S.C. § 2153
18 U.S.C. § 2511
18 U.S.C. § 3056
18 U.S.C. § 3061
18 U.S.C. § 641
18 U.S.C. § 871 & 879
18 U.S.C. §§ 1541-1546
18 U.S.C. §§ 1701-03
18 U.S.C. §§ 2511-21
18 U.S.C. §§ 793-94
18 U.S.C. §§ 876-77
18 U.S.C. Chapter 40
18 U.S.C. Chapter 44
19 U.S.C. § 1589a
19 U.S.C. § 482
20 U.S.C. § 5801 *et seq.*
20 U.S.C. § 6801 *et seq.*
20 U.S.C. § 8001
20 U.S.C. § 8003
20 U.S.C. § 8003
22 U.S.C. § 211a

22 U.S.C. § 2707
22 U.S.C. § 2709
22 U.S.C. § 2778
22 U.S.C. § 3244
22 U.S.C. § 401
22 U.S.C. §§ 2711-12 & 2708
22 U.S.C. Chapter 64
23 U.S.C. § 125
25 U.S.C. §§ 2101-09
25 U.S.C. §§ 2801-2809
25 U.S.C. §§ 311-328
25 U.S.C. §§ 3801-3804
25 U.S.C. §§ 4001-4061
28 U.S.C. § 533
30 U.S.C. § 202
30 U.S.C. § 185
30 U.S.C. § 185(h)
30 U.S.C. § 185(o)
30 U.S.C. § 226(g)
31 U.S.C. § 3121 *et seq.*
31 U.S.C. § 5317(b)
31 U.S.C. § 5318
33 U.S.C. § 1225, 1231
33 U.S.C. § 467h
39 U.S.C. § 2005(a)
39 U.S.C. § 401(2), (10)
39 U.S.C. § 401(3)
39 U.S.C. § 404(a)(7)
39 U.S.C. § 407
39 U.S.C. §§ 2003(b), 2401(a)
39 U.S.C. §§ 201, 403(a)
40 U.S.C. § 471 *et seq.*
40 U.S.C. § 490(a)(17)-(19)
40 U.S.C. §§ 481 & 491
40 U.S.C. §§ 601-619, 490
41 U.S.C. § 521 *et.seq.*
41 U.S.C. §§ 405 & 421
42 U.S.C. § 1856
42 U.S.C. § 2011 *et seq.*
42 U.S.C. § 2133
42 U.S.C. § 2167

42 U.S.C. § 2201(b)
42 U.S.C. § 2201(i)
42 U.S.C. § 2236
42 U.S.C. § 2271 *et seq.*
42 U.S.C. § 2280
42 U.S.C. § 2282
42 U.S.C. § 4001 *et seq.*
42 U.S.C. § 5170 *et seq.*
42 U.S.C. § 5171
42 U.S.C. § 5191(b)
42 U.S.C. § 5196(c), (g) & (h)
42 U.S.C. § 5196(e)
42 U.S.C. § 5811
42 U.S.C. § 5846
42 U.S.C. § 6201 *et seq.*
42 U.S.C. § 7701 *et seq.*
42 U.S.C. § 8301 *et seq.*
42 U.S.C. §§ 6231-6241
42 U.S.C. §§ 7101-7352
42 U.S.C. §§ 7270a-7270b
42 U.S.C. §§ 7701 *et seq.*
42 U.S.C. § 5195a(a)(1)
43 U.S.C. § 1331 *et seq.*
43 U.S.C. § 1732(b)
43 U.S.C. § 1733
43 U.S.C. § 1762(a)-(b)
43 U.S.C. § 1763
43 U.S.C. § 373
43 U.S.C. § 485h(d) & (e)
43 U.S.C. § 485i
43 U.S.C. § 502
43 U.S.C. § 503
43 U.S.C. § 506
43 U.S.C. § 620 *et seq.*
43 U.S.C. §§ 390b
43 U.S.C. §§ 498-499
43 U.S.C. 1737
46 U.S.C. § 3703
46 U.S.C. § 3306
47 U.S.C. § 1
47 U.S.C. § 1 *et seq.*

47 U.S.C. § 2(43)
47 U.S.C. § 2(b)(2)
47 U.S.C. § 214(d)
47 U.S.C. § 302(d)
47 U.S.C. § 303(m)(1)(C)
47 U.S.C. § 303(r)
47 U.S.C. § 333
47 U.S.C. § 4(i)
47 U.S.C. § 605
47 U.S.C. § 606
47 U.S.C. § 706
47 U.S.C. § 721(a)(6)
47 U.S.C. § 902(b)(2)(A)
47 U.S.C. §§ 201-276
47 U.S.C. §§ 3(44) & 3(46)
49 U.S.C. § 13931
49 U.S.C. § 5307(d)(1)(J)(i)
49 U.S.C. § 5321
49 U.S.C. § 60102
49 U.S.C. § 60103
49 U.S.C. § 60123
49 U.S.C. §§ 1(5),2,6,15(1) & 15(7)
49 U.S.C. app. §§ 1801-19
50 U.S.C. § 1701 *et seq.*
50 U.S.C. § 1701 *et seq.*
50 U.S.C. § 191
50 U.S.C. § 402(I)(1)
50 U.S.C. § 403
50 U.S.C. § 403-3(c)
50 U.S.C. § 403-3(d)(1)
50 U.S.C. § 403-4(e)
50 U.S.C. § 413b(a) & (e)
50 U.S.C. § 7270b(a)
50 U.S.C. §§ 205-226
50 U.S.C. §§ 219-226
50 U.S.C. §§ 401 *et seq.*
50 U.S.C. §§ 403a *et seq.*
50 U.S.C. app. § 2061 *et seq.*
50 U.S.C. app. § 2071(a)
50 U.S.C. app. § 2071(c)
50 U.S.C. app. § 2158

50 U.S.C. app. § 2160
50 U.S.C. app. § 2401 *et seq.*
8 U.S.C. § 1185
Title 18 U.S.C. Chapter 25
Pub. Law No. 99-399
Pub. Law No. 100-235
Pub. Law No. 101-298
Public Law 101-509
Pub. Law No. 102-486
Pub. Law No. 102-580
Pub. Law No. 103-62
Pub. Law No. 104-104
Pub. Law No. 104-106
Pub. Law No. 104-201
Pub. Law No. 104-208
Title 19, United States Code

Regulations

10 C.F.R. § 1048.5
12 C.F.R. § 201
12 C.F.R. § 229.13(f)
12 C.F.R. Part 216
12 C.F.R. Part 353
15 C.F.R. Part 730
18 C.F.R. § 141.300
18 C.F.R. Part 1b
18 C.F.R. Part 294
18 C.F.R. Part 35
18 C.F.R. Part 37
19 C.F.R. Part 122
23 C.F.R. Part 668
25 C.F.R. Part 169
25 C.F.R. Parts 225-27
28 C.F.R. Part 0.85
28 C.F.R. Part 0.85(a)
30 C.F.R. Part 250
36 C.F.R. § 14
39 C.F.R. § 224.233
41 C.F.R. Chapter 101
41 C.F.R. Part 101-38
43 C.F.R. § 2803.3

43 C.F.R. § 421
47 C.F.R. § 22.317
47 C.F.R. § 25.210(j)(2)
47 C.F.R. § 63.100
47 C.F.R. § 90.16
47 C.F.R. Part 11
47 C.F.R. Part 68
47 C.F.R. Part N, §§ 87.393-397
48 C.F.R. Chapter 1
49 C.F.R. § 192.613
49 C.F.R. § 192.615
49 C.F.R. § 195.402 (c)-(d)
49 C.F.R. § 195.403
49 C.F.R. §§ 192.614 & 195.442
49 C.F.R. §§ 192.705 & 192.721
49 C.F.R. §§ 195.116, 195.254, & 195.258
49 C.F.R. §§ 195.430 & 195.436
49 C.F.R. Part 190
49 C.F.R. Part 192, subpart D
49 C.F.R. Part 193
61 Fed. Reg. 21,737 (May 10, 1996)
62 Fed. Reg. 12,484 (March 14, 1997)
Title 10 of the Code of Federal Regulations

Presidential Authorities

Executive Order 10485

Executive Order 11423

Executive Order 12148

Executive Order 12333

Executive Order 12472

Executive Order 12631

Executive Order 12656

Executive Order 12844

Executive Order 13010

Executive Order 13031

National Security Directive 42

Presidental Memorandum, Facilitating Access to Federal Property for the Siting of Mobile Services Antennas

Agency Authorities

BLM Law Enforcement General Order 23

FERC Order No. 889

1993 Cooperative Fire Protection Agreement in the State of Alaska

DOE Delegation Order No. 0204-1

Electronic Banking Safety and Soundness Procedures and Financial Institution Letter (FIL) 14-97

FCC Instruction 1479.1

FDIC, Financial Institution Letter 68-97

Federal Preparedness Circular 60

Federal Response Plan

H-9260-1 BLM Law Enforcement Management

Inter-Departmental MOU allowing Cross-Deputation of Law Enforcement

Memorandum of Understanding Between the Department of Transportation and the Department of Interior Regarding OCS Pipelines

Memorandum of Understanding Between the Director of the National Institute of Standards and Technology and the Director of the National Security Agency Concerning the Implementation of the Public Law 100-23532

NTIA Emergency Readiness Plan for Use of the Radio Spectrum

OMB Circular No. 1-119

OMB Circular No. A-130

Other Authorities

4 D.C. Code § 201 *et seq.*

Cases

United States v. Ajlouny, 629 F.2d 830 (2d Cir. 1980)

Reference Materials

Argonne National Laboratory, *Overview of the Impacts of Regulatory Agency Practices on Critical Infrastructure Protection*

Critical Infrastructure Working Group Report

Network Reliability Performance (NRIC)

Network Reliability—The Path Forward (NRIC)

APPENDIX
Federal Agency Contacts

CENTRAL INTELLIGENCE AGENCY

Tom Benjamin
Jonathan Fredman

FEDERAL BUREAU OF INVESTIGATION

M. E. Brown, Associate General Counsel
Marion Bowman

DEPARTMENT OF ENERGY

Samuel Bradley

DEPARTMENT OF TRANSPORTATION

Steve Metoyer
David Tochen

FEDERAL EMERGENCY MANAGEMENT AGENCY

Joseph Flynn

DEPARTMENT OF COMMERCE

Roman W. Sloniewsky

DEPARTMENT OF TREASURY

Virginia Rutledge

DEPARTMENT OF JUSTICE

Vincent Garvey, Civil Division, Federal Programs Branch
Chris Kelly, Antitrust Division
Jeffrey Axelrad, Civil Division, Torts Branch
Sarah Locke, Civil Division, Torts Branch
Roger Pincus, Office of Intelligence Policy and Review
Rosemary Hart, Office of Legal Counsel

NATIONAL SECURITY AGENCY

Richard Marshall, Associate General Counsel

U.S. CUSTOMS SERVICE

Ellen Y. McClain, Deputy Assistant Chief Counsel (Enforcement)

BUREAU OF ALCOHOL, TOBACCO & FIREARMS

David Lieberman, Attorney

U.S. SECRET SERVICE

Mark Mulligan, Attorney/Advisor

DEPARTMENT OF EDUCATION

Deborah Friendly

GENERAL SERVICES ADMINISTRATION

Michael Ettner, Senior Assistant General Counsel

FEDERAL COMMUNICATIONS COMMISSION

Jim Keegan, Associate Chief, Networks

U.S. POSTAL SERVICE

Fred Eggleston, Chief Counsel
Andrew German

DEPARTMENT OF INTERIOR

Arthur Gary, Senior Attorney

DEPARTMENT OF STATE

Michael Matheson, Acting Legal Advisor
K. E. Malmborg, Attorney/Advisor
Idris M. Diaz

U.S. INFORMATION AGENCY

Gordon John Dickey, Assistant General Counsel

NUCLEAR REGULATORY COMMISSION

Martin Malsch, Deputy General Counsel
Kathryn L. Winsberg, Senior Attorney

FEDERAL ENERGY REGULATORY COMMISSION

Robert C. Fallon, Senior Counsel to the Chair, FERC
David N. Cook, Deputy General Counsel

DEPARTMENT OF AGRICULTURE

Michael W. Kelly, Assistant General Counsel

SECURITIES AND EXCHANGE COMMISSION

Marva Simpson, Attorney/Advisor

BOARD OF GOVERNORS OF THE FEDERAL RESERVE

Stephen L. Siciliano, Special Assistant to the General Counsel for Administrative Law

FEDERAL DEPOSIT INSURANCE CORPORATION

INTERNATIONAL TRADE COMMISSION

Richard Liebeskind, Deputy Assistant Director, Bureau of Competition

COMMODITY FUTURES TRADING COMMISSION

Harold L. Hardman, Assistant General Counsel

U.S. INTERNATIONAL TRADE COMMISSION

Mark Morin

DEPARTMENT OF DEFENSE

Mary De Rosa
Stuart Aly

CHAPTER 5

THE REGULATORY LANDSCAPE

OVERVIEW OF THE IMPACTS OF REGULATORY AGENCY PRACTICES ON CRITICAL INFRASTRUCTURE PROTECTION

Report to the President's Commission on Critical Infrastructure Protection 1997

This report was prepared for the President's Commission on Critical Infrastructure Protection, and informed its deliberations and recommendations. The report represents the opinions and conclusions solely of its developer, Argonne National Laboratory.

OVERVIEW OF THE IMPACTS OF REGULATORY AGENCY PRACTICES ON CRITICAL INFRASTRUCTURE PROTECTION

TABLE OF CONTENTS

OVERVIEW OF THE IMPACTS OF REGULATORY AGENCY PRACTICES ON CRITICAL INFRASTRUCTURE PROTECTION

1 INTRODUCTION

1.1 Purpose

The purpose of this study is to provide an understanding of the relationship between protection of critical infrastructure and the regulatory environment in which the infrastructure operates. This understanding will allow policy makers to target key areas in the search for solutions to the problem of assuring protection of our nation's critical infrastructures.

Specific objectives of this study are to:

- Provide a general overview of the regulatory practices of federal, state and local regulatory bodies that govern the operation of seven critical infrastructures;
- Identify current trends in regulatory reform as they affect critical infrastructures;
- Profile key representative agencies involved in developing and enforcing regulatory requirements in these seven areas;
- Analyze the effects of current regulatory schemes on critical infrastructure protection in the regulated industries; and
- Suggest approaches and measures to consider in developing regulatory measures for critical infrastructure protection.

1.2 Scope

The scope of this study is limited to seven of the critical infrastructures identified in Executive Order 13010: telecommunications, electrical power systems, gas and oil delivery and storage, banking and finance, transportation, water supply systems, and emergency services. Each of these infrastructures represents an industry that is subject to extensive regulation by federal, state and local agencies.

1.3 Organization of This Report

The remainder of this report is organized into eight sections. Section 2 provides an overview of the entire field of infrastructure regulation, identifying common themes and significant distinctions among the different infrastructure sectors and their regulatory environments. Overall regulatory trends are highlighted, as well as recurrent themes that appear among the suggested measures for improvement. Sections 3 through 9 contain individual treatments of the infrastructures. For each infrastructure, the following items are provided:

- A "snapshot" overview of the current regulatory situation, covering the full regulatory field for that industry at the federal, state and local level.

- An analysis of regulatory trends at work. Most infrastructures have experienced considerable regulatory changes in the past two decades; the effects of these changes in some cases are still playing out.
- Detailed studies of a sample of regulatory agencies. For each infrastructure, a handful of regulatory agencies were selected for more detailed profiling, to provide a feel for how the regulatory system works in that industry. Selections were based on the following factors: (1) spanning the entire field of the infrastructure; (2) covering the spectrum of government levels (federal, state, local); (3) recognizing diversity of approaches; and (4) highlighting innovative or model approaches. For each agency profiled, there is a discussion of its regulatory jurisdiction, authority, and regulatory methods, and a summary of the pertinent regulations.
- An evaluation of the effect of regulatory practices on critical infrastructure protection. Elements of the regulatory systems that appear to enhance or detract from infrastructure assurance are identified. Examples of elements that enhance protection include direct security requirements, such as restricted access, background checks on employees, and emergency planning. Also included are requirements that are aimed at protecting the system against other hazards (e.g., harsh weather) but incidentally have the effect of reducing vulnerability against hostile action; for example, requirements for physical protection of equipment or system redundancy. Examples of elements that may *reduce* protection include requirements that force disclosure of sensitive information that hostile elements might use to identify vulnerable points.
- Suggested strategies to promote infrastructure assurance objectives through regulatory means. This section provides suggestions on strategies for enhancing critical infrastructure protection, without doing harm to the overall objectives of particular regulatory schemes. In light of the current trend toward deregulation, in many cases these strategies focus on cooperative industry standard setting, rather than development of traditional "command and control" regulations.

This report also includes an appendix listing relevant acronyms, agencies, associations and statutes.

2 OVERVIEW OF INFRASTRUCTURE REGULATION

2.1 The Regulatory Landscape

2.1.1 Historical Development of Infrastructure Regulation

Our critical infrastructure is the product of a long process of co-evolution between government and industry. Every aspect of the infrastructure, from its physical configuration down to patterns of ownership, reflects the influence of past and present government controls. These controls in turn originated in response to three main concerns: access to service, market power, and safety. These three issues cut across many different infrastructures.[1]

Access to Service

Promoting universal access to service has long been a feature of utility regulation, and is a stated goal of most utility-regulation statutes.[2] A prominent example of this policy was the establishment of a program of assistance to rural electric cooperatives under the Rural Electrification Act of 1936 (Pub. L. 74-605) for the express purpose of serving rural areas that might not otherwise be economic for utilities to serve. Similar policies have heavily shaped the development of the natural gas, telecommunications, and transportation industries.

Universal access has been perceived as a desirable social goal for several reasons: it provides the basis for economic growth, increases the standard of living, and generally brings the benefits of our social investment in infrastructure to the largest possible population. In addition, for some types of infrastructure, there is a positive feedback mechanism such that the more widespread the system is, the more useful it becomes. For example, as more people and businesses become reachable by telephone, the more valuable a telephone is. People come to rely on the telephone, and the pressure on each household (and especially each business) to have one increases. Similarly, the more extensive the road system, the more valuable a car becomes as a means of getting places.

Market Power

Most of the infrastructures studied involve the transport of something such as electricity, gas, or water from a few production points, through main trunk lines, then through a very finely divided and extensive distribution network to individual users. Setting up such a system involves a huge capital investment, and in most cases also involves a delegation of governmental authority in order to condemn easements that spread across the

---

1 For example, the Illinois Public Utilities Act provides for "adequate, efficient, reliable, environmentally safe and least-cost public utility services. . . ." (220 ILCS 5/1-102). The Communications Act of 1934 (48 Stat. 1064, as amended) established the Federal Communications Commission (FCC) "For the purpose of regulating interstate and foreign commerce in communication by wire and radio so as to make available, so far as possible, to all the people of the United States a rapid, efficient, Nation-wide, and world-wide wire and radio communication service, with adequate facilities at reasonable charges, for the purpose of the national defense, [and] for the purpose of promoting safety of life and property through the use of wire and radio communications . . ." (47 U.S.C. § 151).

2 *See* footnote 1.

landscape. The required investment in physical infrastructure creates a barrier to entering the market. On the one hand, no company would want to lay out such an investment without some assurance that it will pay off. On the other hand, once completed, the infrastructure investment puts the supplier in a position of power with respect to both customers and competition—to compete, another company would have to duplicate the delivery network. It would not be practical, nor desirable from a public policy standpoint, for another company to begin stringing wires or laying pipelines that are redundant with the existing network. But from the customers' standpoint, there is no choice but to deal with the company whose pipe, or wire, or railroad comes to town. The term "natural monopoly" was coined by economists to describe this situation.[3] Regulators have traditionally approached this situation by granting franchises to serve particular areas, thus assuring the company of demand, and at the same time regulating prices so as to protect the captive consumers from the effects of monopoly pricing.

Safety

Another factor that historically drove development of regulation is concern over safety. Power lines, pipelines, water supplies, railroads, airlines, etc. all carry the potential for disaster if not built, maintained and operated correctly. The community has an interest in ensuring that the companies performing these activities have the technical competence to manage their facilities in a safe manner. Thus, specifications for equipment, installation, operating practices, and employee qualifications became a common purview of infrastructure regulation.[4]

Banking and Financial Services

Regulation of banking and financial services has followed a somewhat different path and reflects different priorities than the regulatory systems sketched above. For one thing, it has a much longer history: the government's role in banking long preceded the technologies that gave rise to the oil, gas, electric, and telecommunications industries. Also, the challenges posed by regulation of banking are substantially different than those associated with the other infrastructures. Regulation of banking and financial services is primarily aimed at (1) protecting depositors and investors; (2) providing a steady source of funds for investment in economic ventures; (3) maintaining a stable currency; and (4) providing an atmosphere

---

3 *See, for example,* discussion of natural monopolies in Kenneth E. Train, *Optimal Regulation*, MIT Press, Cambridge, Massachusetts, 1991, pg.1 *et seq.*

4 *See, for example,* the Illinois Gas Pipeline Safety Act, 220 ILCS 20/3(b): "Standards established under this Act may apply to the design, installation, inspection, testing, construction, extension, operation, replacement, and maintenance of pipeline facilities."

of stability, trust and legal enforcement that will foster healthy economic growth. These goals are pursued through regulatory measures designed to ensure full and fair disclosure of information to investors, prevent banks and savings institutions from taking undue risks, and ensure the safety of deposited funds. The government also actively manages financial markets through control of interest rates and money supply, and through budgetary measures.

Emergency Services

Regulation of emergency services also differs significantly from regulation of the other infrastructures, since emergency services are primarily provided by governmental entities. A majority of emergency services are provided by local fire and police departments, public health services, and local medical services. The effectiveness of these organizations is subject to local political review and control rather than the pervasive, nationally based regulatory systems found in other infrastructures. With minor exceptions, only in the last couple of decades has the federal government established a major role in emergency planning and response.[5]

2.1.2 Current Trends in Regulation

All of the infrastructures have undergone major regulatory changes in recent years. The rationales underlying the original regulatory schemes have not been abandoned. However, in most of the infrastructures there have been major efforts to open markets and increase competition.[6] The major trends are summarized below.

Restructuring of Utilities

Starting in the 1970s, the traditional utility regulatory schemes began to be perceived as overly restrictive, stifling competition and in some cases causing artificial shortages of the commodities they were supposed to assure. A series of reforms was initiated, with the primary goal of increasing opportunities for competition and market forces to balance supply and demand. The reform process and its consequences are still playing out today. Although the reform process is commonly referred to as "deregulation," that

---

5 *See* discussion and notes regarding regulatory trends in Section 2.1.2.

6 Although the historical goals of universal service and economics have not been explicitly disowned, some observers have pointed out that they conflict with the current emphasis on market competition. See, for example, Robert M.Frieden, *Dialing for Dollars: Should the FCC Regulate Interstate Telephony?*, 23 Rutgers Comp. & Tech. Law J. 47, note 16: "Presently, the universal service system is incompatible with the procompetitive efforts required by the Telecommunications Act . . . The current universal service system is a patchwork quilt of implicit and explicit subsidies. These subsidies are intended to promote telephone subscribership, yet they do so at the expense of deterring or distorting competition."

is not an entirely accurate characterization. "Industry restructuring" is perhaps a more accurate term.

Industry restructuring refers to the fact that in many areas, including at least telecommunications, electricity, gas and oil, and rail transport, the effect of the recent changes has been to segment the industry into two or three parts. For example, AT&T divested its local telephone companies while retaining its long-distance service. The natural gas industry is being segmented into natural gas producers, interstate pipeline companies, and local distribution companies. Companies that were once "vertically integrated" from production facilities to ultimate consumers, are now broken up into separate units that either produce, transport, or distribute to individual customers. The purpose and effect of this restructuring has been to create separate markets for the bulk production and transport segments. Thus for example a large industrial consumer of natural gas is no longer tied to the local gas company; it can now purchase bulk quantities directly from a supplier and then contract with a pipeline company to have it transported to the factory.

This type of change puts suppliers and bulk transporters in the position of competing for business and lets market forces set prices, driving increases in efficiency and decreases in cost. Bulk pricing is in some sense deregulated. However, there is still a considerable regulatory presence. In fact, some transporters and local distributors might consider themselves to be less autonomous than before, in that they are now obliged to provide their services for the benefit of other corporate entities. For example, electric companies must now make their transmission lines available for other companies (their competitors) to send power to large customers (who were formerly *their* customers). Railroads must allow passage of competing shippers' freight over their tracks. Arrangements, procedures, liabilities, and rates for these transactions are still subject to regulatory control through the filing of tariffs.

Application of Risk Management Techniques

Under recent Congressional direction, some agencies are incorporating elements of risk management technique into their regulatory schemes. For example, the Accountable Pipelines Safety and Partnership Act of 1996 (Pub. L. 104-304) calls for the U.S. Department of Transportation (DOT) to perform risk assessments and cost-benefit analyses when setting pipeline safety standards. It also establishes a voluntary risk management demonstration program. Under this program, participating pipeline operators will have the opportunity to submit risk management plans designed to achieve an equivalent or greater overall level of safety (as compared with strict compliance with DOT standards) at lower cost. Another example is found in water monitoring. The U.S. Environmental Protection Agency

(EPA) has proposed regulations for reform of chemical monitoring regulations under the Safe Drinking Water Act. Under this proposed rule, drinking water suppliers would be able to tailor their monitoring programs to achieve the maximum safety benefit at the lowest cost, on the basis of risk analysis.[7]

Growth of Technological Emergency Response Capabilities

A major development in the regulation of emergency services has been the growth of specialized programs aimed at technological hazards. Governmental interest in this area was stimulated by technological disasters such as the nuclear power plant accidents at Three Mile Island and Chernobyl and the chemical releases at Bhopal, India and Institute, West Virginia. Subsequent regulatory action has included programs to promote radiological emergency preparedness at the state and local level,[8] a national hazardous material-response program,[9] a program to stimulate state and local hazardous material planning,[10] workplace safety rules for emergency responders,[11] and numerous requirements for emergency planning on the part of industrial firms handling hazardous materials.[12]

Reduction of Banking and Financial Service Restrictions

Recent changes in banking and financial service regulations have tended to reduce the degree of compartmentalization that had previously been imposed on the industry. By repealing part of the Glass-Steagall Act, Congress has allowed banks to grow larger and branch across state lines, subject to concentration limits, state laws, and evaluation of the companies. Further changes in the law allowed bank holding companies to own more than one kind of banking enterprise, so that small banks and mortgage lending institutions began to be combined under centralized regional management. However, such combinations are potentially subject to regulation by many

---

7 U.S. EPA, *Drinking Water Monitoring Requirements for Certain Chemical Contaminants—Chemical Monitoring Reform (CMR) and Permanent Monitoring Relief (PMR), Part II*, 62 FR 36100, July 3, 1997.

8 *See* NRC regulations at 10 CFR 50, App. E and FEMA regulations at 44 CFR 350 *et seq.*

9 National Contingency Plan, 40 CFR 300 *et seq.*

10 40 CFR 355.

11 29 CFR 1910.120(q).

12 EPA regulations issued under the Resource Conservation and Recovery Act (RCRA) [applying to facilities that store or process hazardous waste (40 CFR 265.56)], Clean Air Act (CAA) [release prevention and response for hazardous air pollutants (40 CFR 68; 61 FR 31668, June 20, 1996; 61 FR 16958; and 61 FR 31730)], and Clean Water Act (CWA) [oil spill prevention and response (40 CFR 112)] all contain provisions addressing emergency preparedness.

agencies. In addition, banks are now permitted to offer types of financial services and investment opportunities that they previously could not.[13]

Regulation of the Internet

The Internet has experienced exponential growth, and is currently estimated to have at least several tens of millions of users worldwide. At the same time, the types of uses to which it is put are also multiplying. From the initial postal-like exchange of written messages and data, use of the Internet has diversified to include commercial transactions, advertising, financial transactions, and real-time audio and video transmission, among others. The Internet itself is rapidly becoming a vital part of our critical infrastructure.

Regulation of the Internet is still mostly at the discussion stage. Of the steps that have been taken, some have failed in the courts. However, it may be expected that the next several years will bring increased regulatory attention to at least certain types of Internet activities. A few such areas are outlined below.

- *Regulation of moral content.* Congress's attempt to sanction pornography on the Internet was declared unconstitutional by the Supreme Court in *Reno, Attorney General of the United States et al. v. American Civil Liberties Union et al.*, 117 S. Ct. 2329, June 26, 1997. In so doing, the Court identified First Amendment interests in Internet transmissions that will make it difficult to enforce regulations on the moral content of messages.
- *Telephony.* Given sufficient data transmission rates, users with appropriate equipment and software can use the Internet to conduct real-time audio exchanges; i.e. to mimic a telephone. Telephone service representatives have argued to the Federal Communications Commission (FCC) that Internet service providers performing this function should be subject to the same regulatory obligations imposed on telephone companies.[14]
- *Advertising.* Many home pages, particularly the popular search engines, are accompanied by advertising messages. Such advertising bears certain similarities to broadcast or print media advertisements; however, it differs in that the particular ad that appears to the user can be selected based on information about that particular user's preferences, patterns of Internet use, or demographic profile. Application of truth-in-advertising standards and other regulatory

---

13 Riegle-Neal Interstate Banking and Branching Efficiency Act of 1994 (Pub.L. 103-328, 108 Stat. 2338).

14 *See* Frieden, *supra* note 6.

measures to Internet advertising has been approached differently by different jurisdictions.[15]

- *Commercial transactions.* As commerce on the Internet grows, regulations and legal norms will be developed to ensure that the system can be relied on, and that transactions will proceed according to a well understood set of expectations. As stated by one scholar,

> Businesses of all types routinely use the Internet for a variety of commercial transactions, and consumer services have begun to appear. At present, commercial traffic on the network generally culminates in an exchange of physical goods, and it is presently possible to access a variety of mailorder catalogs online, to arrange for purchase of music, books, fast food delivery, even flowers. Unlike transactions involving physical goods, delivery of digitized information products such as music, photographs, novels, motion pictures, multimedia works, and software can be accomplished entirely within the network itself. Such information products already represent a sizeable portion of the gross national product of developed nations. That portion is likely to increase world-wide, and the Internet will facilitate such increases. And, where there is commerce to be had, regulation is sure to follow.[16]

2.2 Opportunities for Enhancement of Infrastructure Security

2.2.1 Current Regulatory Requirements for Critical Infrastructure Protection

A survey of the regulatory environment for critical infrastructure reveals that there are relatively few instances where regulations directly address protection against hostile acts. Nuclear power plant operators are required to be prepared against a "design basis threat" described in some detail in terms of the numbers of persons involved, vehicles and weaponry at their disposal, types of expertise at their command, and cooperation by insiders.[17] These regulations make sense in the context of a nuclear power plant, where the consequences of an incident can be severe, the target is highly

---

15 As discussed by Dan L. Burk in *Federalism in Cyberspace,* 28 Conn. L. Rev. 1095 (Summer 1996), "Of course, even without the enactment of new laws or regulations, there are already on the books plenty of laws that states might apply to the Internet, including consumer protection statutes and other public law to police online behavior and commerce. The Minnesota Attorney General's office in particular has been very aggressive in pursuing what it considers to be online violations of Minnesota law, filing a flurry of lawsuits against out-of-state advertisers and service providers. The Illinois Attorney General's office is equally eager to get into the cyberspace game. By contrast, the Attorney General of Florida, has opined that because of the novel nature of the Net, forays into online enforcement of current law would be premature."

16 *Id.*

17 10 CFR 73(a)(1).

valuable and visible, and the facility is relatively defensible. The result is that nuclear power plants have multiple security perimeters, fencing, monitoring equipment, and round-the-clock security forces. Similar considerations and requirements apply, to varying degrees, at liquefied natural gas (LNG) storage facilities[18] and airports.[19] However, it is questionable whether this type of approach could be practically employed in other infrastructure contexts. Most elements of critical infrastructure are by their nature more spread out and less defensible than these facilities.

Many critical infrastructures are subject to regulatory requirements that have the effect of making the system more resistant to attack, even though they are aimed at reducing vulnerability to natural or accidental threats. For example, natural gas mains are required to be buried below ground. Burial protects them from ordinary threats such as storms or vehicle accidents; and it incidentally makes them less accessible to saboteurs. In fact, the more densely populated the area they go through, the deeper they are required to be buried. Similar examples are found in other infrastructures.

- Electrical substations are required to be fenced and locked, primarily to protect the public from accidental electrocution but also to prevent vandalism.
- Transmission line towers are built to exacting technical standards to prevent wind damage. Their robust construction also makes them generally more difficult to knock over, by whatever means.
- Certain types of protection are required for watersheds and surface water reservoirs that serve as drinking water supplies. These are primarily aimed at preventing contamination of drinking water supplies by pollution from runoff, migration from hazardous waste sites, or other types of conventional industrial pollution.

These and other standards that provide incidental protection from hostile acts are concerned exclusively with physical threats. Protection from cyber threats was not found to be addressed among the regulatory systems surveyed.

2.2.2 Proposals for Regulatory Measures to Enhance Infrastructure Protection

As noted, at present there are few regulatory measures directed at protection of critical infrastructures from hostile actions. Development of such measures will require careful review and analysis of costs, benefits, and options for achieving the desired level of assurance. Any proposal for

---

18 49 CFR 193.2001 *et seq.*

19 14 CFR Part 107.

regulatory action would have to be individually fitted into the existing regulatory system for that infrastructure. However, review of the current regulatory schemes for each infrastructure suggests a small number of themes that are common to multiple infrastructures—areas in which it may be worthwhile to seek regulatory solutions to problems of infrastructure protection. These recurrent themes are described below.

1. *Information protection.* In several areas, the operation of regulatory schemes causes information about the industry (e.g., schematic diagrams outlining networks of pipelines or power lines, or summaries of contingency planning) to be collected at a centralized location. Such concentrated information might be of interest to individuals intending to disrupt the system. Interviews with industry representatives indicated that they are concerned about this possibility. In one area—airport security—regulatory initiative has already been taken to impose controls over sensitive information.

2. *Contingency planning.* Most utilities, and certainly emergency services, already perform contingency planning to prepare for disasters. Such planning is generally supplemented by drills and exercises to hone response capabilities. However, preparedness efforts are based on responding to identified scenarios. The types of scenarios considered are often limited to natural disasters or technological problems of an accidental nature. Preparedness for malevolent acts could be enhanced by including that type of scenario among those considered in developing emergency plans.

3. *Preservation of redundancy.* In several industries, recent regulatory changes may reduce the level of redundancy in the infrastructure. As competition places greater premiums on economic efficiency, existing equipment is being used to capacity. There is more traffic on fewer lines. The result is less margin for dealing with disruptions to the system. Incentives for retaining reserve capacity could enhance the resiliency of infrastructure networks, possibly preventing small disturbances from being amplified into regional or national problems.

4. *Protection from cyber threats.* Reliance on computerized information systems is increasing. This trend is most pronounced in the telecommunications and banking areas, but is also present in all of the infrastructures in varying degrees. Electrical grids, oil and gas valves, and municipal water systems are all operated remotely by means of computerized data transmission. Security measures have been implemented in order to protect critical data and functions; however, there is little regulatory oversight of this area. There is a need for study in this area, to identify vulnerabilities, and determine for each infrastructure whether there may be an appropriate role for government-mandated or industry-based standards regarding cyber security.

2.2.3 Availability of Regulatory Leverage

In order to implement measures for infrastructure protection, it is necessary to have a mechanism for introducing new ideas or practices on a national scale. For some infrastructures, the regulatory field has a strong federal or coordinated industry presence that provides a convenient means for introducing new initiatives. In other areas, however, regulatory control is fragmented among many state and local jurisdictions, making it difficult to promulgate widespread change. For example, the EPA provides strong leadership in regulation of drinking water supply systems. Although most enforcement is conducted by states, they more or less uniformly adopt the EPA standards; any regulatory changes adopted by the EPA are soon echoed at the state level. Other aspects of water system regulation are covered by a set of industry-developed standards (promulgated by the American Water Works Association) that have been widely adopted by state agencies. Among electric utilities, the North American Electric Reliability Council, an industry organization, promulgates standards for contingency planning and operations that are uniformly followed. On the other hand, there does not appear to be any analogous forum in the oil industry. Although there is a substantial federal presence in oil industry regulation, it is split among several different agencies, and in fact much of the detailed control of production and safety standards takes place at the state level. It can be difficult to determine whose regulations apply to certain particular pipes or valves. In considering opportunities for implementing change, it will be important for policy makers to consider what forums are available to disseminate their ideas to the industry.

3 INFORMATION AND TELECOMMUNICATIONS INFRASTRUCTURE

3.1 General Description of Regulation

3.1.1 Current Regulatory Environment

Federal regulation of wireline and wireless communication of voice, data, and image signals within the United States is authorized by the Communications Act of 1934 (the Act) (47 U.S.C.A. 47), as amended by the Telecommunications Act of 1996 (1996 Act) (Pub. L. 104-104, 110 Stat. 56). The Act established the Federal Communications Commission (FCC), as the body to execute and enforce its provisions.

With regard to national infrastructure policy, a role is also performed by the National Telecommunications and Information Administration (NTIA) which was created in 1970 within the U.S. Department of Commerce (DOC). Through its Office of Policy Analysis and Development, it plays the role of "principal advisor to the President on telecommunications policies."[20] As

---

20 47 U.S.C.A. § 901(b)(6) and DOC World Wide Web home page, National Telecommu-

one of its primary policy functions, the NTIA was directed by Congress to foster "national safety and security, economic prosperity, and the delivery of critical social services through telecommunications."[21] The NTIA has filed numerous comments in matters docketed before the FCC to express the Administration's desires with regard to deregulation of the industry, particularly as to the assurance of universal access to telecommunications service at affordable rates.[22]

In addition to the federal presence, all 50 states have some form of regulatory body that addresses telecommunications; generally it is a public utility commission (PUC) or a subpart thereof. For the most part, these state agencies are concerned with facility siting, equipment siting, and rate setting, but a growing number are addressing access and continuity of service.[23] They generally focus on promoting and accommodating commerce while ensuring the most democratic distribution of service economically permissible.

3.1.2 Current Trends in Regulation

At present, the information and telecommunications infrastructure is experiencing deregulation as a result of the 1996 Act. The primary objective of deregulation was to stimulate increased competition in the industry so as to ensure the lowest feasible service prices to individuals and businesses.

The stated intent of the 1996 Act was "to promote competition and reduce regulation in order to secure lower prices and higher quality services for American telecommunications consumers and encourage the rapid deployment of new telecommunications technologies."[24] The deregulatory content of the 1996 Act opens markets by imposing new obligations for

---

nications and Information Administration, Office of Policy Analysis and Development, at http://www.ntia.doc.gov/opadhome/opadhome/html.

21 47 U.S.C.A. § 901(c).

22 See, for example, Reply Comments of the National Telecommunications and Information Agency, May 30, 1996, *In the Matter of the Local Competition Provisions in the Telecommunications Act of 1996*, CC Docket No. 96-98.

23 *See*, for example, (1) Illinois Code of Statutes, Chapter 220 (220 ILCS), 5/13-102; (2) *Implementing the Telecommunications Act '96*, Washington Utilities and Transportation Commission, summary of report findings, November 30, 1996; (3) California Public Utilities Code, Chapter 4, Article 8, Paragraphs 851-856 (Universal Service); (4) New Jersey Board of Public Utilities, Ratepayer Advocate Recommendations, Section V, *Definition of University Service*; (5) Public Utility Commission of Texas, Rule ¶23.97 relating to interconnection for local exchange service; (6) Report of the Public Utility Commission of Texas, references to universal, nondiscriminatory service as defined by the U.S. Telecommunications Act of 1996; and (7) *Texas Register* (20 TexReg 8777), October 24, 1995.

24 *See* Telecommunications Act of 1996, P.L. 104-104, February 8, 1996.

interconnection, unbundling of services, and resale of infrastructure access on existing providers of local exchange services.

Implementation of the provisions of the 1996 Act will require development of new business models for telecommunications to address costing and business practices in an industry that was previously highly oligopolistic. In recognition that local exchange markets will not immediately become competitive, Congress imposed immediate telemessaging and electronic publishing requirements on local exchange carriers aimed at detecting and preventing improper cost allocation, discrimination, or other anticompetitive conduct. The provisions of the 1996 Act are being implemented by the FCC through a series of hearings and subsequent orders.

Simultaneous with deregulation of the industry, there is a continuing and growing wave of new products and services being introduced. Emerging technologies such as synchronous optical networks, new wireless access with personal communications services, satellite nets, and advanced intelligent networks, when coupled with smaller, cheaper, and more powerful computers, will change communication and information access in ways that are not yet fully clear. There will be a greater number of independent service providers, furnishing a greater variety of services and products to the end user, presumably at lower costs. This variety will make any system-wide security initiatives more difficult to develop and enforce. Each new technology will create additional challenges for regulators to balance service and security objectives.

In any communication system, there is a tradeoff between security and access. These conflicting goals are reflected in the statutes that set regulatory policy. For example, the principal objectives of the Act include service to "so far as possible . . . all the people of the United States." The Act also states as primary objectives the provision of communication services for the purposes of national defense and promoting safety of life and property. Most of the state statutes also contain language about the importance of both nondiscriminatory service and public safety. Resolving conflicts between the desire for a highly available, reliable, and affordable system for the general public, and the need for a secure, responsive, and reliable system for national security, law enforcement, and emergency response agencies will be a great challenge for the regulators.[25]

---

25 As was pointed out in an assessment of the impact of these technologies on military operations, "the characteristic that gives any system its potency — that the parts of a system enhance the effectiveness of one another — also make some systems susceptible to catastrophic failure if one of their central parts can be jeopardized." (Admiral William Owen, U.S.N. [retired], former Vice Chairman of the Joint Chief of Staffs, with Martin Libicki, in the introduction to *Dominant Battlespace Knowledge*, the National Defense University, Washington, D.C., 1995, pg. 11).

The greater part of the national telecommunications infrastructure is owned by private companies. These companies also possess the largest portion of the technical and operational expertise required to maintain the infrastructure. Federal and state regulators have long recognized these facts and have used the expertise of the system operating companies to help develop polices and regulations. The disaggregation of the industry brought about by the 1996 Act will bring in many new players at the local service level, with greatly varying degrees of expertise. Few if any of these new players will have the capital or expertise to address infrastructure assurance unless it directly relates to their service delivery. Any successful plan for accomplishing both the service and security objectives of the 1996 Act will require both (1) close coordination among the federal and state agencies involved and (2) the assumption of many of the tasks involved in managing the consequences of failure of infrastructure protection by the newly deaggregated telecommunications industry. Encouraging and embracing this industrial assumption of risk will involve a rethinking of traditional regulatory schemes, which have as one main tenet to keep individual service providers self-sufficient, and which discourage pooling of assets and activities.

3.2 Description of Selected Regulatory Agencies

3.2.1 Selection Method

The FCC is the primary agency for federal regulation of the communications infrastructure. It is the principal interface between the federal government and the many public and private entities that make up the telecommunications industry. Even in areas where it does not exercise direct regulatory authority, the FCC is a "persuasive presence" promoting a coherent policy.

The state agencies selected for analysis represent various approaches to regulation of the telecommunications industry. They are among the states cited as exhibiting "leadership on the primary journey that will lead to the information superhighway."[26]

3.2.2 Agency Descriptions

3.2.2.1 Federal Communications Commission

Authority and Jurisdiction

The Act is the primary source of authority for the FCC. As stated in the Act, the jurisdiction of the FCC pertains to:

> all interstate and foreign communications by wire or radio and all interstate and foreign transmission of energy by radio, which originates and/or is

---

26 *State Report on Citizen Participation*, Center for Policy Alternatives, Washington, D.C., August 1994.

> received within the United States, and to all persons engaged within the United States in such communication or such transmission of energy by radio, and to the licensing and regulating of all radio stations. . .

The term "radio" has long since been interpreted to include all forms of transmission of electromagnetic energy for purposes of carrying information within the United States.[27] The FCC views its mission as being to "encourage competition in all communications markets and to protect the public interest."[28]

Techniques of Regulation

The FCC, like most regulatory bodies, has both quasi-legislative and quasi-judicial authority. It establishes and enforces administrative regulations but can also take testimony, subpoena witnesses and records, and issue decisions and orders. Although it has a direct investigative and enforcement capability, the principal regulatory technique used by the FCC is rulemaking, in which the provisions of applicable federal law are applied to a specific matter brought before the commission. The subject matter of a rulemaking may be a clarification or specification of the law for all parties[29]; establishment of a federal rule for governance of a particular part of the communication spectrum[30]; a ruling on a controversy concerning a regulated company or companies[31]; or any other matter that the FCC feels can be properly brought before it.

3.2.2.2 Other Federal Agencies and Departments

While the FCC has sole authority for regulating the interstate communications systems in the U.S., it is not alone in having responsibilities and duties that affect the operation and security of those systems. Within the federal criminal code, there are now provisions for investigation of fraud and related activity in connection with computers.[32] Among the "related activities" covered by this new statute are those computer crimes generally associated with the term "hacking," which result in threats to national security or financial damage to any person, corporate or real. This statute expanded the role of the Federal Bureau of Investigation (FBI), the U.S.

---

27 Mission statement, FCC home page at http://www.fcc.gov.

28 Ibid.

29 Op. cit., *In the Matter of Implementation of the Local Competition Provisions in the Telecommunications Act of 1996.*

30 For example, FO Docket No. 91-171/91-301, FCC 94-228, establishing portions of 47 CFR Parts 0, 11, 73, and 76.

31 For example, DA Docket No. 97-1019, *In the Matter of Bellsouth Cellular and GTE Wireless, and AT&T Wireless Services, Inc.*

32 18 U.S.C.A. § 1030.

Secret Service, and other federal and state law enforcement agencies with respect to investigating misuse of the national information infrastructure. In addition to these civilian agencies, the U.S. Department of Defense (DOD) has become more active in the last five years in investigating computer crimes involving defense property and personnel.

3.2.2.3 California Public Utilities Commission

Authority and Jurisdiction

The California Public Utilities Code (CPUC) (CalStat, Title 70 *et seq.*) is the foundation of regulation in California. The CPUC was developed from the Railroad Commission Amendment of 1911, establishing a constitutionally mandated Railroad Commission, which became the California Public Utilities Commission (the Commission) in 1946.

The jurisdiction of the Commission has long since been extended generally to water, waste disposal, electrical generation and service, natural gas, and all forms of transportation. Under the CPUC, the Commission has broad powers to regulate safety, standards of service, and rates in regulated industries. Part 2, Chapter 10, of the CPUC, entitled "Telephone Corporations," gives the Commission authority to regulate telephone service and equipment.

Techniques of Regulation

The five commissioners, who make all policy, procedural, and rulemaking decisions, are appointed by the governor of California, with the advise and consent of the state senate, for staggered six-year terms. The Commission has quasi-legislative and quasi-judicial powers over regulated companies. The guiding principal for the Commission is to "benefit ratepayers through lower rates, new and improved utility products and services, and protect consumers where competition otherwise does not."[33] In meeting this objective, the Commission "balances the public interest and need for reliable, safe utility services at reasonable rates with the need to assure that utilities operate efficiently, remain financially viable, and provide their stockholders with an opportunity to earn a fair return on investment."

The Commission has a Telecommunications Division, composed of three branches responsible for advocacy, advisory, and compliance activities. These branches and their specific responsibilities are as follows:

- The Carrier Branch is responsible for processing tariffs of local exchange carriers, competitive local carriers, nondominant interex-

---

33 Unless otherwise noted, all quotes in this section are from "About the CPUC," at ftp.cpuc.ca.gov.

change carriers, and wireless service providers. Thus this branch oversees price setting for all local commercial companies that provide telecommunications services to the public. The Carrier Branch also enforces compliance with Commission decisions and with service and reliability standards. In addition, it reviews general orders of the Commission to ensure that they reflect the competitive environment and changing regulatory structure for the telecommunications industry.

- The Market Structure Branch is responsible for implementation and oversight of Commission rulings regarding local competitive issues. This Branch reviews, analyzes, and advises the Commission about carrier-to-carrier arrangements and interconnecting agreements, competitive costing, pricing, and unbundling of services. This Branch also reviews utility applications for mergers, divestitures, and acquisitions.
- The Public Programs Branch is responsible for oversight and implementation of special issue programs (such as those for the disabled, elderly, and bedridden), service assurance for the needy, and pay telephones. This Branch also develops and implements consumer protection and education programs, and it reviews, analyzes, and implements the Universal Service program for California.

The Commission has also formed a federal/state legislative team to actively coordinate responses to legislation and proposed legislation.

3.2.2.4 Illinois Commerce Commission

Authority and Jurisdiction

The Illinois Public Utilities Act lays out the structure and authority for regulation of telephone public utilities in Illinois (220 ILCS 5/1-101 *et seq.*). Under the Act, the Illinois Commerce Commission (the Commission) has a general mandate to assure safe and efficient operation of utilities and has powers to investigate accidents and enforce administrative orders related to safety equipment and practices. In general, however, the Commission's authority focuses on financial and economic matters, including rate setting, rather than public health and safety. The Commission determines acceptable rates and charges on the basis of a prescribed formula and a hearing process. The Commission has issued rules and regulations concerning the rate-setting process and other aspects of its oversight of utilities.

The Commission is responsible for general supervision of all public utilities and is authorized to inquire into the management of such utilities and to keep itself informed as to the manner and method in which the utilities' business is conducted.[34] This supervision covers the utilities' general con-

---

34 A public utility is defined as anyone that owns, controls, operates, or manages,

dition, capitalization, and the manner in which their plant, equipment, and other property is managed, conducted, and operated. The Commission is responsible for ensuring that utilities provide adequate, reliable service and otherwise comply with the Illinois Public Utilities Act. However, the Act specifically limits the Commission's authority so as not to overlap with that of other agencies. The Commission may not enforce requirements that are vested in another entity.

Every public utility must obtain a certificate of public convenience and necessity from the Commission before transacting business within the state or beginning any construction of any new plant, equipment, property, or facility. A certificate will be issued only if the public utility demonstrates that the proposed construction is necessary to provide adequate, reliable, and efficient service to its customers. Every public utility must furnish to the Commission any information required by the Commission, including maps, profiles, reports, documents, books, account, papers, and records that in any way relate to its property or affect its business, and inventories of its property.

The Commission is authorized to conduct management audits or investigations of utilities if necessary. Upon complaint or during a rate proceeding, the Commission may investigate claims of poor service and reliability, but generally only in connection with fiscal responsibility and to determine the return on investment the company may receive in its rates. Claims for damages resulting from service outages, or other cases where a public utility has failed to comply with any provision of the Illinois Public Utilities Act or any rule, regulation, order, or decision of the Commission, are not within the jurisdiction of the Commission and must be brought in the Illinois Circuit Court.

The Commission also has the authority to investigate all accidents involving utility property or activities that result in loss of life or injury to a person or property. The Commission then has the authority to issue orders or recommendations to the utility concerning the accident circumstances. This provision is rarely used. If the Commission determines, after hearing, that the practices, equipment, or service of any public utility, or the methods of distribution, storage, or supply employed by the public utility are unsafe, improper, inadequate, or insufficient, the Commission may issue an

---

within the State of Illinois, directly or indirectly, for public use, any plant, equipment, or property used or to be used for, or in connection with, the production, storage, transmission, sale, delivery, or furnishing of heat, cold, power, electricity, water, or light, except for those public utilities owned or operated by any political subdivision, public institute of higher education, or municipal corporation of the state. The definition of public utilities also does not include water companies that are purely mutual concerns having no rates or charges for service but that pay the operating expenses by assessment upon the members of such a company and no other person.

administrative order, decision, rule, or regulation requiring corrective action by the utility.

The Commission can issue administrative orders to require public utilities to comply with the Illinois Public Utilities Act or any rule, regulation, or order issued by the Commission. If the utility does not comply with such order, the Commission may file an action or proceeding in the appropriate Illinois circuit court. Any public utility or other corporation, that fails to comply with the provisions of the Illinois Public Utilities Act or Commission rules, regulations, or orders is subject to a civil penalty of not less than $500 nor more than $2,000 for each and every offense. Every person who fails to comply with the provisions of the Illinois Public Utilities Act or any order, rule, or regulation of the Commission is also guilty of a Class A misdemeanor.

The Commission has additional responsibilities for regulation of "telecommunications rates and services" under the Universal Telephone Service Protection Law of 1985 (which amended the Illinois Public Utilities Act, 220 ILCS 5/13-100). The duties of the Commission in this regard, as set out in 5/13-301 *et seq.*, are to participate in all federal programs intended to preserve or extend and monitor universal service. The authorities and powers of the Commission under the Illinois Public Utilities Act, such as issuing certificates of service authority, establishing rates and charges, and conducting general oversight, are applicable to all noncompetitive telecommunications services. Certain sections of the above law also apply to competitive telecommunications rates and services. System security issues are not addressed in this portion of the Illinois law; however, the Telecommunications Facility Fire and Emergency Act (220 ILCS 45/0.01 *et seq.*) requires appropriate measures to prevent major interruptions of telecommunications services. The implementing regulations for this Act are discussed below in Section 3.3.1.

Also of particular interest in the context of the information and telecommunications infrastructure, 5/4-201 of Chapter 220 states that:

> It shall be the duty of the Commission, at the direction and discretion of the Chairman, to assemble and maintain an electronic trespass enforcement assistance staff consisting of experts in computer systems, electronics, and other professional disciplines to aid public utilities, businesses, individuals, and law enforcement agencies in detecting and preventing electronic trespass violations and enforcing the provisions of Section 16-9 of the "Criminal Code of 1961," approved July 28, 1961, as amended or any other relevant statute.

This provision has not been implemented. The Commission has, however, been very active with the Office of the Attorney General in developing the Illinois Commission on Electronic Commerce (ICEC) and the resultant draft Illinois Electronic Writing and Signature Act. At present, the focus of the ICEC and the draft legislation is to further the commercial use of emerging electronic media by legally defining documents, records, and signatures in their electronic form and by specifying how and under what conditions such electronic versions of legal instruments can be used in place of their physical counterparts and how the public can be protected when it relies on such electronic instruments.

Techniques of Regulation

The Commission carries out its duties through (1) hearings to determine compliance with applicable laws and its own rules, (2) publication of reports on the status of the regulated industries, and (3) rate setting procedures for all providers of regulated services in the state. The Commission also conducts and publishes studies on a variety of subjects as directed by the Illinois General Assembly. For the purposes of this report, the most pertinent of these studies are those concerned with the classification of new services and with the implications of regulating multiple certificates of service authority for intermarket service areas and local exchange services. It is in these areas of local service delivery where the emphasis on ensuring universal service and unbundled services will have the greatest impact on infrastructure protection. New technologies are being rapidly introduced into telecommunications, often in the form of new services. If a narrow view of classification of services is taken, complications in providing adequate measures of security will increase. For example, a regulatory viewpoint that fails to accommodate the convergence of video, information delivery and processing, communication, and entertainment product delivery will not have the scope necessary to correctly address system reliability and performance.

3.2.2.5 New Jersey Board of Public Utilities

Authority and Jurisdiction

The New Jersey Board of Public Utilities (NJBPU or the Board) was put in place by then-Governor Woodrow Wilson in 1911 for the purpose of setting rates, approving financing, and setting service standards for all regulated industries. The agency is autonomous in its decision making with regard to the energy, transportation, commercial, and consumer services and telecommunications industries that it regulates.

The Board is in the process of completing a reorganization plan begun in 1995. The reorganization is intended to increase the efficiency of its operations while emphasizing consumer issues such as affordability of service, safety, reliability, and economic issues such as maintaining the value of shareholder investment in utility companies.

As prescribed in New Jersey law, the duties and authority of the NJBPU are to:

1. Maintain universal telecommunications service at affordable rates;

2. Ensure that customers pay only reasonable charges for local exchange telecommunications services, which shall be available on a nondiscriminatory basis;

3. Ensure that rates for noncompetitive telecommunication services do not subsidize the competitive ventures of providers of telecommunications service; and

4. Provide diversity in the supply of telecommunications services and products in telecommunications markets throughout the state.

The law also gives the NJBPU the authority to approve alternative forms of regulation in order to address changes in technology and the structure of the telecommunications industry, modify the regulation of competitive services, and promote economic development.[35]

Nowhere in the Title 48 is system security or protection addressed. Section 2-23, entitled *Safe, Adequate Service,* states that the Board can require a utility to provide such service but limits the objective of such a requirement to environmental protection.

Techniques of Regulation

The NJBPU carries out its duties principally by means of hearings and rulemaking focused on rate setting, tax policy for utility services, industry restructuring, stranded assets, expanding competition, and economic development. There are no records available of any hearings held or studies made by the Board with specific regard to infrastructure reliability or protection.

3.2.2.6 Public Utility Commission of Texas

Authority and Jurisdiction

In 1995, the legislature of Texas adopted a new Public Utility Regulatory Act (PURA95), which directed a substantial reorganization of the Public Utility Commission of Texas (PUCT or the Commission). The Commission as reestablished consists of three commissioners, appointed by the governor for staggered, six-year terms, with the advise and consent of the state senate. The powers and duties of the PUCT include establishing "a comprehensive regulatory system which is adequate to the task of regulating public utilities . . . to assure rates, operations, and services which are just and reasonable to the consumers and to the utilities."

Under PURA95, the jurisdiction of the PUCT is limited to regulation of electrical energy (Title II) and telecommunications (Title III).Within that

---

35 New Jersey Permanent Statutes, Title 48.

context, the PUCT is given significant authority to carry out the objectives of PURA95 with regard to enhancing competition in local service. For example, the provisions of §§ 1.101, 3.001, 3.051, and 3.458 would appear to give the Commission jurisdiction to require all holders of certificates of convenience and necessity to actively work toward redundant interconnection. PURA95 does not make any specific reference to system security or infrastructure protection. The Commission has adopted a policy of monitoring industry compliance with FCC orders with regard to priority of restoration of service after an interruption. The PUCT *Emergency Procedures for Telephone Utilities* requires local exchange companies to report service outages of four hours or more that affect:

- 50% of the toll circuits serving an exchange,
- 50% of the extended area service circuits serving an exchange,
- 50% of a central office, or
- 20% or more of an exchange's wire center access lines.

These procedures also require that in the event of a major emergency (e.g., loss of major portions of local service due to fire, floods, hurricanes, freezing, sabotage, or war), the PUCT is to be notified as soon as possible. However, according to a respondent at the PUCT staff, no specific state response plan standards have been established by the PUCT. The PUCT is relying on the industry to meet emergency response needs by reliability standards and the return-to-service prioritization determined under federal guidelines. The Commission has appointed one of its members to represent Texas on the National Network Reliability Council.

Techniques of Regulation

Under the PURA95, the PUCT carries out its responsibilities principally through the ratemaking process. The PUCT uses economic incentives to gain its objectives of quality and reliability of service, universal availability, and financial stability for the industry. System security has not been a stated objective for the ratemaking process. The PUCT is required to report periodically to the state legislature on a variety of subjects. Infrastructure protection is not included in any of these reports. PUCT staff indicated that assessing and addressing security and infrastructure protection are implicitly made the responsibilities of the industry, not the PUCT.

3.3 Regulations and Critical Infrastructure Protection

3.3.1 Impact of Regulations on Critical Infrastructure Protection

Threats to Information and Communications Infrastructure

To assess the impacts of regulations on information and communications infrastructure assurance, it is necessary to characterize the types of threats being considered. Appropriate protective measures may depend on the type of threat contemplated, although some measures may be common to all threats. Threats to information and communications may be divided into three types:

1. *System destruction.* Hostile elements might attempt to destroy a data bank or communication system so that it cannot be used at all (e.g., bring down a local [big city] telephone network or long distance service). This type of threat might involve either physical assault (e.g. bombing a key facility) or cyber attack.

2. *Corruption or destruction of data (cyber threat).* Hostile elements might attempt to corrupt or delete information in key systems. For example, they might try to disrupt banking, commerce, or utilities by "hacking" into particular systems and introducing false or misleading data, deleting data, or deliberately manipulating the system in certain ways (e.g., turning off power grids or disrupting market trading)

3. *Unauthorized access to information.* Hostile elements might attempt to use information systems to obtain data that is classified, proprietary, or "private" in the personal sense. While potentially harmful, such actions fall more appropriately under the category of national or industrial espionage rather than terrorism.

4. *Theft of service.*

5. *Service abuse.*

Discussions of information security may focus on one type of threat or encompass all of them. "Access" to a computer system is often discussed, on the basis of the assumption that it may lead to any on of the three consequences above.

Impacts of Current Regulations

As indicated by the sample of regulatory agencies discussed in Section 3.2, few regulations directly address information and communication system security. At the federal level, there are criminal sanctions for tampering with the Public Switched Network (PSN) and national security information. These penal statutes promote system protection by deterring some who would intentionally tamper with the PSN.

Law enforcement agencies, intelligence agencies, and the military all have specific regulations regarding operational security (OPSEC), system security (SYSEC), and information security (INFOSEC). However, these regulations address only data and systems under the direct control of the government and only indirectly affect security standards for the PSN.

Apart from standards for systems used by the government and those directly related to national security, the development of standards that specify protection of data is in its infancy. Current regulations emphasize system reliability rather than system protection. To a significant degree, all of the equipment standards, frequency separation specifications, and reliability measures do affect system security, but only tangentially.

At the state level, the regulatory bodies focus primarily on economic considerations. State regulators are principally concerned with consumer protection and investor confidence. They must ensure that telecommunications service is available to the greatest number of people for whom such service can be economically provided. Additionally, they must ensure that investors in utility securities have confidence in the health of their utilities and will thus accept favorable interest rates on the bonds and debentures that must be sold to finance the needed infrastructure. Security per se has not been a major concern. A notable exception is the State of Illinois, which adopted security requirements for telephone switching stations after an accidental fire caused a significant disruption of service.

A fire at a switching facility in Hinsdale, Illinois, resulted in a serious, extended loss of telephone service over a large suburban area. In response, the Illinois Commerce Commission, in conjunction with the Office of the State Fire Marshal and the Illinois Emergency Management Agency, promulgated special rules for telephone utilities (83 IAC 785). These rules require all switching facilities to have physical security in the form of a door lock, card control entry, or security guard to prevent unauthorized entry or malicious disruption of service. Each facility must have a lock box system to facilitate access to the telecommunications equipment during an emergency situation. Each telephone utility must develop procedures providing for the continued operation of its services in the event of natural or human-caused disasters, including priority restoration of critical services such as police, fire departments, hospitals, and 911 service. In addition, each utility must notify the Illinois Emergency Management Agency of any major service outage expected to last 12 hours or more. Each switching facility must have a direct alarm monitoring communication channel to a fire department. All electrical service to switching facilities must meet fire code requirements.

In addition, the ICC requires central offices to have a reserve battery supply of five hours in locations where emergency power generators are not installed and three hours where they are. In central offices that have more than 3,000 working lines, a permanent power generator must be installed. There are also requirements to ensure that emergency systems (911 service) are not interrupted, which include physical security for all vital equipment to prevent malicious disruption of service, and wherever practical, underground service entrances for electric and telephone service.

Effect of Deregulation and Technical Change

In the last decade, there have been a great multiplication of information and communication services, brought about partly through industry deregulation and partly through the development of many new technologies and products. The popularization of cell phones, cable television, satellite communications, home computers, the Internet, distributed computer networks, electronic banking, and many other products and services has greatly increased the role of the information and communications industry in business and personal life. This trend is anticipated to continue for at least the next decade as well.

In the communications industry, one result of this trend may be an increase in the natural resiliency of the infrastructure. Twenty-five years ago, essentially all general-purpose personal and business communications were conducted via a single company's network. This system worked well and incorporated some redundancy and safeguards against breakdown. However, it is far different from the current situation, in which multiple long-distance providers have independent trunk networks, cellular telephones, and widespread access to the Internet. There are now many more ways to get a message from point A to point B. In addition, local telephone service is currently being deregulated, and it is anticipated that this will lead to further redundancy in local networks, as independent local nets and switching equipment are installed. In this respect, the information and communications industry may differ from other infrastructures (such as gas distribution or rail transport), where deregulation appears to be leading to decreased capacity margins and reductions in the physical redundancy of the system.

Another effect of the current regulatory climate is that protection of information and communications is a major private-sector activity, undertaken by commercial and industrial firms. Firms performing information-intensive work have developed methods of data protection and transmission to ensure privacy and have a reservoir of expertise in this area. There are tremendous business incentives to protect vital information, and the market has responded with methods and equipment for doing so. All this has occurred without benefit of regulatory intervention or standards.

3.3.2 Measures for Improvement

Governmental Recommendations

A number of federal-level groups have been studying information and communications security and have developed recommendations on how to proceed. The National Communications System (NCS)[36] has recommended further study to identify and characterize the nature of the threat to our national information infrastructure. In addition, the NCS has recommended that government and industry cooperate to develop:

- Protocols for industry-government information exchange regarding threats and risk;
- Protocols and mechanisms for information protection that are cost-effective and will not unduly impede communications;
- Statutes and regulations to implement such protocols, as needed; and
- Priorities for investment in infrastructure to reduce vulnerability.

The Network Reliability and Interoperability Council (NRIC), an FCC-chartered group, has studied telecommunications network accessibility and security. This group warns that "without a full range of security measures in place, both systems and data are in danger of misuse, corruption, and loss of confidentiality."[37] The NRIC has developed recommendations on government-industry cooperation that are somewhat similar to the NCS recommendations. In addition, the NRIC has recommended expansion of reporting requirements for outages and suspected attempts at intrusion. In interviews, members of the NRIC and FCC staff expressed concern that the reporting standards for outages affecting especially important classes of service such as major airports, military installations, government facilities, nuclear power plants, and 911 special facilities[38] were not rigorous enough for the growing complexity of the system. There was a strong consensus that an improved automated reporting system, with a greater ability to discriminate among kinds of system interferences and outages, was needed. Such a system would need to be augmented with regulatory language requiring additional and more finely focused data capture. Changes of this kind are unlikely to occur if left strictly to the competitive pressures of a deregulated industry.

---

36 The NCS was established in 1963 to provide oversight of systems to support critical diplomatic, military, and intelligence communication. It has assumed a prominent place in discussions of current vulnerabilities of the national communications and information infrastructure.

37 NRIC Focus Group 1, item 3.01.01.04, opened 2/18/97.

38 47 CFR 63.100.

The NTIA has focused on protection of personal information. The NTIA has recommended that national standards be developed for the collection, use, and disclosure of telecommunications-related personal information. Such information includes personal telephone numbers and addresses linked in databases with employment, age, income level, marital status, purchasing habits, health records, and other highly personal data.

Industry Perspective

Interviews with numerous industry executives, cited throughout this section, have shown that the industry is quite cognizant of the need for infrastructure protection. System security is considered vital to the industry's economic interests. However, members of the industry are wary of federal regulation. They consider themselves to be the appropriate entities to address system security issues. They favor innovative and energetic development of new "best practices" (which are industrywide methods of dealing with operational security and reliability issues) and "standards" (which are industrywide technical requirements for dealing with functional performance issues). Representatives of both established carriers and new entrants see the need for incentives that will bring them together in a community of interest to develop standards and best practices.

Recommended Actions

The rate of technological change in the information and communications industry makes it necessary to adopt new approaches to regulation. If federal and state regulators follow the traditional process of hearings, followed by promulgation of rules, followed by amendment of the rules to adjust to emerging needs, they may be caught in a control loop in which their actions will lag behind reality to such a degree as to be ineffective or even harmful.[39] Instead of the usual top-down regulatory approach, the opportunity presented by the 1996 Act should be used to facilitate standard setting on the part of industry. This effort will involve (1) permitting firms to cooperate on security measures without fear of antitrust prosecution and (2) punishing those who abuse the cooperative nature of the system as a means to circumvent competition.

The appropriate roles for the government to take in protecting the emerging telecommunications and information infrastructure are as follows.

---

39 The FCC has always relied on the industry for technical support in crafting federal rules and guidelines. As long as there were relatively few members in the industry and the industry members could stay ahead of the introduction of new technologies that altered the regulatory landscape, there was adequate "slack" in the process to allow thoughtful development of these national rules and guidelines. Both the pace of change and the increasing number of actors have made this process very difficult to manage.

First, it must articulate a clear statement of the priority of such protection over the competing goals of fostering competition, growth, and universal service. Second, it must establish criteria that target public and private development of solutions to technical issues in infrastructure assurance. Third, it must provide a mechanism through which the stakeholders in the infrastructure can exchange information about the nature and extent of threats to the infrastructure.

In developing a national policy that addresses both physical and cyber security, the FCC is the best candidate to act as convener of the community of interest. However, the objective should not be codification of the resulting policy into agency rules, which would then need to be under constant review and amendment. To provide the appropriate guidance with flexibility, the most promising approach is to create a government-industry council, modeled after the NRIC[40] and charged with addressing infrastructure protection issues. This group would be responsible for integrating the recommendations of the federal, state, and industry into measures for improvement, such as best practices, standards, and research and development objectives. The forum body should be organized and staffed so that its activities are ongoing and continuous, with the statements and objectives constantly under review.

Such a council should view its purpose as developing policies aimed at providing the greatest degree of choice among telecommunications services and products, to the greatest number of people possible, with the greatest degree of security and reliability that is technically and economically feasible. This is not an easy task, but it can be done if the federal government regards issuance of rules and regulations as a recourse of last resort, to be used only when community development of standards, statements of best practice, and research objectives prove to be unwieldy, unreliable, or unproductive.

4 ELECTRIC INFRASTRUCTURE

4.1 General Description of Regulation

4.1.1 Current Regulatory Environment

Industry Overview

The U.S. electrical infrastructure is designed and operated to provide electrical power to industrial, commercial, and residential customers. It is composed of power generating facilities, high-voltage bulk power trans-

40 See a similar quasi-regulatory structure and function in the discussions of the role of the North American Electric Reliability Council (NERC) for the electric industry in Sections 4.1.1 and 4.3.1.

mission lines that carry power over long distances, substations that "step down" the voltage for local distribution, and local distribution networks that carry the power to end users. At the level of generating and bulk transmission, the facilities are interconnected to form networks. Electricity generation and distribution are actively managed (in real time) on a regional basis to maintain an even supply of power to all users.

The majority of the U.S. bulk power system has evolved into three major networks (power grids), which include smaller groupings or power pools. The major networks consist of extra-high-voltage connections between individual utilities designed to permit the transfer of electrical energy from one part of the network to another and between one utility and another. These transfers rely on contractual arrangements and adequate transmission capacity. The three networks are (1) the Eastern Interconnected System, consisting of the eastern two-thirds of the United States (to the Rocky Mountains); (2) the Western Interconnected System, consisting primarily of the Southwest and areas west of the Rocky Mountains; and (3) the Texas Interconnected System, consisting of a significant portion of Texas. The Texas Interconnect is not interconnected with the other two networks (except for certain direct-current lines). The Eastern and Western Interconnects have limited connections to each other. The Eastern and Western Interconnects are completely integrated with most of Canada and have links to the Quebec Province power grid. The Western and Texas Interconnects are linked with different parts of Mexico.

The business units that make up the electrical infrastructure include public, private, and cooperative entities. The ownership structure of this infrastructure is in part the result of past and present federal programs. For example, the Rural Electrification Act of 1936 (Pub. L. 74-605) set up the mechanism for creating rural electric cooperatives. Rural cooperatives are consumer-owned and operated nonprofit utilities. They operate only in rural areas and are generally unregulated for the production and transmission of electricity within their own territories. These rural cooperatives can be small concerns serving a farming community or can be large companies providing electric service to consumers in several states. The federal government also has established a number of power administrations that generate and sell electricity. Federal generating facilities are normally hydroelectric dams, but the Tennessee Valley Authority (TVA) also operates other types of power plants. In addition to the federal power administrations, there are numerous state and local government power systems, primarily municipal systems. Some of these do not generate their own power but instead purchase it from another utility and perform only local distribution. The remainder of the infrastructure is owned and operated by investor-owned utilities. Investor-owned utilities constitute the majority of the infrastructure. The business structure of these entities has been shaped by legal constraints. The Public Utility Holding Company Act of 1935

(PUHCA) (15 U.S.C.A. §§ 79-79z), enacted to eliminate holding company abuses, required them to be composed of a single integrated public utility system located in and serving a single geographic area.

Overview of Current Regulatory Scheme

Electric utilities are regulated by both federal and state agencies. The Federal Energy Regulatory Commission (FERC) regulates the interstate wholesale transmission and sale of electricity, and issues construction permits for power plants and transmission facilities.[41] This permitting process takes into consideration both the public need and the utility's ability to meet public demand for electricity and comply with all applicable state and federal regulations. These permits are issued in the form of FERC Orders, which are enforceable in the U.S. district court of appropriate jurisdiction. FERC also exercises its economic regulation of electric utilities through approval of electricity rates and corporate structure (18 CFR 35). FERC has the authority to order interconnections between utilities when it is determined to be for the public good (18 CFR 32).[42] FERC also exercises plenary policy-setting authority through the issuance of general orders. A recent example is the orders implementing wholesale market deregulation (FERC Orders 888 and 889). Those orders are discussed in more detail below.

State public utility commissions (PUCs) have jurisdiction over the intrastate retail sale of electricity. Usually, a state PUC must issue a permit, license, or certificate to an electric utility before it can begin construction and operation of a generating facility or transmission line. This permitting process is economic in nature and usually involves a determination that there is a public need for the electricity to be generated and that construction of the plant is a prudent investment by the utility. The order may set out specific directions concerning the financing of the construction and the treatment of the completed plant and the related construction expenses in the utility's ratemaking dockets. The order may require that certain construction standards and operational procedures be followed.[43] Many states regulate the erection of transmission lines, including the height, tower construction, and spacing of the lines along the utility's easement. In some cases, municipalities may have additional regulations that must be met in erecting or burying electrical distribution lines within the city limits.

---

41 Federal Power Act, 16 U.S.C.A. §791 et seq., and implementing regulations at 18 CFR Part 4.

42 The Department of Energy (DOE) also has authority (under § 202(c) of the FPA) to order temporary interconnections between electric utilities during emergency situations such as unexpected outages or breakdowns of facilities.

43 Except for nuclear power plants, which receive a separate license from the U.S. Nuclear Regulatory Commission (NRC) and comply with NRC construction standards and operational procedures.

Utilities with nuclear power plants are also under the jurisdiction of the U.S. Nuclear Regulatory Commission (NRC). The Atomic Energy Act of 1954 (42 U.S.C.A. §§ 2011-2296) authorizes the NRC to regulate the safety aspects of the construction and operation of nuclear power plants.[44] The NRC promulgates regulations, issues construction and operation licenses, maintains on-site inspection teams, and issues enforcement actions when license conditions or its regulations are violated.

Although not a regulatory body, a major player in electric reliability is the North American Electric Reliability Council (NERC). NERC is a voluntary association of utilities formed in 1968 in response to the Northeast blackout of 1965. NERC's mission is to promote the reliability of the electric supply for North America by reviewing lessons learned; monitoring the existing system for compliance with NERC policies, standards, principles, and guidelines; and establishing forward-looking policies, standards, principles, and guidelines to assure the future reliability of bulk power systems. NERC is proactive in assuring continued reliability of the power system. As discussed below, despite the presence of federal and state regulatory bodies, NERC has become the major influence on electric system security.

4.1.2 Current Trends in Regulation

Federal regulation of electric utilities dates back to 1935, with the enactment of PUHCA and the FPA. The Federal Power Commission (FPC) was created to oversee interstate transmission and wholesale of electricity. Its authority was later transferred to FERC.

The recent changes to electric industry regulation started with the Public Utility Regulatory Policies Act of 1978 (PURPA). Passed in response to uncertainty in the energy market in the 1970s, it allowed nonutility facilities to generate and sell electricity on the wholesale market. These qualifying facilities (QFs), either cogeneratorsor small power producers, sell their excess energy to public utilities pursuant to contracts that by law set the rate at the purchasing utility's "avoided cost." QFs are exempt from rate and accounting regulation by FERC, regulation by the Security and Exchange Commission, and rate, financial and organizational regulation by the state.

---

44 The Atomic Energy Act of 1946 (60 Stat. 755) established the Atomic Energy Commission (AEC) to administer the transfer of nuclear power development from the military to the civilian government. The 1946 Act was changed substantially by the Act of 1954. In 1974, Congress abolished the AEC and replaced it with two agencies: the NRC, which is responsible for safety and licensing, and the Energy Research and Development Administration (ERDA), which was responsible for promotion and development of nuclear power. ERDA was later absorbed by the U.S. Department of Energy (DOE).

The Energy Policy Act of 1992 (EPACT) amended the FPA and further expanded competition within the wholesale market by allowing exempt wholesale generators (EWGs), which, once they are so-designated by FERC approval, are exempt from PUHCA's corporate and geographic restrictions.[45] Unlike QFs, EWGs' rates or charges are unregulated. EPACT also mandated FERC to open up the national electricity transmission system to all wholesale suppliers on a case-by-case basis, thus allowing QFs, EWGs, and others to transmit their power over other utilities' transmission lines to wholesale customers.[46] This mandate significantly affected FERC's jurisdiction regarding wholesale electricity.

The recently issued FERC Orders 888 and 889 have been instrumental in achieving the goals mandated by the EPACT. Order 888 addresses the issues of open access (opening the transmission system to new generating entities) and stranded costs (recovering embedded costs of facilities that may no longer provide the specified rate of return). Order 889 mandates the formation of utility-owned information systems and requires them to publish information on the available transmission capability of all utility-owned and operated primary transmission lines (Open-Access Same-time Information System [OASIS]).

The electric industry is changing from a fully regulated, vertically-integrated structure to a competitive structure via unbundling of services and disaggregation of functions. These changes may create uncertainty as to who is responsible for reliability of the power system. Mergers and divestitures will occur as companies adjust to enter the competitive markets. Generation will be separated from transmission and marketing functions. State legislatures are grappling with new legislation to guide restructuring of the intrastate, retail electric industry. Most of the debate, however, focuses on the economic impacts of such deregulation on the local utilities, particularly the recovery of stranded costs, rather than on system reliability.

California was the first to enact a plan for restructuring. California's plan provides for recovery of stranded costs through "competition transition" charges: a current customer surcharge, and a severance fee for customers leaving the utility system to seek competitive rates elsewhere. Rates for residential and small commercial customers will be reduced beginning January 1998. This reduction will be funded by revenue bonds to finance the competition transition charges over a ten-year period. Thus California utilities should enter the competitive market without undue financial burdens.

---

45 Both QFs and EWGs, however, are only power producers and do not own transmission facilities.

46 The law specifically prohibits FERC from ordering wheeling of the power to a final customer (retail).

The California plan also established an independent system operator (ISO), as recommended in FERC Order 888, to operate the transmission grid, thus eliminating preferential treatment by transmission line owners along grid corridors. The plan addresses reliability through the ISO. The ISO is to seek FERC authorization to perform its system functions and to secure the generation and transmission resources needed to achieve specified planning and operational reserve criteria. It must have standards no less stringent than those of the Western Systems Coordinating Council (WSCC) and the NERC. It may develop inspection, maintenance, repair and replacement standards for transmission and distribution systems, specifically aimed at preventing system outages. The ISO, in consultation with the California Public Utilities Commission, the California Energy Commission, and other Western state agencies, would have the power to conduct a reliability study of the California interconnected transmission and generation system. The National Association of Regulatory Utility Commissioners (NARUC) adopted principles at its 1996 Summer Meeting asking legislatures to ensure reliability is maintained in the restructured industry.

As the changes resulting from these legislative and regulatory actions work their way through the industry, NERC has also had to change its focus. NERC has revised the composition of the Board of Trustees and its committees to include two new seats representing the "customer" sector. One of the new seats will likely include an industrial customer representative. The other seat has not yet been defined. This modification to the Board composition improves the representation of customer concerns. NERC also decided to sponsor an independent organization to review its present managerial and technical structure. The "Future Role of NERC" Task Force will provide its findings to the review team and oversee the review program. Recommendations are expected to be made available to the board in January 1998.

On January 6, 1997, the NERC Board of Trustees unanimously voted to obligate Regional and Affiliate Councils and their members to promote, support, and comply with all NERC reliability policies. This recent change in position from voluntary to obligatory compliance places NERC in a central position with respect to establishing national standards for system reliability.

The NERC Reliability Compliance Task Force and Future Role of NERC Task Force-II have issued a report[47] about options to ensure compliance with NERC and Regional Council policies. The report recommends that each NERC member organization establish formal coordination agreements with other relevant NERC member(s). Such agreements are to

---

47 *Options to Ensure Compliance with NERC and Regional Reliability Council Policies, Standards and Criteria,* October 12, 1996, NERC, Princeton, New Jersey.

(1) specifically state the intent of the organization and its members to comply with NERC and NERC member reliability criteria and (2) establish authority to impose sanctions to ensure compliance with mandatory protocols. These agreements will then be filed with NERC. In addition, all entities performing primary reliability functions of generation control (control areas) and system security coordination must be members of the Regional Reliability Councils in which they carry out their business. To ensure membership and compliance, the report suggests that regulatory authorities require mandatory compliance with NERC reliability protocols (NERC certification) as a condition for approval of tariffs, contracts, licenses, and other instruments. The report also proposes that FERC ensure that all expenses incurred to meet reliability criteria established by NERC are fully recoverable through rates.

The NERC task force report also recommends the formation of regional security coordinators that will be equipped to evaluate regional security plans, perform transmission system assessments, and start control actions if and when reliability is jeopardized. NERC concludes that the traditional voluntary compliance approach, with the obligation and cost of providing a reliability safety net imposed on utilities entering the competitive environment, will no longer serve the needs of the new, restructured industry. The U.S. Department of Energy (DOE) Task Force on Electric System Reliability, which is to provide advice, information, and recommendations to the Secretary of Energy Advisory Board on issues related to the reliability of bulk electricity systems in the United States, has completed a draft report entitled *An Organizational Framework for Bulk Electric System Reliability: Functions and Interrelationships* (May 1997). It basically reiterates the NERC mandatory compliance suggestions.

4.2 Description of Selected Regulatory Agencies.

4.2.1 Selection Method

The following regulatory and nonregulatory agencies were selected because they represent the significant entities involved in electric utility regulation: FERC, NRC, DOE, Illinois Commerce Commission, the City of Chicago, and NERC. FERC is the primary federal regulatory body overseeing interstate wholesale electricity transactions. The NRC was chosen because it has sole jurisdiction over matters of nuclear safety at nuclear power generating stations, which account for approximately 20% of the electricity generated in the nation. The Illinois Commerce Commission was selected as an example of a state PUC in a state with a well-developed generating and transmission infrastructure. The City of Chicago was chosen as an example of local safety or reliability requirements for electric utilities operating within city limits. NERC, although not a regulatory agency, was chosen because of its historical role in assuring reliable power system operation and its current influence over the continued reliable operation of the electrical infrastructure.

4.2.2 Agency Descriptions

4.2.2.1 Federal Energy Regulatory Commission

FERC is an independent regulatory commission within DOE. Under the FPA, FERC has jurisdiction over interstate, wholesale sales of electricity by privately owned utilities.[48] FERC has the power to conduct formal and informal investigations (18 CFR Part 2). FERC is authorized to issue permits for the construction, operation, and maintenance of powerhouses, transmission lines, or other works for the development and utilization of power. It is also authorized to establish, review, and enforce rates and charges and prescribe a uniform accounting system for the transmission or sale of electric energy. Therefore, every power company transmitting power in interstate commerce must file with FERC for new construction or modification of its power generation facilities and the construction of any interstate transmission lines. They must also file their rates and charges for FERC review. FERC also has jurisdiction over corporate restructuring of electric utilities, including mergers and divestitures, and has the power to order an electric utility to provide transmission services once a good faith request has been made by another utility. In addition to reviewing and approving individual projects and rates, FERC also implements policy by issuing general orders that are applicable to all regulated power companies. DOE can suggest areas for rulemaking to FERC, however, FERC is not required to adopt DOE-suggested rules or policies.

FERC also holds hearings and resolves requests for emergency interconnections made to the DOE under Section 202(c) of the FPA. FERC has specific regulations pertaining to operations during a national emergency, including moving its offices and delegating emergency powers to its staff (18 CFR Part 376). FERC regulations also provide specific exemptions from certification for certain auxiliary installations constructed solely for use during a national emergency (18 CFR 2.60).

FERC has the power to maintain and investigate compliance with its permits, and may revoke a license for non-compliance. After notice and hear-

---

48 The line between federal and state jurisdiction is not always easy to determine, but the federal government's jurisdiction over electric utilities has been interpreted broadly by the courts. In *FPC v. Florida Power & Light Co,* (404 U.S. 453 [1972]), the Supreme Court found that a Florida utility's sales to another Florida utility through interconnection transmission lines, together with the simultaneous transfer of electric power from the second Florida utility to a Georgia utility, was an interstate sale within the jurisdiction of the FPC (FERC's predecessor). Thus, only clearly intrastate transmission of electricity remains under state jurisdiction (i.e., the local distribution).

ing, it may levy civil penalties of $10,000 per day for non-compliance. An order for civil penalties is enforceable in any appropriate district court of the United States.

4.2.2.2 Nuclear Regulatory Commission

The Atomic Energy Act of 1954 (42 U.S.C.A. §§ 2011-2296) authorizes the NRC to regulate construction and operation of nuclear power plants. The NRC promulgates regulations, issues construction and operation licenses, maintains on-site inspection teams, and issues enforcement actions when its regulations or license conditions are violated. The NRC's jurisdiction is limited to the safety aspects of nuclear power and does not address economic regulation. In *Pacific Gas & Electric Co. v. State Energy Resources Conservation & Development Commission* (461 U.S. 190 [1983]), the Court found that the NRC regulatory scheme occupied the entire field of nuclear safety concerns, except the limited powers expressly ceded to the states, and therefore states did not have jurisdiction over the safety aspects of nuclear power plants. Conversely, the only economic considerations by the NRC are those that would affect the ability of the utility applicant to meet safety and health requirements. Therefore, all safety, construction, and operation regulations applicable to nuclear power plants are under the jurisdiction of the NRC. Because of the severe health risk and significant economic loss that could result from nuclear power plant incidents, the NRC has promulgated unusually detailed and comprehensive regulations applicable to the construction and operation of commercial nuclear power plants.

4.2.2.3 Illinois Commerce Commission

Under the Public Utilities Act and the Electric Supplier Act (220 ILCS 5/1-101 *et seq.* and 220 ILCS 30/1 *et seq.*), the Illinois Commerce Commission (Commission or ICC) has authority to supervise the generation, transmission, and distribution of electricity in the State of Illinois, including construction permitting, rate setting, and reliability standards. In general, the ICC regulatory supervision of electric public utilities is the same as that for other types of public utilities; see the detailed discussion of ICC powers and authority in Section 3.2.2.4. Generally, the ICC's focus is the economic regulation of the power industry in Illinois; however, it has issued general standards of service for electric utilities (83 IAC 410) and more specific construction standards for electric power lines (83 IAC 305).

The ICC has a general reliability policy (83 IAC 410.410–410.490). Under the policy each utility must provide reliable service and strive to prevent interruption of electric service. When interruptions occur, the utility must restore electric service within the shortest reasonable time. Utilities must maintain records of service interruptions, including the time, duration, and cause of each interruption. The Commission has the power to adopt electric service reliability standards.

4.2.2.4 City of Chicago

Under the Electric Supplier Act, a public utility supplying electricity within an incorporated municipality is not subject to the ICC requirements for establishing service territories and supplying service to other utility service territories. Usually, the utility operates under a franchise with the city that gives it exclusive rights to serve the city's citizens and to use the city's rights-of-way to extend and construct its facilities and lines. Commonwealth Edison has a 29-year franchise with the City of Chicago, effective until the year 2021. Commonwealth Edison is still subject to the jurisdiction of the ICC for reliability, construction permitting, and rate setting.

The City of Chicago has no specific safety or security requirements for the electric lines Commonwealth Edison maintains or constructs within the city limits. The City of Chicago Municipal Code sets out very specific requirements for building wiring, meter installation, and in-building voltage with regard to protecting the public health and safety. The City's Department of Environment, Energy Management Division, supervises Commonwealth Edison's construction and repair activities to ensure they comply with the National Electric Code, as required by the ICC.

4.2.2.5 North American Electric Reliability Council

The members of the 10 coordinated NERC Regional Councils and the one Affiliate Council represent all segments of the electricity supply industry, including investor-owned, federal, and rural electric cooperatives; state/municipal and provincial utilities; independent power producers; and power marketers. These entities account for virtually all the electricity supplied in the United States, Canada, and a portion of Baja California Norte, Mexico. Members from these entities provide unmatched expertise in the planning, engineering, and operating aspects of electric system reliability for NERC's Engineering and Operating Committees. These committees and their subgroups serve the collective needs of the industry by researching and resolving North American operational and planning issues.

NERC members have historically volunteered to comply with established NERC policies, procedures, principles, and guides to ensure a reliable power system.[49] More details on specific operating policies and NERC's move toward obligatory compliance follow in the next section.

---

49 *Policies, Procedures, and Principles and Guides for Planning Reliable Bulk Electric Systems*, June 1995, and *NERC Operating Manual*, December 1996, NERC, Princeton, New Jersey.

4.3 Regulations and Critical Infrastructure Protection

4.3.1 Impact of Regulations on Critical Infrastructure Protection

System Vulnerabilities and Threat Scenarios

The loss of a power system component (generator, substation, or other piece of equipment) threatens the generation capacity available to serve the loads attached to the transmission network. If sufficient generation and transmission reserve margins are available, system disturbances may be minor and sags may not be noticeable, because other online system components respond to the system changes. Continued system operation under these conditions relies on adequate generation and transmission capacity within and between the areas surrounding the affected component. In more extreme cases, inadequate generation and/or transmission capacity can cause electricity from the main power system to be cut off (or shed load). Isolated islands may form that are left either without a power source or with a power source that is extremely vulnerable to additional problems. This situation could affect a large number of customers connected to the isolated transmission system and reflects an undesirable system emergency state that calls for careful system restoration procedures.

The following scenarios describe possible threats to or system conditions of the electrical infrastructure. Some situations describe a general concept that characterizes a particular operating phenomenon. Other situations highlight the potential for direct sabotage. For all situations, possible remedies to alleviate these conditions or threats or practical ways to improve current methods of evaluation are discussed. The discussion also addresses contingency planning, which can mitigate the impact of sabotage on the reliability of the infrastructure if preventive measures prove inadequate.

An operating failure can result from either natural causes or deliberate means. The utility's response, however, is independent of how the component is taken out of service. As a result, a system designed to operate under a given set of system disruption contingencies will remain operable and reliable. Contingencies represent a transmission line, generating facility, or some other piece of equipment that is taken out of service (becomes inoperable) because of a natural disaster (e.g., lightning), equipment failure, or malicious act.

In most respects, isolated sabotage events (single instances of deliberate attack) can be regarded as single contingency events. These are usually addressed in utility/system contingency planning. On the other hand, coordinated sabotage events representing multiple contingencies are very difficult to predict and therefore may not always be adequately addressed in developing a system contingency list to be considered for emergency planning. The effects of multiple sabotage events pose the greatest threat to the electrical infrastructure.

Physical Threat

The loss of a power plant can be accomplished in several ways. Access to plant surroundings provides a way to modify fuel supplies or disable primary transformers, generating equipment, or system controls. Each scenario results in a different degree of plant malfunction; for example, altering the fuel supply has less severe results than bombing the turbine-generator. However, the same system effect is achieved—the plant's generation capacity is removed from service. Risk of this type of occurrence could be reduced through use of fencing, closed-circuit TV, or security forces, although these measures would involve costs.

Most substations are unmanned and therefore face an increased risk of sabotage. Like most substations, transmission lines are located in remote areas, and a person may have air or ground access to utility rights-of-way with little or no chance of being spotted. In addition, transmission towers can be felled. It should be noted, however, that they are built to specifications designed to ensure survival in violent weather. Some instances of attempted sabotage were unsuccessful because detachment of one or even two legs of a four-legged tower did not bring it down.

An easier way of removing a line from service is to shoot the insulators suspending one of the transmission conductors. The result is a short circuit when the line comes in contact with the ground.

Cyber Threat

Cyber threats to control systems are also a concern. Sophisticated electronic systems are used in operating large electric utility generation and transmission facilities. These systems rely on large databases, remote sensoring, and telecommunications between NERC members to maintain control of the energy flow and to ensure system balance. Malicious action against system components controlled by Supervisory Control and Data Acquisition (SCADA) is possible. Hackers could, for example, record control signals that were sent over conventional communications channels and play them back later, causing unscheduled system interference. Research and development activity to address this problem was recently a topic at the Institute of Electrical and Electronics Engineers (IEEE) 1997 Winter Power Engineering Society Meeting ("Information Security Issues in Utilities"). Encryption methods are available; however, utilities would have to upgrade older equipment to accept the additional protocol.

Regulatory Security Requirements for Nuclear Power Plants

All commercial nuclear power plants must obtain a license from the NRC. NRC regulations specifically state an applicant is not required to provide for design features or other measures for the specific purpose of protection against attacks, including sabotage, directed against the facility by an enemy of the United States, whether a foreign government or other person (10 CFR 50.34). However, the regulations provide for design criteria and performance standards for protection against radiological sabotage. Under NRC regulations (10 CFR 73.1(a)(1)), the definition of radiological sabotage includes deliberate acts directed against a nuclear power generating station that could endanger the public health and safety by exposure to radiation. In establishing physical security requirements, the NRC defines radiological sabotage as (1) a determined violent external assault, attack by stealth, or deceptive actions of several persons who are dedicated and well-trained (including military training and skills), (2) possibly using inside information or assistance, (3) possessing suitable weapons, up to and including hand-held automatic weapons, equipped with silencers and long-range accuracy, (4) possessing hand-carried equipment, including incapacitating agents and explosives for use as tools of entry or for otherwise destroying reactor, facility, transporter, or container integrity, and (5) using a four-wheel-drive land vehicle used for transporting personnel and their equipment. It can also include an internal threat of an insider, including an employee and a four-wheel-drive land vehicle bomb. A successful act of nuclear power plant sabotage would likely cause significant damage to the plant and possible permanent shutdown. There would also probably be a significant temporary or permanent loss of electrical generating capacity and economic loss to the operating utility.[50]

The NRC regulations cover design criteria that specifically address security (e.g., built-in redundancy, physical separation of independent circuits to minimize the likelihood of simultaneous failure, and control and monitoring equipment necessary to maintain a radiologically safe condition during emergencies [10 CFR 50, Appendix A]). These design features must ensure that the plant is able to withstand and recover from a station blackout and be capable of shutting down the reactor and maintaining it in a safe condition. In addition, each application for a permit must include a physical security plan that addresses the identification of vital equipment, vital areas, and the security systems and subsystems designed to protect them. It must also include a safeguards contingency plan for dealing with threats, thefts, and radiological sabotage in conformance with the requirements

---

50 NRC regulations also contain detailed specifications for emergency response capabilities on the part of nuclear utilities and nearby communities; these are discussed in Section 9.

found in NRC regulations for physical protection for plants and materials (10 CFR 73).

NRC regulations for the physical protection for plants and materials require the capability to prevent entry of unauthorized persons, vehicles, and materials into material access areas and vital areas. This capability must include a (1) physical barrier system for all vital areas, such that access to them would require passage through at least three barriers, and (2) perimeter protected by two separate physical barriers, with an intrusion detection system between the two. The barrier must incorporate features to prevent forcible vehicle entry. Isolation zones must be maintained adjacent to all outdoor physical barriers to allow observation of all persons or activities on either side of the barrier. The protection system must be able to detect any attempts to gain unauthorized access or to introduce unauthorized material across the plant boundary by stealth or force. Access to vital areas is limited to those employees whose duties require them to enter the area.

The plant must have an established, trained, and qualified security organization that can execute planned responses to emergency and safeguard contingency events (tactical response team). The regulations recognize internal and external threats. For instance, no plant can assign a member of the security organization to have direct operational control over more than one of the redundant elements of a physical protection subsystem if it could result in loss of effectiveness of security performance. All security personnel must meet stringent performance standards concerning physical health and weapon proficiency and undergo an FBI criminal history investigation.

These physical protection systems must be maintained and tested periodically; tests must include real-time exercises. The plant cannot make changes to its safeguards contingency plan that would decrease the effectiveness of the security plan or guard training and qualification plan without prior approval of the NRC. Therefore, sabotage against a nuclear plant would require an extremely well-trained, well-equipped terrorist force or the criminal intent of a very high-placed, knowledgeable employee. Other components of the power system are not so well-protected.

Regulatory Security Requirements for Nonnuclear Facilities

Policy 4 of the *NERC Operating Manual* requires only minimal physical security for electric systems. This policy requires monitoring equipment that will bring important deviations in operating conditions to the system operator's attention and will indicate, if appropriate, the need for corrective action. Each control area must have sufficient metering to ensure accurate and timely monitoring of operating conditions under both normal and emergency situations, including evaluating the loss of significant transmission or generation facilities. Policy 4 also provides that, where practical, critical unmanned facilities should be monitored for physical security, and, at a minimum, scheduled inspections and preventive maintenance should allow the utility personnel to monitor the conditions of these facilities.

The Illinois Commerce Commission regulations do not require specific physical security for any component of an electric utility's system. They do provide that all electric supply and communication lines and equipment must (1) be defined, constructed, and maintained to meet those portions of the National Electric Safety Code adopted in the regulations (83 IAC 305) and (2) generally be able to provide safe, adequate, and dependable service. All distribution lines must have protective devices to effect shutdown in case of a power surge.

Ensuring Reliability after an Interruption

No federal or state regulatory body or legislation provides a concise definition of reliability; therefore, it is appropriate to adopt the definitions and standards of NERC as the basis for discussion. NERC defines *reliability* as the "degree to which the performance of the elements of that system results in power being delivered to customers within accepted standards and in the amount desired."[51] A reliable power system gives customers what they want, when they want it. The "ability of the bulk power electric system to withstand sudden disturbances such as electric short circuits or unanticipated loss of system components" refers to the *security* of the power system. There is a relationship between reliability and system security in that a secure system assures a reliable system. A secure system remains operable after system emergency contingencies occur; it also successfully adjusts for transmission or generating capacity that was removed because of the loss of a critical component. A reliable system provides the adequacy and security needed to react to emergency contingencies. The ability of a system to adjust and continue service after loss of a critical component is what ensures the viability of the electric infrastructure when preventive measures have failed.

The NERC regions all practice methods to measure and model system security.[52] Models are created to simulate the system impacts of individual and multiple contingencies. A single contingency represents a single component failure. Considering the size of a power system, there are literally millions of possible contingencies. Model simulations are used to understand the system response to these contingencies and to assure continued system operation. Operational guidelines that direct how system operators

---

51 *Reliability Concepts*, NERC, Princeton, New Jersey, February 1985.

52 *Overview of Planning Reliability Criteria of the Regional Reliability Councils of NERC*, North American Electric Reliability Council, Princeton, New Jersey, April 1988.

are to adjust the system if such events take place result from these simulations. The practice of modeling these system contingencies greatly contributes to the reliability of the electric power infrastructure.

A critical element of these studies is the accuracy of the selected contingencies. A great deal of system expertise, effort, and advanced modeling software is expended in defining the contingencies represented on the contingency list. If a specific contingency is not explicitly modeled, the system response to it cannot be examined and anticipated. The NERC Reliability Assessment Subcommittee (RAS) reviews and assesses the overall reliability of the regional bulk electric systems to ensure that each Region conforms to its own planning criteria. The planning reliability criteria of the Regional Reliability Councils are published by the Reliability Criteria Subcommittee (RCS). Thus the RCS would be a logical group to consult regarding inclusion of sabotage-based scenarios in contingency planning.

In situations of direct sabotage, a piece of equipment is "prematurely" taken out of service. Regardless of whether a component naturally fails or is deliberately made to fail, the event is viewed as a single contingency that may be represented in the contingency list. This concept is important to understand, since standard modeling practices apply equally to deliberate events and natural events. As long as the contingency is a member of the contingency list, the event can be simulated, system security can be evaluated, and impact mitigation can be planned. In a similar manner, selected multiple contingencies are included in the contingency list, although they pose a greater threat to continued system security. This information must be kept in mind when considering the impact of deliberate power system attacks.

The steps involved in selecting, simulating, and evaluating the impacts of contingencies is considered a system planning activity. In the event that contingencies do occur, NERC policies dictate how the industry is to respond to these events and report them. An example of these policies are found in the *NERC Operating Manual.* It presents specific criteria, requirements, and guidelines with regard to particular system operating characteristics and procedures. Guidance is provided for generation control, transmission, system coordination, emergency operations, operations planning, telecommunications, and operator training issues.

The *NERC Operating Manual* addresses the need for each utility in the control area to maintain a level of operating reserve sufficient to account for system equipment forced outage rates. Following loss of resources or load, a control area must take appropriate steps to return the system load to its predisturbance level within 10 minutes after the start of the disturbance. When an operating emergency occurs, a prime consideration is to maintain parallel operation throughout the interconnection. Each system and control area must promptly take appropriate action to relieve abnormal conditions. A system, control area, or power pool that experiences or anticipates an operating emergency must communicate its current and future status to neighboring systems throughout the interconnection. Systems able to provide emergency assistance then make their capabilities known to the burdened system. Other systems may be notified through predetermined communication paths whenever the system suspects or has identified a multi-site sabotage occurrence or a single-site sabotage of a critical facility. If necessary, the system has a predetermined methodology for shedding load to accommodate the entire interconnected system.

An issue recently arose concerning the original NERC System Coordinators Committee proposal for prioritizing customers, which allowed a utility to interrupt nonutility firm customers before utility firm customers. The proposal was in contravention of FERC's *pro forma* tariff regulations, which prohibit discrimination between similarly situated customers (i.e., all firm customers). NERC and FERC worked together informally to redraft the proposal to meet FERC nondiscrimination requirements, using a *pro rata* interruption methodology for all similarly situated customers whose load was determined to be affected by the contingency.

All disturbances or unusual occurrences suspected or determined to be caused by sabotage must be reported to the appropriate governmental agencies, including the FBI and regulatory bodies. Each control area must have procedures for recognizing sabotage events on its facilities and multi-site sabotage affecting larger portions of the interconnection and for making its system operators aware of them.

A set of plans must be developed and implemented for each system, control area, pool, and region within NERC to cope with operating emergencies. These plans must enable the system or pool to mitigate, to the fullest extent possible, the effect of a capacity or energy emergency on its customers. These must include operating and coordinating agreements between neighboring systems or pools to provide emergency assistance. Such assistance may include supplying reserve generating capacity for a limited period of time. The system operators also must have authority to implement manual load shedding when necessary.

Reports of any emergency conditions or abnormal events, which in the opinion of the reporting utility could constitute a hazard to maintaining the continuity of the bulk electric power supply system, must be reported to the DOE Emergency Operating Center (EOC) within 24 hours. The *NERC Operating Manual* provides for regional and interregional emergency telecommunications networks in Appendix 7A. Each network consists of a preset conference call that interconnects coordination centers in the regions within each interconnection. The *NERC Operating Manual* even sets out call procedures. There are hotlines for aid in emergency or near-emergency situations, which cannot be solved by normal contiguous interconnected system communications. The telecommunication networks are to be tested weekly to assure operation during emergency conditions.

NERC consistently modifies and updates these policies as specific events reveal policy deficiencies. Considerable practical experience has been gained from analyzing actual system disturbances.[53] These studies provide insight on whether to reenforce or change existing policies. Most of these events represent localized problems that affect a small number of customers, if any. In the event of a large-scale incident, NERC assigns focused task forces to study and publish a detailed accounting of the events leading up to the specific situation.[54] Both report formats typically summarize the situation, describe the sequence of events, describe power restoration procedures, draw a series of conclusions, and identify a set of recommendations intended to prevent similar incidents. Similarly, specific changes to NERC policies may be proposed to reduce the likelihood of reproducing similar situations in the future. Again, NERC's methods provide a proactive means of acquiring and including lessons learned into operational procedures to increase overall system reliability.

Availability of Sensitive Data

FERC Form 715, *Annual Transmission Planning and Evaluation Report,* provides NERC regional load flow data to the general public. This data could be used by an experienced energy system modeler to pinpoint potential weak network facilities or interconnections. This poses a risk to system security by providing information on the system topology and likely vulnerable areas. Unrestricted access to this data was a controversial issue before Form 715 was ordered. Now the FERC bulletin board system (BBS) offers the data free of charge. Anyone can log onto the BBS by typing a name and password; there is no mechanism to verify the name or company affiliation. The system provides downloadable data files, including annual reports, rates, planning area reports, and other documents. In addition, similar information, including the date, time, duration and cause of outages experienced by every NERC member, is available on the NERC Internet site (http://www.nerc.com/dawg/database/dawg-96.html). The exact potential for abuse is difficult to determine, since some amount of system expertise is still required to derive a contingency set that causes trouble. However, it is possible that access to this data could enable an intelligent hostile element to establish a contingency list, prioritize the effects of represented contingencies, and produce a multiple contingency

---

53 *System Disturbances,* NERC, Princeton, New Jersey, July 1996.

54 For example, see *Report on Electric Utilities' Response to the Cold Wave of January 1994.* NERC, Princeton, New Jersey, April 1994.

disturbance that could bring down the system in some manner. Access to the contingency set, whether derived from the FERC Form 715 data or purchased from a disgruntled utility employee, poses a significant risk to electrical infrastructure protection.

In addition, with the development of the OASIS, even more information will be easily available concerning the routing of interstate, wholesale electric sales throughout the nation. Also, to facilitate the wheeling ordered by FERC Order 888, independent system operators (ISOs) (centralized control points) may be developed for many power pools and interconnections to broker the movement of electricity from control area to control area.[55] This again will tend to concentrate information on generation and transmission adequacy in one database.

4.3.2 Measures for Improvement

From discussions and professional interactions at conferences and meetings, one can sense that utilities sincerely want to provide a reliable product to their customers. As it does in any business, risk plays a role in both reliability and earnings. If utilities had the option, they would install state-of-the-art equipment designed to provide the most reliable operation for their given transmission and distribution network configuration. However, in the deregulation climate it is necessary to closely monitor costs. The open market will send the proper price signals when transmission bottlenecks become a limiting factor in assuring economical power.

One simple improvement in the present regulatory structure would be to reduce the red tape associated with recovery of costs incurred to improve reliability. The NERC report, *Options to Ensure Compliance with NERC and Regional Reliability Council Policies, Standards and Criteria*, recommends that FERC ensure that all expenses incurred to meet reliability criteria established by the NERC are fully recoverable through rates. After short-run improvement costs are recovered, the long-run benefits of relieving these bottlenecks are as follows: increased transfer capability, increased transmission margins, improved operating flexibility in the event of natural or deliberate interruptions, and reduced customer costs.

The analysis of regulatory impacts on critical electrical infrastructure protection in Section 4.1 suggests four areas for possible initiatives to improve security. Three of the four are aimed at preventing attack, and the fourth is aimed at preventing widespread consequences from an attack.

---

55 In the California electric utility restructuring scheme, when the ISOs are formed the utilities will transfer their transmission grids to the ISO and either sell or lease their control centers (computers and all) to the ISO.

1. *Develop higher standards for physical security at vulnerable nonnuclear facilities.* As discussed, security standards for nuclear power plants are extensive and apparently effective. Security standards for physical protection of nonnuclear facilities such as power plants and substations are much narrower in scope. To improve overall system security, one approach would be to research cost-effective measures to enhance security at such facilities. A model standard for implementing such measures could then be developed for adoption by NERC and state regulatory bodies. Uniform application of requirements would help ensure that these measures do not interfere with the market by favoring some firms over others.

2. *Develop methods for protecting against cyber attack on control systems.* As discussed, network control systems are considered vulnerable to intercept and interference. An obvious response would be to develop methods or protocols for protecting this information, either through the use of physically secured transmissions or through encryption or other data-protection methods. Regulatory action to require use of such protections may be appropriate since they might be cost-effective from an overall system standpoint, yet not so from the standpoint of any given company.

3. *Control key information that identifies system vulnerabilities.* Information on system configurations that might benefit hostile elements is currently available over the FERC BBS, on the Internet, and via Freedom of Information Act requests. Placing controls on such information might serve a protective function. However, research would be needed to determine whether (1) information can practically be controlled without substantially interfering with normal or emergency operations and (2) control of this information would run counter to any current regulatory disclosure requirements.

4. *Add more sabotage-based scenarios to NERC contingency modeling and planning.* Modeling and planning for system contingencies by NERC is the main method by which the electric industry assures that local emergencies do not become regional or national emergencies. Some sabotage-based scenarios are already part of this planning. It might be worthwhile for a government-industry group to examine this planning and determine whether other plausible scenarios should be considered.

5 OIL AND NATURAL GAS INFRASTRUCTURE

5.1 General Description of Regulation

5.1.1 Current Regulatory Environment

Natural Gas Industry Profile

Natural gas is produced at privately owned wells located throughout the United States. Some wells are located on private lands. Others are located on federal lands leased from the U.S. Bureau of Land Management (BLM) or under leases of rights on the Outer Continental Shelf (OCS), which is under the jurisdiction of the U.S. Department of Interior (DOI), Minerals Management Service (MMS).[56] The gas is pumped, either as a by-product of oil production or as a singular product, and transported along gathering lines to a consolidation/compression facility. Natural gas is also imported from (and exported to) Canada and Mexico along international pipelines. Liquefied natural gas (LNG) is also imported via tank ships to port facilities where it is stored, then converted to the gaseous form for pipeline shipment. The safety regulation of LNG facilities is under U.S. Department of Transportation (DOT) jurisdiction. However, states may have requirements for liquefied petroleum gas (LPG) and compressed natural gas (CNG) facilities.

The gas is then transported along a network of interstate and intrastate pipelines that traverse the 48 contiguous states and Alaska. These pipelines are owned and operated by pipeline companies. Pipeline companies are engaged in the transportation of natural gas; however, these companies may have sister divisions or companies engaged in production or distribution operations. The gas can be delivered to the ultimate consumer either directly from the pipeline, as is the case with some industrial or commercial facilities, or to an interconnection (city gate) with a local distribution company (LDC) for transportation along distribution lines to homes and businesses.

Overview of Natural Gas Regulation

As discussed below, each of the natural gas system components (production, transportation, and distribution) is regulated by different regulatory bodies.

- Production facilities are subject to both economic and environmental regulation. Economic regulation of natural gas production addresses

56 MMS has jurisdiction over areas except those specific areas under state jurisdiction that are delineated by using a formula. For Texas and the Gulf Coast of Florida, it is three marine leagues (approximately nine nautical miles) seaward from the baseline from which the breadth of the territorial seas is measured.

well drilling, well management, and management for gathering, treating, and compressing gas for pipeline transmission. In general, privately owned production facilities are under the jurisdiction of the state in which the facility is located. Production facilities located on federal lands are regulated via the terms of their lease by the BLM. The MMS exercises control via lease terms of facilities located on the OCS. Production-related facilities are also subject to environmental regulation by the U.S. Environmental Protection Agency (EPA) and its authorized state counterparts. Applicable environmental regulations include those pertaining to waste disposal and air emissions.

- For interstate pipelines, the economic aspects of wholesale transportation and sales of natural gas are within the jurisdiction of the Federal Energy Regulatory Commission (FERC). Pipeline safety and construction standards for interstate transportation pipelines are under the DOT.
- For intrastate pipelines and LDC distribution lines, pipeline safety and construction standards are within the jurisdiction of the state agency (certified by DOT). The economic aspects of intrastate, retail transportation and sales of natural gas (intrastate pipeline companies and LDC public utilities) are regulated by the state public utility commission (PUC).

Oil Industry Profile

Like natural gas, oil is produced at privately owned facilities located throughout the United States. These facilities may be located on private land or on leased federal lands, including the OCS. Although there are small independent producers, the major oil companies (the 20 largest, vertically integrated, multinational oil companies) predominate the oil industry. The independents vary in size and generally are not involved in every stage of the oil industry. Most engage only in production and sell their crude product to the major oil companies or to smaller independent refineries. In the case of gasoline, the oil companies and refiners sell to other wholesalers or sell at retail from their own gas stations or under contract with independent gas stations. Other grades of oil are also sold to end users or to retailers (e.g., fuel oil suppliers). In some cases, oil or oil products are shipped via pipeline directly to large industrial customers. Most wholesale transportation is by pipeline or truck to a local end terminal, where the product is stored and then transported by truck to individual customers such as gasoline stations.

Overview of Oil Regulation

The regulation of oil production facilities (including treatment facilities such as refineries) is generally under state jurisdiction, except for facilities operating under lease on federal lands or the OCS. Oil producers are generally subject to state regulations concerning well drilling, well management (including maximum production limits), gathering equipment, storage, and refining.

They are also subject to environmental regulations (including requirements for spill prevention and emergency response) administered by the EPA and its authorized state counterparts. Emergency response and personnel safety for off-shore production facilities on the OCS come under the jurisdiction of the U.S. Coast Guard and MMS. However, pipelines running from production facilities to on-shore facilities (that are not gathering lines) are under the jurisdiction of DOT.

Safety and construction standards for interstate oil transportation pipelines are under the jurisdiction of DOT. Intrastate pipelines are similarly regulated by state agencies under DOT-certified programs. Most economic regulation of oil pipeline construction, both interstate and intrastate, ended in 1994. However, oil pipelines must still file simplified rates and tariffs for transportation services with the FERC (18 CFR Parts 340-342).

Complexity of the Oil and Gas Regulatory Scheme

The regulatory scheme for oil and gas is complex and divided among many agencies. There are numerous provisions and exceptions to particular parts of the scheme for particular types of equipment or situations. In some cases, the scope of a particular agency's jurisdiction may extend only to production equipment or only to transmission or distribution equipment, but the real world of oil and gas systems does not always fall neatly into these categories. Therefore, determining who has jurisdiction and what rules apply to a given location is sometimes difficult. For example, DOT has issued regulations prescribing minimum safety standards for pipeline facilities, covering the construction, maintenance, and operation of gas and oil interstate pipelines. The scope of these regulations includes the gathering, transmission, and distribution of gas by pipeline and the storage of gas, but does not include the gathering of gas through unregulated gathering lines in rural locations located outside the limits of any (1) incorporated or unincorporated city, town, or village; (2) any other designated residential or commercial area (including a subdivision, business, shopping center, or community development); or (3) similar populated area. DOT-regulated transportation of oil includes the movement of petroleum or petroleum products by pipeline or the incidental storage of such products but does not include oil moving through gathering lines in a rural area; on-shore production, refining, or manufacturing facilities; or storage or in-plant piping systems associated with on-shore production, refining, or manufacturing facilities. The safety aspects of gathering lines and distribution lines in rural areas or within production facilities are generally under the jurisdiction of state pipeline safety regulation.

The complexity of the regulatory scheme may or may not be a problem for the oil and gas industry *per se*; however, it can complicate efforts to institute changes in the regulations to accommodate new policies.[57]

5.1.2 Current Trends in Regulation

Deregulation of Natural Gas

Federal regulation of natural gas began with the Natural Gas Act (NGA) passed in 1938. Regulation of natural gas under the NGA was implemented by the Federal Power Commission (FPC).[58] The NGA allowed pipelines to transport gas from producers to customers across state boundaries. The NGA applied to interstate pipeline companies only; gathering, production, and distribution facilities were not included. However, the FPC's jurisdiction under the NGA was extended to producers by the Supreme Court.[59] After a relatively long quiet period, oil and gas shortages in the 1970s led to a series of legislative changes. In 1978 Congress passed the Natural Gas Policy Act (NGPA), which moved from cost-based, utility-type regulation toward a pseudo-free market pricing scheme. The NGPA had a built-in time schedule for price deregulation, and a substantial portion of gas was deregulated by January 1985. In 1989, the Natural Gas Wellhead Decontrol Act (NGWDA) removed all remaining NGPA wellhead price controls and eliminated all NGA filing requirements for natural gas producers. This situation created a competitive market and the start of market-based wholesale rates. As discussed below, pursuant to FERC Orders 436 and 636, producers and users of natural gas now have open access to the network of pipelines on a contractual basis with pipeline companies. However, FERC still requires pipeline companies to obtain approval for corporate merger and divestiture, obtain permits for constructing interstate pipelines, and file rates for transporting natural gas along those pipelines and for operating storage facilities connected to the pipeline.

The 1980s brought gradual deregulation of natural gas production pricing. In 1984, FERC issued Order 380, which outlawed minimum bills, giving customers the opportunity to purchase low-cost supplies while avoiding

57 *See* Appendix A to 49 CFR Part 195—Delineation Between Federal and State Jurisdiction—Statement of Agency Policy and Interpretation.

58 The FPC was the predecessor to the FERC. Intrastate pipelines (Hinshaw pipelines) are not subject to NGA jurisdiction but to the jurisdiction of the applicable state agency.

59 *Interstate Natural Gas Co. v. FPC*, 331 U.S. 682 (1947) and *Phillips Petroleum Co. v. Wisconsin*, 347 U.S. 672 (1954).

the high cost old contracts. The following year, FERC issued Order 436, which required pipelines to provide separate transportation service to other producers, thus allowing marketers and end-users to compete. Finally, FERC Order 636 (August 1992) completed deregulation by forcing pipelines to fully unbundle their merchant, transportation, and storage services, thereby allowing for market-based prices and creating a secondary pipeline capacity market.

The recent changes in federal regulation of natural gas pricing effectively segment the industry into three parts: producers, interstate pipeline operators, and intrastate pipelines and distribution networks. Members of each of these groups are now more or less free to contract for the sale, purchase, and transportation of natural gas on an as-needed basis and at mutually negotiated prices. However, there is still federal oversight, and the federal government can act to alleviate shortages if they occur.

At the state level, the trend is for state PUCs to follow the federal pattern of deregulating pricing. However, other aspects of state-level regulation, such as well production limits and pipeline safety requirements, are still closely regulated, as detailed below.

Deregulation of Oil

As did natural gas regulation, oil industry regulation underwent significant changes beginning in the 1970s, leading to wholesale pricing deregulation in the 1990s. The Emergency Petroleum Allocation Act of 1973 (EPAA) gave broad authority to the President to regulate oil production and wellhead prices to stimulate domestic exploration and production. The President was also authorized to establish priorities among end users in times of shortage. In 1979, President Carter used his EPAA authority to begin a phased decontrol of oil and oil products regulation. Economic regulation of oil products was essentially ended under Executive Order 12287 (January 28, 1981). In 1992, FERC jurisdiction over oil transportation pipelines was reduced to the acceptance of simplified rate filings, and in 1994 FERC's authority over permitting the siting and construction of oil pipelines was repealed altogether. Although the federal economic regulation of oil and natural gas production has in effect ended, production functions and pipeline construction are still regulated at the state level.

The deregulated oil industry is in an economically challenging situation. The number of refineries has decreased from more than 300 to 165 over the past 15 years. The profit margins have decreased while the competition, particularly from overseas companies, has increased. There has been an increase in environmental regulation, in part from the closing of loopholes and exemptions that once protected the industry. Recent environmental trends include (1) producers having to report data for the EPA's Toxic Release Inventory, (2) compliance with air quality standards, (3) an increase in safety programs, and (4) restricted exploration. All of these result in increased costs to the oil industry. The EPA's Toxic Release Inventory requirements are projected to cost producers more than $100 million per year. Increased air quality standards and reduced emissions require major capital investments in environmental technologies. In addition, because of environmental concerns, some potential oil-producing areas, such as the Arctic Refuge and recent finds off the coasts of Florida and North Carolina, are not being opened up to producers. These regulatory changes and access restrictions will ultimately increase the cost of production and lower the amount of domestic production.

In addition, the construction of new natural gas and oil pipelines must now confront the opposition of environmental groups and communities. Recently in Illinois, a request for permission to build an oil pipeline through a suburban area was denied by the Illinois Commerce Commission for failure to demonstrate a public need. The Commission found the desire of one refinery to obtain lower-cost, additional oil supplies was not sufficient to establish a public need and invoke the Commission's power of eminent domain, when compared with the concerns of affected communities.[60]

Decreased Margins in Capacity

Economists link competitive markets and efficient markets together. The assumption is that the gas and oil suppliers will meet the demand, provided consumers are willing to pay to maintain adequate supplies. However, there may be some potential pitfalls to consider. Under pure competition, pipelines must operate at high utilization rates to keep costs down. Each pipeline will optimize its own deliveries without consideration for the network system or for establishing emergency supplies. The excess capacity on the existing pipeline infrastructure will decrease, thereby reducing the current transportation system redundancy.

Use of Risk Analysis in Setting Requirements

DOT, EPA, the Coast Guard, and state agencies still strongly regulate the safety of pipelines and production and storage facilities. The recent reauthorization of the Pipeline Safety Act (Accountable Pipelines Safety and Partnership Act of 1996) does not erode the safety standards set for pipelines. However, for the first time, it does provide for the systematic application of a risk management scheme to its rulemaking process and to the industry's implementation of the regulations. This new approach provides that safety standards prescribed by DOT must be not only practicable

60 *Lakehead Pipeline Company* (Ill.C.C. Docket No. 96-0145). The applicant did not need Commission approval to construct the pipeline but required a Certificate of Good Standing in order to exercise the right of eminent domain.

and designed to meet the need for gas pipeline safety but must also consider the reasonably identifiable or estimated costs and benefits expected to result from the new standard. In conducting the risk assessment, DOT must identify the regulatory and nonregulatory options considered in prescribing the new standard and the associated costs and benefits. However, this balancing is not required when the regulation is the product of negotiated rulemaking or the adoption of industry standards that receive no significant adverse comment after notice in the *Federal Register.*

To ensure continued pipeline safety, under the current rulemaking process, proposed standards are reviewed by the Technical Pipeline Safety Standards Committee and the Technical Hazardous Liquid Pipeline Safety Standards Committee. These committees are made up of individuals from the federal government, state government, natural gas or petroleum industry, and general public. They review each proposed standard and publish a committee report on its technical feasibility, reasonableness, and practicability. Although the DOT Secretary is not bound by the conclusions of the committees, if the Secretary adopts regulations in contravention to the report, the reasons for deviation must be published. In addition, the committees may propose safety standards to the Secretary for adoption.

Under the Accountable Pipelines Safety and Partnership Act of 1996, a risk management demonstration program is to be established, with the voluntary participation of gas and oil pipeline operators. The volunteers will submit risk management plans that contain measures designed to achieve an equivalent or greater overall level of safety (compared with DOT standards) with increased efficiency and cost-effectiveness. The expectation is that through the application of some form of risk assessment of the volunteer's systems, areas needing the greatest risk reduction will be identified. The company will have the opportunity to design and propose plans based on risk modeling that may deviate from the current standards but can be demonstrated to lead to greater levels of systemwide safety at lower cost. This program is still in the formative stage.

Increase in Liability Exposure

One other recent legal change that has affected safety practices is the increase in liability exposure provided for in the Oil Pollution Liability and Compensation Act of 1990. Specifically, the Act allows for unlimited liability for negligence or willful misconduct and increased liability for oil spills. Although not regulatory in nature, this change in the legal landscape drove upgrades in many of the transport vessels.

Proposals by Industry Associations

Although they are not regulatory bodies, the Interstate Natural Gas Association of America (INGAA), American Gas Association (AGA), and Gas

Research Institute (GRI) are influential in setting national policy concerning the operation of natural gas production, transportation, distribution, and storage facilities. The INGAA and AGA closely monitor legislation and rulemakings that may affect the natural gas industry. All three associations participate in research and studies on new technologies and economic market conditions. GRI has prepared two topical reports concerning the future of gas safety issues under deregulation.[61] These reports conclude that the trend for deregulation is a positive change and "the increased use of voluntary programs that are co-designed by government and industry can result in compliance with safety goals in a more efficient manner." However, they also find that "more flexible regulation, while introducing opportunities for cost savings, also introduces less certainty, a greater need for decision-making and documentation, and a lengthy learning curve." They also postulate that the shift away from federal regulation to state regulation could result in less uniformity in safety regulation. The reports stress that the cost-cutting requirements associated with the newly competitive market may result in a smaller, untrained workforce that will not have sufficient knowledge to implement the necessary safety procedures. The reports also express concern with the new risk management demonstration program and whether it will result in a perception by the public that safety requirements are being reduced.

5.2 Description of Selected Regulatory Agencies

5.2.1 Selection Method

The following agencies were selected for detailed discussion: DOT, EPA, DOE, Illinois Commerce Commission (ICC), and Texas Railroad Commission (TRC). DOT establishes the basic safety standards for interstate transportation of natural gas and oil via pipeline and for the construction and operation of LNG facilities. These DOT standards are then adopted and enforced for pipelines by the states. The EPA establishes spill response requirements for petroleum and petroleum product storage facilities[62] and environmental standards for the oil production industry. DOE is included as the policy-setting body for all energy industries. The ICC was chosen as an example of a state PUC that regulates natural gas public utilities and gas pipelines. The TRC was chosen as an example of the regulation of oil and natural gas production facilities in a major oil- and natural-gas-producing state.

---

61 *The Future of Gas Safety Issues: Views from Outside the Industry*, GRI-96/0198, June 1996, and *Trends, Issues and Recommendations for Natural Gas Safety: Views from the Industry*, GRI-96/0410, December 1996.

62 End terminal storage facilities are under the jurisdiction of the EPA. Storage facilities incidental to the transportation of oil along oil pipelines are under the jurisdiction of DOT.

5.2.2 Agency Descriptions

5.2.2.1 Department of Transportation

The Pipeline Safety Act (PSA) (49 U.S.C.A. 60101 *et seq.*) authorizes DOT to establish minimum safety standards for oil and natural gas pipelines. DOT has responded by issuing standards that apply to all owners and operators of pipeline facilities. The standards address design, installation, inspection, testing, construction, extension, operation, replacement, and maintenance of pipelines and emergency plans and procedures for them. As discussed above, these standards are promulgated via a cooperative system that allows federal and state government, industry, and public participation through the Technical Pipeline Safety Standards Committee and the Technical Hazardous Liquid Pipeline Safety Standards Committee.

The DOT Office of Pipeline Safety[63] has primary enforcement jurisdiction over interstate transportation pipelines. Primary enforcement responsibility for pipelines is delegated to the states, with DOT supervision. Each state authority must submit an annual certification to DOT, showing that it (1) has regulatory jurisdiction over the pipelines; (2) has adopted regulations at least as stringent as the DOT safety standards and regulations; (3) has the legal power to enforce and is enforcing each adopted standard; and (4) is encouraging and promoting programs designed to prevent damage by demolition, excavation, tunneling, or construction activity to the pipeline facilities under its jurisdiction. In addition, the certification may show that the state requires record maintenance, reporting, and inspection programs substantially the same as those provided in the DOT regulations.

To enforce the pipeline standards, DOT may request the Attorney General to bring a civil action in an appropriate district court of the United States, and that court may award appropriate relief, including punitive damages. In addition, DOT can, after written notice and an opportunity for hearing, find a regulated entity in violation of the PSA and impose a civil penalty of not more than $25,000 per violation or for each day of violation, not to exceed a total penalty of $500,000.

DOT is also authorized under the PSA to conduct research and studies. For instance, under the Accountable Pipeline Safety and Partnership Act of 1996, DOT is to survey and assess the effectiveness of remotely controlled valves to shut off the flow of natural gas in the event of a rupture. DOT will determine whether the use of remotely controlled valves is technically and economically feasible and would reduce risks associated with such a rupture. If so, DOT will prescribe standards for use of these valves by interstate natural gas pipeline operators.

---

63 The Office of Pipeline Safety is located in the DOT Research and Special Programs Administration (RSPA).

5.2.2.2 Environmental Protection Agency

The Clean Water Act (CWA; 33 U.S.C.A. § 1321) authorizes the EPA to regulate owners or operators of any off-shore or on-shore facility that could reasonably be expected to cause substantial harm to the environment by discharging oil into or on navigable waters or adjoining shorelines. These regulations (40 CFR 112 *et seq.*) apply to owners or operators of non-transportation-related on-shore and off-shore facilities engaged in drilling, producing, gathering, storing, processing, refining, transferring, distributing, or consuming oil and oil products.[64] In general, the regulations require regulated facilities to prepare and implement a spill prevention control and countermeasure plan, designed to provide for prompt emergency response and notification to authorities in the event of a spill.

5.2.2.3 U.S. Department of Energy

DOE was created in 1977 by the Department of Energy Organization Act of 1977. The Act consolidated many energy-related functions of the federal government into a single, Cabinet-level organization. DOE sets national policy concerning oil and gas production, supply, and distribution, including response to an energy crisis. It has emergency powers to allocate oil or natural gas to regions affected by a shortage and to increase production. For instance, during the Gulf War, DOE requested production states to act to lessen the impact of any disruption of supplies due to Iraq's invasion of Kuwait. In response, the Texas Railroad Commission issued an emergency statewide rule (Statewide Rule 87, September 1, 1990) to allow oil wells to increase production over the state's normal production allowances up to each well's capacity for the months of September, October, and November.

DOE also operates the Strategic Petroleum Reserve (SPR). The SPR was created in 1975 by the Energy Policy and Conservation Act (EPACT) (Pub. L. 94-163, 89 Stat. 882, 42 U.S.C.A. 6234) in response to the oil crisis of that period. The Act was amended in 1992 to enlarge the SPR to one billion barrels. The SPR is an important resource for managing oil supply. The oil is located at various storage locations throughout the United States. The SPR is designed to minimize the impact of any interruption or reduction in imports of refined petroleum products and residual fuel oil. Congress can

---

64 These regulations do not apply to any on-shore or off-shore facilities that, because of their location, could not reasonably be expected to discharge oil into or on the navigable waters or adjoining shorelines. This determination must be based solely on consideration of the geographical, locational aspects of the facility and shall exclude consideration of human-made features such as dikes, equipment, or other structures. They also do not apply to transportation-related facilities that are subject to the authority and control of DOT, nor do they apply to facilities that do not have at least 42,000 gallons of underground buried oil storage or at least 1,320 gallons of aboveground oil storage. (49 CFR 192.5)

authorize DOE to sell this supply when market conditions dictate such action. During the Gulf War, the DOE did draw down some portion of the SPR to provide additional oil supplies. When the decision is made to use the SPR, DOE sells oil from it according to a formula set forth in DOE regulations (10 CFR 625).

The DOE can conduct sales and begin physical delivery of oil to purchasers in approximately sixteen (16) days following the Presidential decision to draw down. However, it may take weeks for purchasers to transport, process and deliver oil to consumers. Under Section 101(a) the Defense Production Act of 1950, DOE can obtain equipment needed for repair of damaged oil production facilities or expedite delivery of fuel oil to electric utilities, however, the exercise of such authority requires a Presidential declaration of national defense emergency (64 Stat. 798; 50 U.S.C.A. Appx §§ 2061, 2062, 2071. Also see E.O. 12038, February 3, 1978). Under the NGPA, DOE has been delegated the power to make emergency purchases and to allocate natural gas to high-priority users (E.O. 12235, September 3, 1980). DOE can also order pipelines to transport gas or construct emergency facilities. To activate these special powers, the President must declare a natural gas supply emergency. However, this emergency powers legislation may need to be revisited due to a restructured natural gas industry. (See *Energy Emergency Response Activities,* DOE Office of Emergency Management, Washington, D.C., May 1997)

5.2.2.4 Illinois Commerce Commission

In general, ICC (the Commission) regulatory supervision of public utilities focuses on setting the rates for service to customers. Most of the Commission's investigatory powers and prescription of rules and regulations concern the economic efficiency and prudency of the utility rather than public health and safety. (See the detailed discussion of ICC powers and authorities in Section 3.2.2.4 above.) However, under the Illinois Gas Pipeline Safety Act (220 ILCS 20/1 *et seq.*), the Commission is also authorized to implement a program of safety standards for pipelines within the state. Basically, the Commission adopts the minimum safety standards promulgated by DOT (83 Illinois Administrative Code 590). In addition, whenever the Commission finds a particular facility to be hazardous to life or property, it may require the person operating the facility to take steps necessary to remove the hazard.

The Commission also has the authority to investigate all accidents resulting in loss of life or injury to a person or property arising from a utility's maintenance or operation. Under the Illinois Pipeline Safety Act, the Commission requires pipeline operators to file reports of all accidents. Any accident report filed with the Commission is admissible into evidence in any action for damages arising from the accident.

The Commission can issue administrative orders to require public utilities to comply with the Public Utilities Act or any rule, regulation, or order issued by the Commission. If the utility does not comply with such an order, the Commission may file an action or proceeding in the appropriate Illinois circuit court. Any person who violates any provision of the Gas Pipeline Safety Act or any order, rule, or regulation of the Commission is subject to a civil penalty of not more than $1,000 for each violation for each date the violation persists. The maximum civil penalty may not exceed $200,000 for any related series of violations. In addition, the Commission may request an injunction to secure compliance. The Illinois Pipeline Safety Act does not specifically provide for criminal penalties.

5.2.2.5 Texas Railroad Commission.

The TRC has jurisdiction over any operation that involves:

- Drilling, operating, or producing oil, gas, or geothermal resources; brine mining injection; fluid injection; or oil and gas waste disposal wells;
- Transporting, reclaiming, treating, processing, or refining crude oil, gas and its products, or geothermal resources and associated minerals;
- Operating gasoline plants, natural gas or LNG processing plants, pressure maintenance or repressurizing plants, or recycling plants;
- Operating a pipeline; or
- Operating an underground hydrocarbon or natural gas storage facility.

The TRC exercises its economic regulation on the production of natural gas by setting production rates based on priority categories and the ratable share of market demand (16 Texas Administrative Code [TAC] 3.34). These production rate regulations are complex and established to prevent waste, including producing natural gas in excess of market demand, protecting correlative rights, preventing discrimination between different wells in the same field, and conserving the natural resources of Texas. In addition, the TRC requires each oil and gas producer to obtain approval for drilling wells, operating pipelines, operating LPG or CNG facilities, or constructing and operating underground hydrocarbon storage facilities.

The TRC is the agency certified by DOT to regulate the safety of natural gas and oil pipelines within the state. The TRC receives all reports of accidents and safety-related conditions for each facility under its jurisdiction.

The TRC has regulations providing for emergency response to shortages, including a means for each producer or purchaser to increase production from wells in a field in response to an increase in demand caused by unforeseen events. As discussed above, this was done in Statewide Rule 87 in response to Iraq's invasion of Kuwait. If economic inequities occur as a result of the production authorized in an emergency, the TRC may make an adjustment after a hearing.

5.3 Regulations and Critical Infrastructure Protection

5.3.1 Impact of Regulations on Critical Infrastructure Protection

Oil and Gas Pipelines and Equipment

The oil and natural gas supply systems may be vulnerable to criminal tampering at any point in the system: production wells and off-shore platforms, gathering lines, treatment facilities (*i.e.*, refineries), transportation pipelines, compressing and routing stations, storage facilities, and consumer distribution mains. Because production facilities are numerous and widespread, it is unlikely that destruction of any particular facility would create a significant crisis in the oil or natural gas industry. For example, the destruction of a natural gas production well field could result in a serious fire emergency situation requiring immediate response and possible danger to adjacent properties; however, it would not create a national supply crisis. As another example, damage to an off-shore oil production well could result in a significant environmental emergency requiring immediate response and might result in public health, environmental, and economic impacts on local shorelines and the fishing/shellfish industries. But again it would not cause a significant supply crisis.

Most oil and gas pipelines are substantially protected from damage by being buried underground. In fact, the more populated the area, the deeper pipelines are required by DOT to be buried to prevent accidental harm to public health or property. The most vulnerable components of the oil and natural gas infrastructure would be aboveground compression and routing stations, large pipeline intersections, large pipeline interfaces (*e.g.*, natural gas city gates), or large storage facilities. Of particular vulnerability are aboveground interconnections between interstate pipeline supplies and LDC distribution systems (city gates). Often only a small number of interconnections serve to supply gas to the whole distribution system, and without significant LDC storage facilities, the destruction of these city gates could disrupt gas service to a large number of residences and public buildings.

The following discussion addresses regulatory requirements for security at vulnerable facilities, along with requirements for emergency response in the event of spill, equipment failure, or fuel shortage.

There are clear, direct requirements for security systems or restricted access to certain vulnerable components of the oil and gas infrastructure. Although not designed for the prevention of intentional sabotage, the EPA regulations on spill prevention for oil storage and production facilities address facility security (40 CFR 112). For instance, the EPA regulations require all plants that handle, process, and store oil to be fully fenced and all entrance gates to be locked and/or guarded when the plant is not in production or is unattended. In addition, all flow and rain valves and any other valves that would permit the direct outward flow of a tank's contents must be securely locked closed. Starter controls on oil pumps must be locked. In addition, the facilities must have lighting that is commensurate with the type and location of the facility, specifically considering the prevention of spills occurring through acts of vandalism. The general requirements to prevent accidental spills, including secondary containment, leak detection systems, and automatic shut-off systems, may also serve to minimize the impact of an intentional release of oil to the environment or creation of a fire hazard.

Although designed for the prevention of accidental releases, pipeline safety regulations enforced by DOT and certified states provide some physical protection for pipeline operations from possible criminal intrusions. As mentioned above, DOT regulations for natural gas pipelines classify areas according to population density.[65] Pipelines constructed through more densely populated areas must meet stricter standards with regard to construction materials, depths, and other design factors (49 CFR 195.248). For oil pipelines, the regulations require pipeline rights-of-way to be selected to avoid, as far as practicable, areas containing private dwellings, industrial buildings, and places of public assembly (49 CFR 195.210). No pipeline can be located within 50 feet of a private dwelling, industrial building, or place of public assembly unless it is provided with at least an additional 12 inches of cover over the normal cover requirements. These construction and siting requirements make the pipeline less vulnerable to destruction in these areas where the greatest harm could result.

The regulations require more vulnerable aboveground natural gas compressor stations to be (1) located on property under the control of the owner/operator of the pipeline, (2) far enough away from adjacent property to minimize the possibility of fire moving between the compressor building and adjacent buildings, and (3) fenced (49 CFR 192.163). In addition, the buildings must be equipped with adequate fire protection facilities and appropriate shutdown or alarm devices in the event of equipment failure. Any aboveground oil facility (e.g., pumps) must be constructed in areas under the direct control of the operator of the pipeline and in an area

---

65 Under DOT regulations, a Class 1 location has 10 or fewer buildings intended for human occupancy; a Class 2 location has 10-46 buildings; a Class 3 location has 46 or more buildings or is situated where the pipeline lies within 100 yards of an area where 20 or more persons gather (i.e., playground, outdoor theater, recreation area); and a Class 4 location is any location where buildings of four or more stories aboveground are prevalent (49 CFR 192.5).

inaccessible to the public (49 CFR 195.254). Valves must be installed so as to be protected from damage or tampering (49 CFR 195.258). Aboveground oil breakout tanks must be adequately protected against unauthorized entry, and oil pipeline pumping equipment must be equipped with emergency shutdown devices (49 CFR 195.264 and 195.436).

DOT requirements also require each operator of a regulated natural gas or oil facility to have a manual of written procedures for conducting emergency response, including procedures for handling abnormal operations (49 CFR 192.605 and 195.402). Abnormal operations can include unintended closure of valves or shutdowns; increases or decreases in pressure or flow rates outside normal operating limits; loss of communications; and operation of safety devices. All of these abnormal operations could be caused by normal equipment failures, natural disasters, or criminal activity. The operator's response, however, would be the same. Under the requirements, the operator must investigate and correct the cause of the abnormal operation, including checking sufficient critical locations in the system to determine the continued integrity and safe operation of the pipeline, even after the abnormal operation has ceased.

LNG Facilities

Special DOT regulations govern the operation of LNG facilities used in the transportation of gas by pipeline (49 CFR 193.2001 *et seq*). LNG facilities could possibly be vulnerable to hostile action because of the volatile nature of LNG and its ability to disperse in a dangerous (explosive) vapor-gas cloud. DOT regulations require all LNG container and transfer system facilities to use a specific formula to determine a dispersion exclusion zone. Within this zone, there may be (1) no outdoor area occupied by 20 or more people during normal use (e.g., beaches, playgrounds, other recreations areas), (2) no residence or other building occupied by 20 or more persons in normal use (schools, commercial establishments, etc.), and (3) no facility containing explosive, flammable, or toxic materials in hazardous quantities or that could be an additional hazard if exposed to a vapor-gas cloud. In addition, the facilities must be equipped with separately routed or underground local, remote, and redundant signal lines for all control systems that affect any operational component that is not otherwise fail-safe.

There are very specific security requirements for LNG facilities. Security personnel must be trained in their duties, including actions to be taken and notification to be made whenever there is any indication of an actual or attempted breach of security. Access to the plant must be restricted, and positive identification of all persons entering is required. Vulnerable facilities (e.g., storage tanks, transfer system, process equipment, control rooms and security communications systems) must be surrounded by a protective enclosure, and the enclosure must be located such that features outside the facility (e.g., trees, poles, buildings) cannot be used to breach the security of the enclosure. All enclosures must be locked unless continuously guarded and sufficiently illuminated. The protective enclosure must be of sufficient strength and configuration to obstruct unauthorized access to the facilities, including 7-foot-high fencing with at least one foot of barbed or similar topping. Security procedures must include the use of direct communications between all on-duty security personnel and all control rooms and stations. Each enclosure must be monitored for the presence of unauthorized persons, either by visual observation on a schedule or by security warning systems that continuously transmit data. In addition, security equipment must have an alternative power source.

The regulations also require emergency planning, including emergency evacuation plans. Emergency response procedures must consider the worst-case discharge volume and the potentially affected environment.

Role of Planning and Procedure

Emergency planning is not a way to prevent attack on the oil or natural gas infrastructure but is critical in minimizing the impacts of such attacks. Planning is performed for short-term, immediate response and for longer-term cleanup. Probably the most critical emergency response planning is for the resumption of natural gas or oil service after an equipment or facility failure. The ability to bypass vulnerable facilities and still move the product to the end user will minimize or eliminate any significant impact on the infrastructure from the destruction or immobilization of a critical facility.

As discussed above, both oil and gas pipeline operators must have a manual of written procedures that address the response to abnormal operations (loss of important facilities or equipment). In addition, oil pipelines must have an oil spill response plan to reduce the environmental impacts of an oil discharge (49 CFR Part 194). The response plans must determine the worst-case discharge for each response zone (49 CFR 194.105). This planning must include procedures for repairing pipeline facilities and starting up and shutting down any part of the pipeline in a manner designed to assure operations within the maximum allowable operating pressure limits prescribed in DOT regulations. It requires that personnel have the ability to isolate portions of the system to allow for repairs. The regulations require the operator to periodically review the response of operating personnel to determine the effectiveness of the procedures for controlling abnormal operations and taking corrective actions when problems are found. Procedures are in place to ensure the continued operation of a critical facility, even in a diminished state. As long as the natural gas or oil can be transported to the end user, the infrastructure continues to function.

In addition, under the NGPA, the President has authority to declare a natural gas supply emergency upon finding that there is a shortage threatening high-priority uses (e.g., space heating in winter). Pursuant to Executive Order 12235, the President has delegated emergency powers under the NGPA to DOE. (September 3, 1980, 15 U.S.C.A. § 3363 note). Under the emergency powers, once the President has declared an emergency, DOE may authorize any interstate pipeline or LDC served by any interstate pipeline to contract, upon such terms and conditions as DOE determines to be appropriate, for the purchase of emergency supplies of natural gas. DOE may also require any interstate pipeline to transport the natural gas for the contract purchaser. DOE may also allocate natural gas supplies to any interstate pipeline or LDC served by a interstate pipeline for the provision of high-priority gas uses. The Executive Order provides that DOE must consult with the EPA and the Federal Emergency Management Agency (FEMA) in exercising these emergency functions.

It should be noted that although this authority exists, the subsequent deregulation of the natural gas industry, resulting in greater pipeline utilization, may make it difficult to redirect gas on an emergency basis.

5.3.2 Measures for Improvement

Review of the current regulatory scheme for oil and natural gas production, transport and distribution suggests the following areas for consideration in terms of improvements to infrastructure security.

1. *Develop systemwide contingency planning.* Deregulation has created a competitive interstate market for producers and transporters, who must now become efficient and cost-effective to meet competitive pressures. This effort may include filling pipelines to capacity with consumer shipments, thereby reducing excess capacity for use during an emergency when rerouting is necessary. It may be worthwhile to encourage industry members to use the existing associations to develop a reliability plan, after the pattern of the North American Electric Reliability Council (NERC). Possibly FERC or DOE, as energy-policy-making bodies, could work with the National Petroleum Council (NPC), American Petroleum Institute (API), Interstate Natural Gas Association of America (INGAA), American Gas Association (AGA), and Gas Research Institute (GRI) to establish an integrated, cooperative contingency planning system.[66]

   It may also be worthwhile to work with these groups to develop contingency procedures for quick access to the SPR. With the current trend

---

66 See Natural Disasters and the Gas Pipeline System, GRI-96/0385, Gas Research Institute, Chicago, Illinois November 1996. The API already issues recommended standards in various areas, including safety and fire standards, environmental and health issues, measurement standards, and other categories of interest to its members.

toward fewer production facilities and reduced domestic production, the United States will become more dependent on foreign oil. This reduction in domestic facilities may mean that sabotage to a small number of domestic production facilities will have a greater impact on infrastructure and continuous delivery. Although the SPR provides backup reserves in case of shortage, due to the regulatory requirements concerning the sale process for these supplies, it may be difficult to quickly disburse these reserves in emergency situations. The SPR was established to react to gradually increasing supply shortages (e.g., the discontinuance of foreign supplies due to political difficulties) rather than an immediate loss of supply to a region (due to the immobilization of the production or transportation infrastructure).

2. *Include sabotage scenarios in the development of risk management plans.* New legislation contemplates using risk management plans to examine pipeline vulnerabilities. Under this system, for infrastructure security concerns to be addressed, hostile action scenarios must be considered in the risk analyses. DOT should work with industry so that the requirements for these risk management plans are clearly laid out and that planning for sabotage is considered in the initial development of the risk models.

3. *Consider effect of fragmented jurisdiction.* Finally, it should be noted that the fragmentation of regulatory authority in this area may create obstacles to implementation of any new initiatives. Federal regulation of the industry is split among DOE, FERC, DOT, EPA, and the Coast Guard. Moreover, many safety requirements are enforced at the state level. Except for DOT pipeline safety standards, which are adopted by state agencies for implementation on the state level, changes in state regulation of the oil and natural gas industry would also require reforms addressed to PUCs, pipeline-safety-certified agencies, and other oil or natural gas production regulatory bodies in 50 states.

6 BANKING AND FINANCIAL SERVICES

6.1 General Description of Regulation

Our economy depends on a healthy banking and financial services sector. Regulation of this sector is intended to promote economic stability and growth by: (1) protecting depositors and investors; (2) providing a steady source of funds for investment in economic ventures; (3) maintaining a stable currency; and (4) providing an atmosphere of stability, trust, and legal enforcement that will foster the conclusion of financial transactions. These have always been the goals; the methods of achieving them have evolved over time. The overview below provides a brief outline of the financial services sector, the types of regulation that are applied to it, and how the regulatory system has developed in response to various crises.

6.1.1 Current Regulatory Environment

Description of the Banking and Financial Services Sector

Banking

Banks are the primary institutions for holding, lending, borrowing, and circulating money. The essential economic characteristic of banks is the creation of credit. In the United States, banking is generally thought of as consisting of two parts; commercial banks, whose main functions are accepting deposits and creating credit through short-term loans, and investment banks, which primarily underwrite and sell new issues of stocks and bonds to investors. When acting in the role of underwriter, a bank guarantees to furnish a definite sum of money on a specific date in exchange for a given number of bonds or shares of stock.

From the 1930s until recently, the United States has not allowed banks to perform both commercial and investment banking functions under one charter. The functions had to be managed under separate ownership and control. Additionally, banks have not been allowed to provide other types of financial products or services, such as insurance or brokerage. Provision of these products and services has been restricted to the financial services houses, traditionally outside the sphere of banking in this country.

Financial Services

Brokerage houses and financial services institutions deal in the transfer of risk, equities, and expectations from one holder to another. The major types of transactions are:

- *Insurance.* The most obvious risk-transfer transactions are those involving insurable risk. Daily, billions of dollars discount to other parties willing to purchase and hold them.

  These transactions cover far more than the familiar fire, flood, theft, or life insurance; they also include the vast sums needed to reduce the risk of commerce, such as business interruption, goods-in-transit, accounts receivable collection, and so on. This is a fiercely competitive industry, and the timeliness and accuracy of information is crucial. The entire industry has placed as many of its activities on-line as can be accomplished within the boundaries of safety and security.

- *Equity.* Equity transfers in the United States are best exemplified by the trading on the major exchanges. While the law still requires that shares of stock, bonds, and debentures have a physical manifestation, trading in the *value* of these pieces of paper occurs without the items being present or even visible. One exchange, the National Association of Securities Dealers' Automated Quotation System (NASDAQ), is a virtual exchange, in that even the traders are present only in cyberspace. Trillions of dollars are represented only by the contents of computer files. The timeliness, accuracy, and security of these files is the basis of trust on which the industry operates.

- *Commodities.* While the trading of commodities is often thought of as buying and selling large quantities of goods and materials, it is actually a means of transferring expectations about the future from one holder to another. These transfers entail agreements as to the quality, quantity, time, and location of delivery of grain, lumber, building materials, petroleum and petroleum products, and financial obligations—all items for which actual delivery is to occur in the future. Many farmers, bankers, insurers, manufacturers, wholesalers, and retailers use this industry to "insure" against the uncertainties of life and commerce. Huge sums are traded daily from those who believe that the subject items will increase or decrease in the future to those who hold opposite expectations. All of these transactions are virtual, in that very few participants in the trading of commodities ever expect to actually make or take delivery. As these markets have become global, with trading continuing around the clock and around the globe, the timelines, accuracy, and security of the information flow are of paramount concern.

Settlement of transactions conducted in any of these industries is conducted primarily by privately owned and operated entities, such as the National Securities Clearing Corporation, Depository Trust Company, Government Securities Clearing Corporation, and International Securities Clearing Corporation. While these entities maintain high standards for security and accuracy, determined individuals have penetrated their databases for brief periods. Although no major losses of funds have occurred, the industries and the attendant settlement entities must be constantly vigilant as new technologies are available to a wider number of users.

As with other firms dealing in electronic funds transfer, a significant issue for the members of these industries is the authentication of who owns the value represented and who has authority to make changes in that ownership.

Current Regulatory Techniques

Techniques of Banking Regulation

Banking regulation in large part consists of measuring the financial health of banks. When a bank is identified as being at risk of failure, generally regulators first attempt to encourage a private industry solution, such as acquisition by another (stronger) bank. As a last resort, the Federal Deposit Insurance Corporation (FDIC) provides protection for depositors (see Section 6.2 below). To make their financial health measurements, regulators have adopted rules, generally centered on the elements of what has become known as the "CAMEL" rating. CAMEL is an acronym for the accepted indicators of performance of commercial banks:

- Capital adequacy,
- Asset quality,
- Management competency,
- Earnings, and
- Liquidity.

Banking examiners have published rules establishing ratios by which an institution is rated. A CAMEL rating of 1 is good and 5 is bad.

Banking regulation is also an essential part of the management of the national economy; regulations on banks are the tools used by the government to regulate the availability and cost of credit. Such regulatory tools include regulations on (1) the portion of deposits held on demand by banks, which must be retained in available cash ("reserves"); (2) the discount that the central bank (the Federal Reserve System) requires from banks that must sell securities to cover demands for cash that cannot be met immediately from reserves ("rediscount rate"); and (3) the interest rate that the central bank charges other banks for short-term loans to ensure liquidity ("funds rate").

Techniques of Financial Services Regulation

For the investment side of the financial services industry, statutes, case law, and regulatory practice have resulted in a standard that is balanced between protecting the investor from fraud and deceit and promoting the free exchange of value among ready, willing, and able buyers and sellers. The underlying principle is that of full disclosure. Regulators cannot directly set the exchange rate for representations of value[67]; the investor or purchaser must make that decision. Regulators can act only to insure that no relevant facts are intentionally withheld from buyers and sellers. This balance between supporting the market and protecting participants has been maintained by focusing on the following elements of financial services:

- *Products.* Representations of value that are offered to the public must be registered with federal agencies, state agencies, or both.

---

67 A "representation of value" is something *other than* cash, coin, or bullion that can be converted, at a rate determined by comparison with a known numneraire, into another such representation, or exchanged for goods, services, or for a claim on goods or services. Such a representation can be physical (e.g., paper or plastic) or nonphysical (e.g., electronic).

- *Practices.* Regulators attempt to prevent (1) misleading or false inducements to buyers and sellers and (2) manipulation of value by persons with fiduciary or insider status superior to that of those to whom financial products are sold.
- *Prices.* Regulations directed at limiting excessive or inflated pricing of representations of value (which set limits on interest rates, service fees, and other transaction costs) are intended to ensure that the price paid for a representation of value accurately reflects the real underlying value and is not inflated by fraud or excessive hyperbole.

Evolution of Banking and Financial Services Regulation

When the United States was formed, there was no constitutional provision for a national banking system. The founders of the new nation were very much aware of how the European nations had suffered from the inflationary practices of their national banks in financing wars and colonial expansion. They believed that leaving all banking activities in the hands of private merchant-bankers would prevent those excesses.

During the War of 1812 and the War between the States, the nation faced serious difficulties in obtaining needed credit at reasonable rates. By 1863, the need for a national banking system to provide a means to discipline the many state-chartered banks and satisfy the credit needs of a growing nation was obvious. Congress studied the crises that arose after the charters for the First and Second National Banks were allowed to expire. The result was legislation[68] that attempted to balance the demonstrated need for a national banking system and the fears of a populist minded Congress of placing too much power in the hands of a few financial czars. This legislation created the Office of the Comptroller of the Currency (OCC), which was made responsible for chartering and supervising national banks owned by private parties. Those forming a national bank were required to purchase U.S. government bonds in amounts equal to one-third of their capital and place these bonds on reserve with the Treasury Department. This meant that the bank's reserves would be backed by the full faith and credit of the federal government.

In 1907, the nation faced another banking crisis. Many of the national banks under the supervision of the OCC and most of the state-chartered banks were caught with seriously short cash reserves when railroad and mining stocks fell. When one bank was unable to meet a depositor's demand for cash, other banks faced a loss of confidence, and they too were subject to a run on their reserves. This chain reaction led to many bank failures and a general loss in confidence in the banking system. Following a public outcry

---

68 The Currency Act of 1863 and the recodification in the National Bank Act of 1864 (Chap. 106, 13 Stat. 99).

to address this crisis, Congress studied the situation for five years, then decided that the national system of supervision needed to be buttressed with a national system of bank reserves that could react quickly to expand or contract the supply of money (credit) to stabilize the need for cash and commerce. Beginning with a Congressional Resolution in 1912 and legislation in 1913, the National Reserve Association (later the Federal Reserve System[69]) was created. It had 12 banks and district branches around the country and authority to mobilize cash reserves, issue credit to national banks, and state-chartered banks who choose to become members, and oversee the required Treasury balances. This legislation also added an incentive for state-chartered banks to join the Federal Reserve System (the Fed) as an alternative to seeking a federal charter. Membership in the Fed allowed state banks to get the same discounts for short-term loans as did federal banks. State banks that joined the Fed had to submit to reserve limits and supervision overseen by the Federal Reserve Board. National banks were automatically members of the Fed, but their supervision remained with the OCC. So in 1913, Congress had dealt with the problems behind the politically sensitive issues of the day, but in 15 years, another crisis was to bring banking back to the front political burner.

After the 1913 act, banks were still allowed to act directly as investors and underwriters of commerce and business expansion. As the fervor of investment in railroads, mines, and the like began to be franchised among banks across the country, control of regional investment potential became a hot political subject. In 1927, smaller Western bankers were successful in getting a bill passed that barred interstate banking.[70] This left the states with the onus of controlling the rampant speculative investments by nonfederal banks. When more than 9,000 commercial banks failed between October 1929 and late 1933, pressure was again put on Congress to do something. Again, Congress studied the situation through a series of investigations and hearings, followed by passage of legislation that:[71]

1. Created a national system of insurance on deposits;

2. Prohibited commercial banks from engaging in offering securities to their depositors, and

3. Prohibited all commercial banks from engaging in investment banking activities.

The Securities and Exchange Act of 1933 established a mechanism for supervising the noncommercial banking institutions now permitted to act

---

69 The Federal Reserve Act of 1913 (Pub. L. 63-43, 38 Stat. 251, 12 U.S.C. 221).

70 The McFadden Act of 192, (Pub. L. 69-639, 44 Stat. 1224).

71 The Glass-Steagall Act of 1933.

as underwriters and offer securities to the public; the goal was to ensure informed investment and prohibit any fraud in inducing people to purchase securities. In 1934, another Securities and Exchange Act broadened the mandate for disclosure of information pertinent to investors, by forcing brokers to disclose who was "behind" each deal and what stake they had in it and to provide some statement about risk factors that should be considered by investors. The temporary depositor insurance was made permanent and placed under the supervision of the FDIC in 1935. The FDIC was also given the additional responsibility of supervising those state-chartered banks that received FDIC insurance coverage but did not join the Fed. Thus, by the beginning of World War II, the nation's laws controlling banking and financial services were as follows:

- National Banking Act of 1864 (established federally chartered banking)
- Federal Reserve Act of 1913 (established the Federal Reserve System),
- McFadden Act of 1927 (prohibited interstate banking)
- Glass-Steagall Act of 1933 (established the FDIC and separated types of banking),
- Securities and Exchange Act of 1933 (established the Securities and Exchange Commission disclosure requirements), and
- Securities and Exchange Act of 1934 (broadened disclosure requirements).

By the end of World War II, the nation had a bank regulatory system organized as shown in Table 6.1.[72]

Federal regulation of the issuance, sale, and trade in securities was solely the province of the Securities and Exchange Commission.

All activities involved in banking and financial services at that time were based on paper, gold, silver, and direct exchange between human agents. The only electronic media involved were the ticker machine for stock quotes and telephones for making deals.

This structure prevailed into the 1980s, even though the adoption of electronic banking and trading had already vastly changed the rate at which this industry was evolving. Banks and bank holding companies had begun to acquire competitors and consolidate operations that could be done more efficiently by the mainframe computers that were rapidly becoming available. The early versions of today's electronic funds transfer systems were established as part of this expansion. Although Congress did pass legislation in 1956 to require the Fed's approval to establish a bank holding

---

72 This chart and the following discussion of the three regulatory agencies is drawn from Anne M. Khademian, *Checking on Banks: Autonomy and Accountability in Three Federal Agencies*, The Brookings Institution Press, Washington, D.C., 1996.

company and prohibit such holding companies from interstate branching,[73] the electronic "cat" was out of the bag. A system that was developed to deal with crises occurring during the first half of the 20th century was faced with new and different problems in the second half.

Table 6.1 Bank Regulatory Agencies from the Late 1940s to the 1980s

	Regulatory Agency		
Characteristics	**OCC**	**Federal Research Board**	**FDIC**
Supervisory Jurisdiction	National Banks	State-chartered banks that belong to the Federal Reserve System and bank holding companies	All other state banks that are covered by Federal Deposit Insurance
Agency type	Bureau within the Treasury Department	Independent agency	Government-chartered corporation
Regional and Field organization for supervision	Six districts that oversee duty stations	12 Reserve Banks	Eight regional offices that oversee field offices
Mandates other than supervision	Administration of the national banking system	Regulation of the money supply	Management of the Bank Insurance Fund

Before the laws were revised in the last two decades, the three regulatory agencies took very different approaches to their responsibilities. Each adopted a different blend of *regulation* versus *supervision*. A strict regulatory approach emphasizes compliance with measurable standards of performance and risk management. A supervisory approach emphasizes safety and soundness of banks as determined by the guided judgment of bank examiners.

The Fed relied on its expertise as the nation's central bank and its independence from political demands. Its examiners followed a strict regulatory approach. They conducted a full examination of every bank under the Fed's jurisdiction every year. These examinations were intended to determine the degree to which a bank was in compliance with strict regulatory mandates for capital reserve, risk management, and other quantifiable measures of safety.

Examiners in the OCC, as a part of the Executive Branch, were given far more leeway to use their judgment in taking a supervisory, versus a regulatory approach. The OCC relied on a concept of hierarchy of risk. Examiners focused on practices related to systemic risk, which included activities

---

73 Financial Institutions Regulatory and Interest Rate Control Act of 1978 (Pub. L. 95-630, 92 Stat. 3641).

posing a threat to the entire industry and on large, regionally important banks, whose difficulties have systemwide impact. Not every bank was examined regularly, and only troubled banks were examined every year.

The FDIC fell somewhere in between the patterns of the Fed and OCC. Its examiners applied strict risk management standards for certain classes of bank activities, but conducted much of their examination off-site by review of reports and documentation submitted regularly by the banks. On-site examinations for strict compliance were conducted at about two-thirds of supervised banks.

As the nature of banking began to change with the introduction of new communications and computing technology, Congress tried to keep the regulatory environment in step with those changes. The Financial Institutions Regulatory and Interest Rate Control Act of 1978[74] created major statutory provisions regarding electronic fund transfers and created the Federal Financial Institutions Examination Council (FFIEC) to begin bringing standard practices to the regulatory bodies. The Depository Institutions Deregulation and Monetary Control Act of 1980 eliminated interest rate ceilings for savings and time deposits and raised the level of FDIC insurance coverage to increase the competition among depository institutions. The three regulatory bodies were, however, given much freedom to conduct their business as they saw fit until the system faced another crisis.

The number of bank failures suddenly jumped from the 5 per year level that occurred in the period 1933-1986 to 200 per year in 1987. The Federal Deposit Insurance Fund showed a negative balance of $7 billion in 1991. Once again, Congress addressed the crisis through a process of hearings and legislation.

The Financial Institutions Reform, Recovery, and Enforcement Act of 1989[75] gave thrift institutions (savings and loans) coverage under FDIC insurance and created the Federal Housing Finance Board (FHFB) and the Office of Thrift Supervision (OTS) to oversee these now federally insured institutions. This act also created the Resolution Trust Corporation (RTC) to manage failed or troubled institutions that regulators took over under the insurance provisions. The Federal Deposit Insurance Corporation Improvement Act of 1991 limited the discretion that federal supervisors from the Fed, OCC, and FDIC had previously exercised in assessing the condition of banks and determining what enforcement actions were needed. Congress wanted certain responsibility for ensuring that the national system of banking is well-managed, and they wanted the federal regulators to shoulder that responsibility.

---

74 Pub. L. 95-630, 92 Stat. 3641.

75 Pub. L. 101-73, 103 Stat. 183.

In addition to responding to the bank failure crisis, Congress also had another specter to cope with: competition from huge foreign banks for financing new ventures and industrial growth. Restrictions on interstate banking and vertical integration put U.S. banks at a disadvantage.[76] Congress eased those restrictions by repealing part of the Glass-Steagall Act and allowing banks to grow larger and branch across state lines, subject to concentration limits, state laws, and evaluation of the companies.[77] Further changes in the law allowed bank holding companies to own more than one kind of banking enterprise, so that small banks and mortgage lending institutions began to be combined under centralized regional management. However, such combinations are potentially subject to regulation by many agencies. This has led to situations, for example, where the various parts of a small regional bank holding company must adhere to the banking rules of the Fed and FDIC, the banking rules of four states, and the mortgage lending rules of at least five states. As one banker put it, "I see more teams of examiners in my office every year than I do baseball teams on television."[78]

6.1.2 Current Trends in Regulation

Electronic commerce is a reality in banking and financial services, and it poses some serious problems for regulators. The following brief discussion is intended to provide an overview of the major vehicles of electronic commerce, and the new challenges they pose to regulators. The impact of these systems on critical infrastructure protection is discussed in Section 6.3.1.

Electronic Funds Transfers (EFT)

This term refers to the movement of funds from one bank account to another by means of electronically communicated payment instructions. It has been estimated that each day, banks, global financial markets, and the U.S. Federal Reserve transfer more than $35 trillion within, into, and out of the United States. Citibank alone transfers $500 million daily, including domestic transfers. Recently, Citibank worked with domestic and foreign law enforcement agencies to track and trap a group of criminal hackers who, before they were caught, had transferred $400,000 into several foreign accounts. The crimes went unnoticed for several months because the transfers were kept small and below set limits for investigation. It was only when the crooks got greedy that they were found out.

---

76 See, for example, BankNet Electronic Banking Service at http://mkn.co.uk/bank.

77 Riegle-Neal Interstate banking and Branching Efficiency Act of 1994 (Pub. L. 103-328, 108 Stat. 2338).

78 Remark made not-for-attribution during an interview.

Electronic Checks

Several major banks have fully instituted paperless checking as an option for payments over the Internet. Basically, this system transmits the same data that is contained in paper checks, but in machine-readable format. Businesses and individuals use this system just as they do paper checks, except that each check bears a digital rather than manugraphic, signature for signing, endorsing, and authentication. The effective and safe use of these systems will require highly secure and reliable management of the electronic signatures.

Digital Cash or Tokens

Any electronic representation of money in digital form can be negotiated and transmitted among payers, payees, depositors, and depositories. Stores of such digital cash or tokens–held in on-line accounts, smart-cards, or "electronic wallets" (palm-sized PCs)—can be presented, transferred, and negotiated just like currency. There is no need for an intermediary, as there is for electronic checks or credit cards. The repository can be "re-charged" on-line, at an ATM, or by a teller. Digital cash can be "identified" and carry data about the person to whom it belongs, or it can be anonymous. The same caveats regarding electronic signatures that apply to electronic checks apply to digital cash. In addition, a major threat resulting from widespread use of anonymous digital cash or tokens is the ease by which money can be laundered in this format. As the U.S. Treasury Financial Crimes Center has stated:

> These systems are designed to provide the transacting parties with immediate, convenient, secure and potentially anonymous means by which to transfer financial value. When fully implemented, this technology will impact users worldwide and provide readily apparent benefits to legitimate commerce, however, [it] may also have the potential to facilitate the international movement of illicit funds.[79]

A significant issue for financial institutions in coping with all of these computer-based systems is authentication of who owns the value represented, and who has authority to make changes in that ownership. What is really needed is a way to identify those who transact business using electronic money, just as we attempt to do with those using paper-based currency—have the transactor show some identification.

There is currently a struggle between elected officials, particularly those in Congress, and the regulatory agencies. The elected officials are again under pressure to do something, and this time their answer is to make regulatory agencies more accountable for the health and safety of the banking system.

---

79 "Money in cyberspace," at http://www.ustreas.gov/treasury/bureaus/fincen/cybpage.html #concern.

Their prescription is to consolidate the regulatory agencies and strictly codify the standards and procedures, reducing the numbers of both regulations and regulators. Those who manage the regulatory system have a different solution: greater autonomy for examiners and less reshuffling and reorganization of authority and responsibility. The trade-off is between accountability and flexibility.

This struggle over accountability versus flexibility is taking place just as the nation faces the potential for another crisis in banking and finance. The lines between banking and non-banking financial services such as investment counseling and brokering, insurance, and investment banking are under attack. The attack is driven not merely by those who stand to gain from consolidation of such services; it is also the result of demands from the public, who want to have access to a full range of financial services and products with as little change in venue and "user interface" as possible.[80]

With substantial changes in banking infrastructure due to increase competition, new technology, and the amendment and revision of the law in the latter part of the 1990's, the banking environment of the United States is entering a period of dynamic change. The regulatory scheme that was present in 1975 will probably not be evident in the system of 2010. While the same federal agencies have the same compartmentalized responsibilities today, it is likely that dramatic changes will be made as changes in the industry render old regulatory methods obsolete.

6.2 Description of Selected Regulatory Agencies

6.2.1 Selection Method

The federal agencies profiled below represent the primary agencies involved in regulation of banking and financial services. Following the discussion of federal agencies is an overview of the general practices among state regulatory agencies. The state overview is not intended to represent any specific state but rather to illustrate the range of regulatory goals and techniques applied at the state level.

6.2.2 Agency Descriptions

6.2.2.1 Federal Agencies

The roles of the federal agencies that oversee the banking and financial services industry all have their origins in previous crises in commercial or investment banking.

---

80 *See*, for example, 1997 Senate Bill 377, "Promotion of Commerce On-Line in the Digital Era."

Bank Regulation

The roles of the three basic federal banking regulatory agencies are shown in Table 6.1 above. Table 6.2 below summarizes key features of three additional agencies that have been created or whose roles have expanded since 1987 in response to the crisis in thrift institutions.

Table 6.2 Recently Created or Expanded Bank Regulatory Agencies

CHARACTERISTICS	FFIEC	OTS	FDIC
Supervisory jurisdiction	National Banks	All thrift institutions insured by FDIC	Also insures deposits of thrift institutions
Agency type	Independent agency	Office within Treasury	Government-chartered corporation
Regional and field organization for supervision	Works through FED districts and state regulators	Works through FED districts and state regulators	Eight regional offices that oversee field offices
Mandates other than supervision	Helps oversee systems for electronic funds transfers	None	Management of the Bank Insurance Fund.

A brief summary of the responsibilities of the selected federal agencies follows.

- *The Fed.* The Fed is an independent agency with the responsibility for regulating the nation's supply of money, and supervisory jurisdiction over all state-chartered banks that belong to the Federal Reserve System. The Fed can mobilize cash reserves, issue credit to national banks, and oversee the required balances.
- *OCC.* The Office of the Comptroller of the Currency (OCC), a bureau in the Department of the Treasury, is the administrator of the national banking system. As such, the OCC has supervisory jurisdiction over all federally chartered banks.
- *FDIC.* The Federal Deposit Insurance Corporation (FDIC) is a federally chartered corporation that manages the Bank Insurance Fund. In that role, the FDIC insures deposits in all federally chartered banks, state-chartered banks that join the Federal Reserve System, and all thrift institutions eligible for the insurance fund. The FDIC also has supervisory responsibility over state-chartered banks that choose to be insured by the Bank Insurance Fund but not to join the Federal Reserve System. The FDIC also resolves and liquidates failed banks covered by the Bank Insurance Fund.

- *OTS.* Another agency of the Treasury Department, the Office of Thrift Supervision (OTS) oversees all thrift institutions insured by the FDIC. The OTS resolves and liquidates all failed thrifts under its supervision.
- *FFIEC.* The Federal Financial Institutions Examination Council (FFIEC) is a coordinating body, created to standardize practices among the regulatory agencies regarding national banks. It has a strong role in overseeing the systems for electronic funds transfers.

Investment Regulation

The Securities and Exchange Commission (SEC) oversees the persons and institutions that offer stocks and bonds to the public. This oversight includes licensing and registration of dealers and brokers, registration of securities, investigation of alleged fraud or misleading practices, and requirements for internal controls for institutions engaged in investment banking.

6.2.2.2 State Agencies

The state agencies go by varied and sometimes confusing names. Some states combine all financial services and banking under a State Banking Commissioner; others have separate offices to supervise banks, thrifts, and securities firms. All supervise credit unions separately, although sometimes under a broader umbrella. A summary of the stated purposes and goals of state regulation follows.

Source of Authority

The basis for most state banking and financial services regulation is constitutional authority implemented by legislation. The legislation often includes many pages defining the exact nature of the things to be regulated. Most of these definitions are concerned with the institutions and their products and activities that are to be the subject of state scrutiny. There is wide overlap with federal language regarding these same entities.

The major difference between federal regulatory language and that of the states is in the purposes for regulation. While language in both systems covers the goals of (1) maintaining the stability of the banking and financial services industry, (2) providing convenient services, and (3) concern for the public interest in safety and security of deposits and investments, the state statutes have one striking characteristic that differs from the federal. State regulators are specifically charged to promote "economic development," "economic growth," "growth of investment opportunities," or some similarly phrased mission to bring new and expanding business to the subject state.

Jurisdiction

Most state regulatory agencies have authority to examine all depository and non-depository institutions for compliance with both state *and* federal laws. Such authority typically includes state chartered banks, national banks, state savings banks, federal savings banks, branches of out-of-state banks of all kinds, trust companies (special entities established to hold property subject to a trust, to buy and sell securities for others, and to offer advice on these matters), securities and commodities dealers, and building and loan associations. Many states are expanding the authority of their regulators to include entities such as personal loan vendors and makers, personal property finance companies, mortgage brokers, check sales and cashing services, and companies engaged in transportation of money and valuables. All of this latter category of entities will become very active in the electronic movement of value. This extension and broadening of jurisdiction at the state level is particularly important as the number of multistate holding companies grows.

Techniques of Regulation

Discussions with senior managers in national banks and multistate holding companies indicated that the techniques used by state regulators and the competency of the regulators vary widely. In the past, some states tended toward the strict regulatory paradigm of across-the-board standards, while others tended toward forbearance and mediation of difficulties that a state-charted bank, for example, might be undergoing. This stems, apparently, from the state agency goal of maintaining and encouraging growth and economic activity in the state.

6.3 Regulations and Critical Infrastructure Protection

6.3.1 Impact of Regulations on Critical Infrastructure Protection

In view of the industry's current and future reliance on electronic data transfer, the security of critical infrastructure in banking and financial services is primarily a function of its vulnerability to electronic manipulation or attack. In general, the reality of current practice has outstripped the regulatory system in this area. As one expert in the field put it:

> Without certainty in the legal rules, electronic commerce will not reach its . . . potential . . . Yet the law always lags behind technology.[81]

Banks and other financial institutions, acting in their own self interest, apply various mechanisms and methods to ensure the security of transmitted information. However, these decisions are not generally driven by a

---

81 Smedinghoff, Thomas J., *Online Law*, Addison-Westley, 1996, p.4.

need for regulatory compliance. Vulnerabilities and the regulatory situation for major electronic vehicles are summarized below.

Electronic Funds Transfers (EFTs)

This term generally applies to the movement of funds from one bank account to another by means of electronically communicated payment instructions. A major departure from traditional practice is that an EFT for which a bank and its customer have established an authentication protocol is effective as an order of the customer, whether or not it was authorized by the customer, if the bank shows that it fully complied with the agreed-to protocol. According to the banking community, the major threat posed by EFT is from organized crime. The movement of value at light speed makes detecting and tracking money laundering and other illegal practices very difficult. Another threat could come from intentional manipulation of certain markets at levels that are too low to detect, but that yield great profit to the criminal element. If criminal hackers could get into the EFT system and manipulate the apparent market for U.S. Treasury products for half a day, they could "earn" hundreds of millions of dollars from straddles that they created.

EFTs are handled by only a few systems in the United States. These are listed in Table 6.3.

Table 6.3 Electronic Funds Transfer Systems

System	Users	Uses
Automated Clearing House (ACH)	>22,000 financial institutions, >160,000 corporations,, and several million private parties	Payment to vendors and suppliers, consumer payments, direct deposits
FedWire	Federal Reserve Banks	Intermediation, settlement, and closure
Clearing House Interbank Payments System (CHIPS)	New York Clearing House Association (12 Money Centers and 125 associate banks)	International funds transfers among members
Society for Worldwide Interbank Financial Telecommunications (SWIFT)	1,700 banks in 80 countries	Transmittal of payment orders and tenders; settlement is via one of the other systems.

Responsibility for the security of these systems lies with the operators. There is no direct security supervision or regulation by any governmental authority. State statutes and the federal Electronic Fund Transfer Act[82] are directed at consumer protection and elimination of fraudulent practices in electronic commerce, but they do not address system security. Standards and best practices are based on business needs and the result of after-the-fact legal action.

Electronic Checking

As with other forms of electronic commerce, this electronic analogue of paper-based checking relies on use and security of electronic signatures. Protocols for transmission and authentication of these electronic signatures are still under development, and regulation is changing as rapidly as the technology. At present, all states have some kind of legislation that attempts to define what must be protected and what standards of protection should apply. Federal legislation on the same theme is pending. The principal threat is, again, that technology and attendant practice is changing faster than the regulatory environment can be adapted.

Digital Cash or Tokens

As described above, this technology is considered vulnerable to misuse for money laundering. There is growing concern about the potential for illegal use of this technology to support the world traffic in illegal drugs. Current proposals center on reporting money movements in excess of $750 to the Internal Revenue Service and analyzing these reports for suspicious patterns.[83] This approach will work only so long as there is some exogenous source of information that will focus the attention of the detection and tracking technology. The cost of any other means to limit the search space would be astronomical.

6.3.2 Measures for Improvement

It is evident that federal and state regulatory practices can have a significant impact on the safety and security of the banking and financial services industry in terms of balancing economic risks and opportunities. It is not at all clear what impact regulation can have on protection of the industry from intentional acts by hostile elements.

Security and safeguard measures taken by this industry have been more directly related to industry concerns for protection of assets than in response to any regulatory pressure. The impact of infrastructure protection is actually felt more directly through regulation of the telecommunications industry than through regulation of banks and financial services firms. That will not be the case in the future.

---

82 15 U.S.C. § 1693.

83 "Treasury Seeks Tighter Rules on Wire Transfers," *The Washington Post*, May 20, 1997, Section 1, pg.1.

Historically, major reforms in the regulation of banking and financial services have followed in the wake of crises. That pattern may continue. Alternatively, regulatory systems might be adapted to provide the flexibility necessary to achieve public policy goals, such as infrastructure protection, without unduly compromising the health of the industry. Consideration of the following issues may serve to anticipate problems before they become severe.

1. *Rethinking oversight structure.* The lines between federal and state regulatory practices and structures must become as transparent as the lines between local, state, national, and international commerce in financial products and services are becoming. This will require rethinking the point of balance between supervisory and regulatory approaches to practice, and assignment of oversight responsibility to the level closest to the commerce being conducted. Research is needed on the appropriate regulatory structure for federal and state oversight of the emerging banking and financial services infrastructure.

2. *Development of On-line Techniques for Oversight.* Responses to problems must be much more immediate than has been the practice in the past. Regulators and examiners must begin to monitor the activities of the industry by observing on-line traffic, rather than depending on periodic reports generated after-the-fact. Research is needed on how federal and state regulators can conduct on-line supervision and oversight. This research should include techniques such as pattern recognition, expert systems, and risk-assessment tools available on-line to both the industry and regulators, so that problems can be quickly identified and acted upon.

3. *Development of On-line Training.* The need for timely and economic training and continuous education for examiners will become greater as changes in the industry broaden and accelerate. The use of the new tools mentioned above and the nature of the emerging and evolving infrastructure that they are intended to address are complex and not familiar to most examiners and regulators. Research is needed to develop distance learning capabilities and on-line expert systems to help prepare and train examiners and regulators.

4. *Development of Standard Definitions of Terms.* The definitions of products, practices, and prices in the industry must be harmonized between state and federal statutes and regulations. It must be clear to all concerned what is being regulated when transactions are taking place among billions of actors in cyberspace 24 hours a day. Research and model law development are needed to clarify the subject matter of regulation in the emerging infrastructure and harmonize the laws and regulations among agencies and between levels of government.

7 TRANSPORTATION INFRASTRUCTURE

7.1 General Description of Regulation

7.1.1 Current Regulatory Environment

The U.S. transportation industry consists of a wide and diverse network of carriers and supporting infrastructure for highway, rail, air, and water transport. Regulation of the industry reflects this diversity. At the federal level, regulatory authority over transportation infrastructure is organized on a modal basis. These modal authorities, now almost exclusively vested within the U.S. Department of Transportation (DOT) are as follows, with some modest overlap:

- Federal Highway Administration (FHWA): highways and bridges (funding for maintenance and replacement; structural and technology research);
- Federal Aviation Administration (FAA): airport and aircraft safety (including navigation facilities for commercial and general aviation);
- Federal Railroad Administration (FRA): railroad safety law administration and enforcement;
- Surface Transportation Board (STB): mergers and infrastructure modifications of commercial carriers;
- Federal Transit Administration (FTA): mass transit (funding for capital investment and operating subsidy, administration and enforcement of federal safety requirements);
- Research and Special Projects Administration (RSPA): safe transport of hazardous materials;
- U.S. Coast Guard (USCG): Administration and enforcement of safety laws pertaining to domestic waterway transport; and
- Maritime Administration: Administration and enforcement of safety laws and multilateral compacts pertaining to international marine commerce using U.S. ports.

Each of these agencies enforces safety and reliability of service through its own regulations. In addition, some of the agencies set priorities for infrastructure maintenance and upgrades by serving as conduits for federal funding.

The mode-based arrangement is generally echoed in the structures of state government transportation agencies, with separate bureaus for roads, railroads, airports and so on. Road construction and maintenance is generally the largest dollar-volume activity. Many state departments of transportation have regional offices to provide close local supervision of transportation infrastructure.

7.1.2 Current Trends in Regulation

Governmental decisions at both the federal and state levels continue to play an important role in the use and fate of transportation infrastructure. Governmental agencies retain both an outright proprietary interest in certain infrastructures (e.g., highways, seaports and airports) and an indirect interest in others through the direction and administration of funding for infrastructure addition, enhancement, and safety.

With the exception of waterway transport, all modal areas have experienced considerable federal deregulation over the past few decades. The role and presence of governmental regulation in transportation activities has diminished, thanks in part to several (again generally mode-specific) pieces of federal legislation:

- The Airline Deregulation Act of 1978 (Pub. L. 95-504, 92 Stat. 1705, October 24, 1978; codified and subsequently revised in 49 U.S.C. Section VII) removed pricing and service provision requirements for commercial carriers, and opened the market to much easier access by start-up operators.
- The Staggers Rail Act of 1980 (Pub. L. 96-448) basically declassified railroads as fully regulated carriers and empowered them to enter into rate and service contracts with shippers in a manner unconstrained by minimum level-of-service or fixed commodity rate requirements.
- The Intermodal Surface Transportation Efficiency Act of 1991 (ISTEA)(Pub. L. 102-240), declared the Interstate Highway System complete and thereby decentralized much of FHWA's direct authority over highway and bridge project funding.
- The Interstate Commerce Commission Termination Act of 1995 (Pub. L. 104-88), in recognition of the modern success of intermodal freight competition, terminated the ICC. The ICC had been established during the era when railroads were the only source of quick, reliable transport in many areas, to mitigate the excesses of the railroads' monopoly power.

One state-level regulatory area that has shown recent growth is the designation of routes for shipping hazardous materials. States designate particular routes for transport of hazardous cargoes to reduce accident risks, for example limiting them to bypass routes rather than through downtown areas.

In general, far more actual decision-making authority is now vested in private ownership of transportation facilities than at any time since the late 19th century. However, relationships between the former regulators and those they regulated persist in many manifestations both formal and informal, and to a considerable degree color perceptions on both sides about responsibility for the maintenance of system integrity and security.

7.2 Description of Selected Regulatory Agencies

7.2.1 Selection Method

Good examples of current regulatory relationships and their effects can be found by studying the FHWA, FAA, FRA, STB, and various large-state transportation agencies, among which is the Illinois Department of Transportation (IDOT). A review of the activities of these five agencies and how their regulatory authority affects critical transportation infrastructure is presented below.

7.2.2 Agency Descriptions

7.2.2.1 Federal Aviation Administration

The Federal Aviation Administration (FAA) certifies aircraft and aircraft manufacturing (through issuance of "airworthiness directives"), regulates airport security standards, and is the lead agency for rulemaking on commercial aviation practices. It was established by the Federal Aviation Act of 1958 and is charged with promoting the growth and safety of civil and commercial aviation through constructive regulation and oversight. Today the FAA has a well-publicized role in overseeing airline passenger safety and airport security. It also regulates commercial air freight transport and civil aviation (private aircraft).

In addition to its regulatory role, the FAA owns and is responsible for much of the *in situ* infrastructure that supports air navigation, including control towers and regional control centers, very-high-frequency omni-range (VOR) radio navigation beacons, Doppler radar installations, and visual approach slope indicator (VASI) and instrumented flight rule (IFR) landing lights and structures. Commercial and most civil aviation could not function without these facilities.

7.2.2.2 Federal Highway Administration

The FHWA was created in the 1966 Department of Transportation Act.[84] It is charged with implementing "highway safety programs, research, and development related to highway design, construction and maintenance, traffic control devices, identification and surveillance of accident locations, and highway-related aspects of pedestrian safety. "

Despite considerable delegation of authority to states under ISTEA (*see* above), FHWA retains major funding responsibility for capital improve-

---

84 *See* 49 U.S.C. § 104.

ment and augmentation projects on the designated National Highway System (including the Interstates) and provides technical guidance to states on road and bridge construction standards and safety. FHWA maintains the inventory of the over 575,000 bridges and structures on U.S. highways, whether on or off the federal-aid system. Probably FHWA's most important fiscal function is apportioning and distributing the respective states' shares of Congressionally authorized fiscal-year funding for highway projects. However, the FHWA has relatively little discretion in allocating this funding; most is dictated by formulas incorporated into the authorizing statutes.

A second key function of FHWA with regard to the security and integrity of national highway infrastructure is research and development. The FHWA performs research and also sponsors research by other organizations pertaining to materials science. Much of this activity is performed and coordinated through FHWA's Turner-Fairbank Highway Research Center in McLean, Virginia. Ongoing projects related to concrete performance and deterioration modes, advances in structural reinforcement mechanisms, and optimal channeling of traffic flows (leading, for example, to better designs for freeway interchanges) keep Turner-Fairbank prominently involved in road and bridge improvement efforts.[85]

7.2.2.3 Federal Railroad Administration

Safe practices on U.S. railroads are regulated and enforced by the FRA. The FRA was created in the 1966 Department of Transportation Act, and its primary function is spelled out in 49 U.S.C. Section 103: "to carry out all railroad safety laws of the United States . . . [the FRA is] responsible for all

85 A specific example of a project responsive to a potential threat to infrastructure integrity is the ettringite distress program, initiated by the Turner-Fairbank Center in 1996 in response to a request from FHWA Region VI. The Texas department of Transportation had discovered that a calcium aluminum sulfate hydrate material called ettringite had formed (possibly due to unusual water and heat exposure conditions) from the concrete in pre-cast bridge support beams awaiting installation at a highway construction site. Expansion cracks had appeared where the ettringite was forming, and the beams were written off as unusable. Unfortunately, ettringite was later discovered in beams that had already been installed at several other locations around the state. Pursuant to the request of Region VI, Turner-Fairbank's Exploratory Research Team developed chemical thermodynamics models to study ettringite formation, evaluated samples of damaged material, and applied an advanced impact-echo ultrasonic system to evaluate the nature and cause of the microcracking of the concrete. Effort and expertise from the National Institute of Standards and Technology (neutron diffraction instruments) and the University of Hawaii (laser scanning microscopes) were also brought into play. The overall objective was to isolate the exact damage mechanism and the conditions under which the damage occurs, and to evaluate whether concrete specifications for future jobs must be changed to mitigate the threat and if damage to existing structures is repairable in place. (Source: FHWA Web page announcement, 1997).

acts taken under those laws and for ensuring that the laws are uniformly administered and enforced."

The FRA railroad safety rules are spelled out in 49 CFR Part 209 *et seq.* Track, structures, rolling stock, and locomotives are subject to unannounced inspections. There are also numerous regulations pertaining specifically to rail shipment of hazardous chemicals; the FRA is empowered to levy fines and other penalties on both shippers and carriers for failures in marking and labeling, inspection and handling procedures, crew training, or other requirements.

Under 49 CFR Part 240, FRA also spells out certification requirements for individuals authorized to operate locomotives. Included in these requirements are (1) written evidence of positive prior safety conduct, including operation of a highway motor vehicle (i.e., driving record); (2) no current record of substance abuse; (3) ability to meet specific vision and hearing standards; (4) qualification for road service by means of a written examination covering personal safety practices, operating practices, equipment inspection practices, train handling practices, and compliance with federal safety rules, as well as by means of an operational test at the controls of either an actual locomotive or locomotive simulator; and (5) periodic performance monitoring by qualified supervisory personnel.

7.2.2.4 Surface Transportation Board

The STB, successor to the ICC under Title II, Sec. 201(a) of Pub. L. 104-88 (the ICC Termination Act of 1995), reviews and approves proposed railroad and motor carrier mergers and the resulting trackage rights (authorization to move cargo with its own rolling stock and motive power) that will be granted to competing carriers over certain lines of the merge partners. It also authorizes return to service (following appropriate safety checks) of railroad lines that have been inactive. Its specific mission, as articulated in 49 U.S.C. § 702, is to carry out activities of the ICC "not otherwise abolished." As discussed below, activities meeting this definition are rather limited.

ICC regulation of rail transport was largely concerned with protecting shippers from potential price gouging or service cutbacks by railroads. This led to curtailments on railroad restructurings such as mergers, acquisitions, and line abandonments. It is clear that the STB endorses the notion that, because shippers and carriers are now fully able to negotiate both short-and long-term haulage rates on a one-to-one basis, and the trucking industry is always standing by to offer "just-in-time" service when needed, shippers no longer require protection from predatory pricing or poor service. All they need is the opportunity to negotiate for better terms from other carriers if dissatisfied with the current provider. Thus, much of the U.S. rail network, formerly tightly controlled by owning carriers capable of restricting access by other haulers unless they agreed to often stringent terms, has been thrown wide open to access by multiple providers of rail services. It is not now unusual to see the locomotives of three of more railroad companies pulling trains along the same stretch of railroad line, and while the "guest" carrier must still pay access charges to its "host," it retains far more control, from origin to destination, over its originating commodity shipments than in the ICC days.

7.2.2.5 Illinois Department of Transportation

IDOT, like many of the large-state departments of transportation, retains substantial authority over intrastate road and transit project funding and contracting, airport planning, and railroad crossing protection standards and equipment. IDOT also participates in *inter*state coordination of infrastructure management (e.g., multistate "smart" highway corridors). IDOT's mission is to provide cost-effective, safe and efficient transportation for the people who live, work, visit and do business in Illinois, and to ensure that the system supports the state's economic growth. The IDOT modal divisions and their respective responsibilities are as follows:

- *Division of Highways.* Directly manages or oversees the design, construction, operation and maintenance of the 17,000-mile state highway system (including the state's federal-aid highways: the Interstates and the Illinois portion of the National Highway System) and administers the state's local roads and streets program. Total road mileage in Illinois is approximately 138,000, making it the third largest system in the nation.
- *Division of Aeronautics.* Works with local airport agencies to maintain and upgrade the state's airport system (at 138 airports, 280 heliports, and nine balloon ports the second largest in the nation), and plans and develops new airport capacity as needed. This division is also responsible for developing aviation safety and education programs and assisting the Civil Air Patrol.
- *Division of Public Transportation.* Provides technical assistance and administers state and federal funding (including capital grants for purchase and upgrading of track) to 50 public transit systems in Illinois, serving about 600 million passenger boardings a year.
- *Bureau of Railroads.* Administers rail service programs that (1) supplement the passenger service provided by Amtrak, and (2) preserve rail freight service critical to keeping and expanding industry and employment in the state. Also coordinates with and may directly participate in the periodic inspections performed by agents of the FRA to determine compliance with railroad safety rules and operating practices.

- *Division of Traffic Safety.* Compiles crash data; evaluates and analyzes information used to identify highway improvement needs in problem areas; enforces and provides education on safety belt and DUI laws; inspects school buses, trucks and ambulances; and oversees the transport of hazardous materials.

Thus, IDOT exercises authority across the entire spectrum of Illinois transportation activities, and is available to consult with shippers, carriers, operators, passengers, and the general public on matters relating to security and safety of transportation properties and physical plant.

Illinois is one of several states that stands to gain a substantial increase in funding authority for transportation projects if Congress directs more federal highway fuel tax revenue back to its originating states. A number of candidate reauthorization bills currently under Congressional scrutiny incorporate this provision. Such a development would be in keeping with the trend to decentralize administrative power in recent years, and would constitute a windfall for IDOT in terms of both resources and influence. Whether a more pronounced regulatory posture for IDOT would also ensue remains open to question.

7.3 Regulations and Critical Infrastructure Protection

7.3.3 Impact of Regulations on Critical Infrastructure Protection

The transportation industry in the United States is well developed, well interconnected, and generally robust. It is a strong and resilient system. In addition, different methods of transportation are fungible to a considerable degree; that is, if one transport system is unavailable, another may be substituted. In short, the potential for major disruption of our ability to move people and goods from one part of the country to another appears relatively small. However, as recent terrorist incidents have shown, there is a potential for loss of life and significant disruptions on a more local scale.

With over 575,000 bridges and 4 million miles of road in the nation's highway system, as well as some 135,000 privately-held railroad miles and about 200 commercial airports (25% of the domestic airport total), continuous security monitoring of all of our transportation infrastructure is a logistical impossibility. The primary—and certainly a very productive—contribution that regulatory bodies make in the area of transportation infrastructure security is promoting a reasonable level of vigilance on the part of the owners and operators of the infrastructure. The regulatory agencies generally have appropriate powers that can be used to achieve this end, as illustrated by the following examples.

Curtailment of Service

Federal and state transportation agencies are generally empowered to curtail service whenever demonstrable threats to public safety exist and are not corrected by the responsible operator. A recent example of this was Emergency Order 19, issued by the FRA in February of 1996, which halted operations of the Tonowanda Island Railroad over a bridge on its property that crossed a navigable waterway (a tributary of the Niagara River). The bridge had repeatedly been cited by FRA inspectors as structurally unsafe. Use of the bridge was halted because of (1) the railroad owner's chronic failure to respond to prior citations, (2) the potential danger posed to recreational boaters on the river, and (3) the danger of violating international pollution accords with Canada in the event that a train should break through the bridge and spill debris and effluent into the waterway. Another, better known, example was the suspension of ValuJet commercial flights (on grounds of not protecting passenger safety) pending the results of the investigation of the 1995 crash near Miami International Airport.

Training directives

A regulatory agency will sometimes uncover potentially serious gaps in the content of training given to private sector employees charged with vehicle upkeep, passenger and cargo screening, scheduling and logistics of operations, structural maintenance, and the like. Under most circumstances, the agency is then empowered to issue rules about the content of training as well as certification of those who have completed such training (including on-the-job experience). One example is FAA's advance notice for a proposed rule on uniform certification of airport security company employees.

Structural Inspection and Maintenance Protocols for Public Roads

Preservation of the structural integrity of the public roads system requires a certain amount of uniformity in inspection and maintenance procedures. IDOT is one of many state Departments of Transportation that, together with their contractors, follow a fixed set of procedures for structural inspections and improvement designations. In order to qualify for state and federal-match funding, an upgrading or replacement project must appear on an approved project list, called the one- or five-year Traffic Improvement Program (TIP). An inspection report indicating priority need for repairs is one of the key ways of getting on the TIP list. As a result, roadways are periodically inspected for signs of trouble, particularly at vulnerable points such as bridges.

Airport Security

The FAA has become increasingly concerned about security practices and procedures at airports. In the wake of the crash of TWA Flight 800, scrutiny of the background, credentials, and training of ground airport staff, especially those working in passenger and cargo screening areas, has moved to center stage. In April 1997, the FAA implemented a new rule, applicable to air carriers and airport authorities, designed to increase airport security by controlling information about airport security procedures.[86] Pursuant to 14r carriers are responsible for screening passengers, carry-on baggage and checked baggage with approved screening devices; controlling access to aircraft and air carrier facilities; and reporting and responding to bomb threats and hazards discovered during screening. Under the new rule, they are not to disclose information about these functions except on a "need-to-know" basis. Similarly, employees of airport authorities having information about the air operations area (control tower and related command centers) may no longer share this information freely. Specifically, the following types of information are considered "sensitive security information" (SSI):

- The content of any approved or standard security program for an air carrier, airport operator, foreign air carrier, or the air transport of United States mail;
- The content of FAA Security Directives and Information Circulars;
- The description of screening profiles used for persons, baggage, or cargo;
- The implementing guidance or other information, comment, or instructions pertaining to a security contingency plan;
- The technical specifications of any device used to detect weapons, explosives, or destructive substances; the descriptions and specifications of objects used to test screening equipment; and the technical specifications of security communications equipment;
- Information that would reveal a potential systematic vulnerability to attack of the aviation system or facilities (such as a specific location where a security violation occurred);
- Information about threats against civil aviation;
- Information about assignments and logistics of Federal Air Marshals; and
- The content of proposed changes to security procedures, information, or records.

FAA is also proposing a rule (Docket No. 28859, posted March 11, 1997) that would strengthen requirements for background checks on the employment and criminal history of those who perform checkpoint screening functions

86 *See* 62 FR 13736, March 21, 1997, rule amending 14 CFR Parts 107, 108, 109, 129, and 191.

at airports. These procedures would now include fingerprint checks for those holding unescorted access privileges to airport security areas. It is believed these additional safeguards have been rendered necessary by the proliferation of contract screening companies, whose employees are not under direct control of the air carriers or airport authorities. The air carrier(s) for whom the screening is being conducted would have the responsibility to assure that the background checks are performed.

The FAA, in an Advanced Notice of Proposed Rulemaking issued March 11, 1997, is also seeking to establish a uniform certification protocol for airport checkpoint security screening companies in order to assure full compliance with the Air Carrier Standard Security Program set forth in 49 CFR Part 108. At the moment, the Agency is only soliciting comments on how this should be done.

7.3.4 Measures for Improvement

A survey of transportation regulatory arrangements reveals a number of areas where regulatory adjustments might be appropriate in the name of increased infrastructure security. In addition, there is one newly emerging practice—centralized traffic management in urban areas—that may lead to a new set of vulnerabilities. There is the opportunity now to incorporate security concerns into the development stage of this technology. Detailed below are the current and future issues identified for enhancement of transportation infrastructure security.

1. *Airport security.* As described above, the FAA is in the process of implementing new rules to increase security at airports. The now-final rule, restricting information about airport security procedures, will presumably make it more difficult for terrorists to circumvent security. However, it is also important that the rule be interpreted and applied in such a way that it does not stifle debate regarding the efficacy of current procedures. For example, the television show *60 Minutes* recently highlighted alleged shortcomings in airport security. Public discussion of such issues may lead to the sort of scrutiny and pressure that uncovers problems and makes resources available to solve them.

   The proposed rule on background checks for security personnel is aimed at maintaining the effectiveness of passenger and baggage screening procedures. The growing use of independent contractors to perform these functions has complicated the process of maintaining security requirements; an insufficiently trained or conspiratorial employee might allow a harmful device to pass. Implementation of the proposed rule has the potential to reduce this possibility.

2. *Protection of aviation control hardware.* As noted in section 7.2.2.1 above, the FAA owns and is responsible for much of the infrastructure that supports air navigation, including airport control towers, landing lights

and other structures, plus regional control centers, VOR radio navigation beacons and Doppler radar installations. This physical infrastructure is essential for air navigation and damage to it could shut down an airport. Some of the key items are within the perimeter of airport security, but others, especially the regional control centers and VOR beacons, are located at either decentralized or remote rural sites with little or no active security present. A review of security requirements for these facilities should be considered.

3. *Protection of key rail facilities.* The Arizona Amtrak derailment in 1995 indicated the potential for terrorist attack directed at rail infrastructure. The track was damaged sufficiently to cause a derailment, and the damage went undetected because the perpetrators tampered with the electronic track monitoring system. FRA standards cover electronic railway track communications and signaling system standards but do not provide for controlling or mitigating *override* of these systems by unauthorized intervention. Clearly, most rail carriers want to have appropriate security and inspection measures in place on their major lines, but resource constraints may preclude such provision on secondary lines (as in the Arizona case) unless a specific regulation were to require it.

Another issue relating to key rail facilities is the presence of a relatively small number of "choke points" in the rail network that connects the eastern and western U.S. There are six locations (mostly major rail junctions) where east-to-west rail lines converge; simultaneous disruption at those six points would sever all U.S. rail connection between the coasts.[87] These six rail funnel points may deserve consideration for upgraded security in the national interest.

4. *Protection of rail passenger terminals.* In some locations, rail passenger facilities are located with or nearby freight facilities. This configuration represents a vulnerability in that hazardous materials or explosives could thus be brought into proximity with a crowded area. No regulation requires complete (lateral) physical segregation of freight cars from passenger equipment in use; a regulation that stipulates internal inspection and locking of freight cars prior to their storage on tracks near rail passenger loading facilities may merit study.

5. *Highway funding formulas and FHWA discretion.* Under various sections of Title 23 of the U.S. Code, FHWA is given a degree of license in including demonstrated need as a criterion in computing funding allocations to states (although admittedly the final allocation remains based on established formulas). However, it is not specifically within FHWA's purview

---

87 It should be noted, however, that traffic could still be rerouted through Canada, unless rail tracks there were also attacked. In addition, highway truck transport could be substituted for most commodities.

to attach, as appropriate, an *infrastructural security or safety* dimension to the concept of need. Any proposed infrastructure improvements of a capital nature might require legislative adjustments to allow use of security considerations in determining funding.

6. *Security of advanced traffic management systems.* Plans are being developed to establish advanced traffic-management systems to relieve congestion on urban roadways. At such facilities, personnel will monitor roadway conditions in real time and implement (or monitor) computer-controlled modification of intersection and freeway access ramp signal timings and variable message signs (hazard warnings, potential reroute guidance). Such systems have the potential to make traffic flow more smoothly—or, if misused, to tie it up, impeding passage of emergency vehicles as well as ordinary traffic. Efforts are already underway to privatize the operation of the these new traffic management centers: while the roadway system itself remains under public jurisdiction, private entities will in many cases be contracted to operate the facilities. State departments of transportation and local authorities will oversee these privatized installations in terms of contractual compliance. Study of methods to include security measures in the oversight agenda may be appropriate.

8 DRINKING WATER INFRASTRUCTURE

8.1 General Description of Regulation

8.1.1 Current Regulatory Environment

Most drinking water is supplied to the public by municipal water systems. There are also private systems for subdivisions or large commercial or governmental facilities. A water system consists of a water supply (either groundwater wells or surface water), a treatment facility, treated water storage facilities, and a distribution system. The configuration of a water system varies depending on the source of water, type of treatment applied, and number of connections to be served. Smaller systems may have minimal treatment facilities in each well house, while larger systems may have a centralized treatment facility. Small systems can distribute the water directly from the source or treatment facility, while larger systems require treated water storage facilities to meet peak demand and maintain system pressure. Some systems consist only of distribution lines and purchase their water from another water system. In general, these municipal distribution systems are exempt from regulation other than local interior or connection plumbing requirements.

The regulation of drinking water systems is conducted primarily by state agencies, but according to federally established standards. The Safe Drinking Water Act (SDWA) (Public Health Service Act, 42 U.S.C.A. §§ 300f to

300j-26) authorizes the U.S. Environmental Protection Agency (EPA) to promulgate regulations applicable to water supply systems; perform studies on drinking water sources, technologies, and health risks; and assist states in implementing drinking water programs. In general, the EPA drinking water regulations address water quality standards, treatment methods, system operations (i.e., system water pressure, operator certification, and engineering requirements), and construction standards. EPA also protects drinking water sources via its underground injection control program, watershed demonstration program, sole source aquifer protection program, and wellhead protection program. These latter regulations, however, are not directly applicable to drinking water suppliers but are instead concerned with identifying potential sources of contamination (landfills, surface impoundments, underground storage tanks of special waste, the application of pesticides and fertilizers, etc.) near public drinking water sources or over areas that recharge drinking water sources.

The EPA Office of Ground Water and Drinking Water directs the National Drinking Water Program, including the establishment of drinking water maximum contaminant standards, monitoring requirements and techniques, and groundwater protection programs. All state programs must adopt regulations at least as stringent as those promulgated by the EPA. If a state does not have an EPA-approved drinking water program, the EPA itself implements and enforces the SDWA in that state. States may also enforce their own additional regulations to address economic matters (such as rates and access to public rights of way), or local hydrological conditions. Generally the SDWA standards are enforced by the state environmental protection agency. Other regulations, such as those for rate approval, may be performed by a state public utilities commission (PUC) or similar body. In any state that has not applied for and obtained approval of an enforcement program, or whenever the EPA Administrator finds an authorized state is not duly enforcing the national primary drinking water standards, the EPA Administrator may take enforcement actions.[88]

Local plumbing ordinances are enforced by municipalities or counties. These ordinances pertain to plumbing inside buildings and the connections to any public water system but not to the operation or construction of the public water system itself. Plumbing ordinances are enforceable against anyone constructing or modifying a building within the municipality's or country's jurisdiction. Often, if the municipality or county is also the owner/operator of the public water supply system, it is required to promulgate and implement such local plumbing regulations and standards to

---

88 In these cases, the EPA may issue administrative orders against any water company not in compliance with the national primary drinking water standards and may enforce those orders in the federal district court.

comply with state water supplier laws and regulations. Non-municipal water companies, which must comply with the state water suppliers regulations but do not have legal authority to promulgate their own ordinances, can still require any applicant for water service to comply with their plumbing and connection requirements and standards.

Usually these plumbing regulations are designed for the protection of the water system from improper piping systems and illegal connections to the potable water supply. They may include requirements for (1) installing backflow prevention devices for certain kinds of equipment and appliances, (2) using particular materials for building internal piping systems (*e.g.*, no lead pipe or soldering or PVC piping), (3) providing venting, and (4) applying to connect to the public water service main.

8.1.2 Current Trends in Regulation

The SDWA was amended in 1996, but there has been no trend toward significant deregulation and no shift in regulatory jurisdiction as a result. The amendments eliminate the automatic requirement that the EPA list additional contaminants for monitoring and treatment every three years. The updated process for setting drinking water standards focuses on contaminants known to pose greater health risks (such as *Cryptosporidium*). The amendments require a new regulation for radon in drinking water and studies of the health effects of sulfate and arsenic in drinking water. The amendments also allow for limited alternatives to current filtration and groundwater disinfection regulations. Other provisions focus on assurance that water supply companies can provide sufficient capacity to meet future needs and on the ability of small water companies to meet the current regulations.

Some of the provisions in the 1996 amendments may affect the security of the drinking water infrastructure. These are discussed below.

The disinfectants and disinfection by-product rule was first proposed in July 1994. It is currently in the negotiated rulemaking process. Public water systems use disinfectants to kill harmful microbial contaminants. However, disinfectants and their by-products may also pose risks, including potential increases in rates of cancer, liver damage, and kidney damage. In Stage I, the standard for certain disinfectant by-products (e.g., trihalomethanes) would be lowered. However, any rule to control these disinfectant by-product health risks must also consider maintaining a level of disinfectant that will protect the population against microbial risk. There is currently a regulatory requirement to maintain a specified level of chlorine residual in all parts of the distribution system. This requirement is intended to eliminate chlorine-sensitive contaminants that may be introduced into the water distribution system. However, if the new rulemaking lowers the level of chlorine residual, although the disinfectant level may be considered safe for normal microbial risk, it may no longer protect against the introduction of a more potent microbial agent.

Although many authorized state programs already include water plant operator certification requirements, the amendments require EPA to promulgate operator certification guidelines that must be adopted by authorized states or the states will lose a portion of their State Revolving Fund allocations.[89] There is no indication what types of certification requirements are being considered.

The amendments also give a public water system a shorter time to notify its customers of violations of SDWA requirements. Currently, the SDWA mandates that a notice of a violation of a maximum contaminant level (MCL) or of any other violation designated by the EPA as posing a serious potential health effect must be given as soon as possible, but in no case later than 14 days after the violation. The EPA regulations, which are generally adopted by the states, mandate that when a water supplier has exceeded the MCLs of contaminants that may pose an acute risk to human health, it must furnish a copy of the notice to the radio and television stations serving the area as soon as possible, but in no case later than 72 hours after the violation (40 CFR 141.32(a)(1)(iii)). The amendments shorten this time period to 24 hours when there is potential for serious adverse health effects from short-term exposure. It is up to the state implementing agency to determine the form, content, and manner (e.g., broadcast media, newspaper, or door-to-door) of the notice.

The amendments also require states with primary enforcement responsibility to conduct an assessment, coordinated with existing information and programs, to determine the vulnerability of sources of drinking water within the state's boundaries . By August 1997, the EPA will issue guidance for these state source water assessment programs to ensure that they will delineate protection areas and assess contamination risks to source waters. The vulnerability discussed in these programs is the susceptibility of a water source to contamination from either industrial activities or natural conditions. The amendments provide for loans to public water systems to acquire land or conservation easements for source water protection. New subsection 1428(1)(5) of the SDWA requires the EPA to conduct a demonstration program on the most effective means of assessing and protecting source waters serving large metropolitan areas. However, this program is aimed at protecting water sources from the effects of development and industrial activities rather than intentional contamination.

---

89 The EPA maintains the State Revolving Loan Fund, which provides funds to state agencies to be used for making loans or loan guarantees to water systems to facilitate compliance with applicable drinking water regulations.

The EPA has also proposed a Chemical Monitoring Reform Program that would allow states to target systems at risk of contamination for increased monitoring and reduce the sampling requirements for systems not at risk. This program would focus available public resources on the highest priority water systems. Although reduced monitoring may affect the ability of a water system to detect the introduction of a contaminant into the system, contaminant monitoring required by current regulations is not designed to detect intentional contamination, and increased monitoring would not be an effective method of deterring or detecting intentional contamination. (See discussion in Section 8.3.)

8.2 Description of Selected Regulatory Agencies

8.2.1 Selection Method

The following agencies were selected for detailed discussion: the EPA, Illinois Environmental Protection Agency (EPA), Illinois Commerce Commission (ICC) and the City of Chicago. The EPA was selected because it is the paramount authority on the regulation of public drinking water under the SDWA.

In general, most state programs follow the EPA program closely with few significant state-specific requirements. Illinois is a representative state, and the IEPA and ICC together regulate drinking water supply systems in the state. The IEPA has primary enforcement responsibility for public water systems in Illinois. Its regulations incorporate the federal drinking water regulations, as required by the SDWA, and add a few Illinois-specific regulations. Non-municipal water supply systems in Illinois are regulated as public utilities by the ICC. ICC authority focuses on reliability of service and rates setting.

The City of Chicago Public Works, Department of Water, maintains the public water system for all citizens within the Chicago city limits. It is the largest public water system regulated by the IEPA. It is representative of a large urban system, which is assumed to be a more likely target of terrorist activity. The City of Chicago promulgates its own plumbing ordinances regulating interior building piping and customer hook-up to the potable water system.

8.2.2 Agency Descriptions

8.2.2.1 Environmental Protection Agency

The SDWA (42 U.S.C.A. §§ 300f to 300j-26) authorizes the EPA to establish national drinking water regulations applicable to all public water systems.[90]

---

90 Under EPA regulations, a public water system is a system for the provision to the

The SDWA applies to all public water systems in each state.[91] Primary enforcement of these drinking water regulations, however, is delegated to states that have approved drinking water programs. To obtain approval, the state must adopt drinking water standards at least as stringent as the federal (EPA) standards (40 CFR 141) The EPA directly regulates drinking water systems in states that do not have approved state programs or where state enforcement has broken down.[92]

The EPA also has the following authorities and functions under the SDWA:

The EPA also has the following authorities and functions under the SDWA:

- Establishes the types and extents of variances and exemptions to national drinking water regulations that can be obtained. Such limitations must be adopted by states with approved programs.
- Establishes the regulations and approves state programs for underground injection control and wellhead protection programs.
- Conducts research, studies, and demonstrations related to (1) the causes, diagnosis, treatment, control, and prevention of physical and mental disease, or (2) the provision of a dependably safe supply of drinking water.
- Provides technical assistance to the states and municipalities to establish and administer public water system supervision programs. The EPA may also make monetary grants to states with approved primary enforcement responsibility to carry out public water system supervision programs and underground water source protection programs.
- Provides technical assistance and monetary grants to states or publicly owned water systems to assist in responding to and alleviating any emergency situation affecting public water systems that the EPA determines present substantial danger to the public health.

---

public of piped water for human consumption, which has at least 15 service connections or regularly services an average of at least 25 individuals daily at least 60 days out of the year. The term includes (1) any collection, treatment, storage, and distribution facilities under the control of the operator of such system and used primarily in connection with such system and (2) any collection or pretreatment storage facilities not under such control which are used primarily in connection with such system.

91 The act does not cover water systems that (1) consist only of distribution and storage facilities (and not collection and treatment facilities); (2) obtain all of their water from, but are not owned or operated by, a public water system to which the regulations apply; and (3) do not sell water to any person. This exclusion is often pertinent to federal facilities which supply municipal water to their employees and residents without charge through their own distribution systems.

92 Currently, only Wyoming and the District of Columbia do not have SDWA primary enforcement responsibility. Therefore, the EPA is the enforcement agency in those two states.

- Is authorized to take action if it receives information indicating that a contaminant may present an imminent and substantial endangerment to public health, and that state or local authorities have not taken appropriate protective action.
- Has primary enforcement responsibility for bringing actions against those who tamper with public water systems.

The SDWA establishes criminal and civil penalties for tampering with public water systems (42 U.S.C.A. § 300i-1). "Tampering" includes the introduction of a contaminant into a public water system with the intention of harming persons or otherwise interfering with the operation of a public water system with the intention of harming persons. Sanctions include imprisonment for up to five years and fines of up to $50,000. Attempting to tamper and threatening to tamper are also subject to (somewhat lesser) criminal and civil penalties.

8.2.2.2 Illinois Environmental Protection Agency/Illinois Pollution Control Board

Under the Illinois Environmental Protection Act (415 ILCS 5/1 *et seq.*), drinking water supplies are regulated by the Illinois Pollution Control Board (IPCB or Board) and the IEPA.[93] The Board issues standard-setting regulations, including those mandated by the SDWA. The Board also conducts hearings upon complaints charging violations of the regulations and determines sanctions. The IEPA conducts the associated administrative functions such as issuing permits, conducting inspections, and sending notices of violation. The IEPA also supervises the state wellhead protection program. The IEPA promulgates its own regulations, setting out how it conducts those duties. Below are summaries of Board and EPA regulations governing various aspects of water supply regulation and enforcement.

Investigation and Enforcement

The IEPA may investigate any drinking water supplier upon the request of the Board or upon receipt of information concerning an alleged violation. The IEPA can then issue a written notice of noncompliance and a formal complaint against the regulated water supplier. The public water system would have to answer the complaint before the Board. In addition, third parties may bring a complaint against any public water system alleging violation of the act or a permit or rule issued under it. After the hearing, the

---

93 The IEPA and Board regulations apply to all public water supplies except for those designated as noncommunity water supplies. A noncommunity water supply is one which serves or is intended to serve fewer than 15 service connections used by residents or regularly serves fewer than 25 residents. These small, noncommunity water systems are regulated by the Illinois Department of Public Health under its own regulations (77 Illinois Administrative Code 900 *et seq.*)

Board issues an administrative order that may (1) direct the public water system to cease and desist from violating the Act, rule or permit condition; (2) revoke the system's permit; or (3) impose civil penalties. All such orders are enforceable by injunction, mandamus, or other appropriate remedy in the appropriate state court.

Emergency Authority

When an episode or emergency condition exists, the IEPA so declares, and if the emergency condition creates an immediate danger to health, the IEPA can seal any equipment, vehicle, vessel, aircraft, or other facility operated in violation of regulations or contributing to the condition.

Compliance Monitoring

State regulations specify the frequency and method of sampling drinking water for each of the drinking water primary and secondary standards; they even designate from which taps samples are to be taken. The supplier must file monthly monitoring reports with the IEPA as demonstration of compliance with the MCLs for the listed contaminants. If the supplier has exceeded any MCL for that month or failed to perform the required monitoring, it must notify the IEPA and make public notice to all its customers in a local newspaper or other form of written announcement.

Construction Standards for New Systems.

Under EPA regulations, a state must have a program that assures that the design and construction of new or substantially modified public water system facilities will be capable of complying with all state primary drinking water regulations. This program ensures that the state enforcement agency maintains up-to-date records of the structure of each drinking water system under its jurisdiction. Also, the state must demonstrate that it has a program that includes maintenance of an inventory of all public water systems in the state. Therefore, the state enforcement agency will have a working knowledge of all drinking water systems in the state, including the structure of their production, treatment, and distribution systems.

The Illinois Act requires all owners of public water supplies to submit plans and specifications to the IEPA and obtain written approval before constructing, adding to, or changing any public water supply installation. The IEPA has adopted very specific regulations governing such permitting and the design and construction of public water systems. The regulations for issuance of a construction permit also require the construction be in accordance with the American Water Works Association Standards (Standards) and the Recommended Standards for Water Works—Policies for the Review and Approval of Plans and Specifications for Public Water Supplies of the Committee of the Great Lakes-Upper Mississippi River Board of State Public Health and Environmental Managers (Recommended Standards). The Recommended Standards are used by regulators in Illinois, Indiana, Iowa, Michigan, Minnesota, Missouri, New York, Ohio, Ontario, Pennsylvania, and Wisconsin to draft and grant permits for the construction of public water systems.

8.2.2.3 Illinois Commerce Commission

In addition to the Board and IEPA, water utilities operating in Illinois are also subject to regulation by the ICC. The general powers and authorities of the Commission are set out in Section 3.2.2.4 above. Water utilities must obtain certificates of public convenience and necessity and meet all Commission requirements for service and information filings. The Commission also has jurisdiction over the rates and charges of water utilities.

8.2.2.4 City of Chicago

The City of Chicago is a public water system regulated under the IEPA jurisdiction. It operates its system in compliance with those Illinois laws, regulations, and standards. The City Council has also issued ordinances concerning the standards for plumbing within buildings and requirements for connecting to the City of Chicago public water system (Chicago Municipal Code § 11-8-010 *et seq.*). Under these local ordinances, the Chicago Department of Water has free entry and access to every part of any building, structure, or premises and the authority to inspect the plumbing.

Permits must be obtained for the installation of any piping in any building that receives its service from the Chicago Waterworks System. Permits must also be obtained from the Department of Water to open any paved street for the purpose of tapping mains or laying service pipe. Department of Water approval is necessary before a service pipe can be connected to any water supply pipe that is part of the Chicago Waterworks System.

8.3 Regulations and Critical Infrastructure Protection

8.3.1 Impact of Regulations on Critical Infrastructure Protection

The current regulatory regime for water supplies is not specifically designed to prevent hostile acts. The EPA's regulations do not contain any clear, direct requirements for security systems or restricted access to any parts of the water supply system. Nonetheless, the regulations do have a number of features that affect the vulnerability of supply systems to attack. Several such features are discussed below. To put this discussion in perspective, a brief analysis of potential threats follows.

Threat Analysis

Water supply systems may be vulnerable to criminal tampering at any point in the system: water sources (surface or groundwater), production facilities (well or intake), treatment facilities (plant or well house), storage facilities (raw water reservoirs, and treated water tanks or towers), and distribution systems (mains, lift stations, sampling points, and customer connections). Tampering could include introducing a contaminant into the drinking water supply, thereby causing illness or death in the population served by the system, or physical destruction of water system facilities.

Destroying a retaining dam in a supply reservoir would cause a physical threat to those located downstream and could affect water pressure and the amount of water available. Taking out a treatment facility or cutting off service from the water source to the treatment facility could result in the loss of treated, safe drinking water to the public. The loss of potable water would cause an emergency situation; however, it would be clearly evident and could be dealt with (e.g., by issuing boil orders and securing alternative water supplies).

Contamination of a large urban drinking water system is a logistically difficult proposition. As discussed in two reports prepared for the President's Commission on Critical Infrastructure Protection (PCCIP), *Identification and mitigation of Threats to Critical Infrastructures in the U.S. Water Supply System*, June 1997, and *New York City's Emergency Services Infrastructure*, May 1997, dilution effects are so extensive in raw water sources (groundwater aquifers, lakes, rivers, or retention reservoirs) that huge amounts of contaminant would be needed for harmful level to be reached at the tap. Introduction of a contaminant near the inlet of a treatment plant would reduce the amount of contaminant needed; however, subsequent treatment would destroy or remove at least some portion of the contaminant, thus rendering it less effective. Therefore, the following discussion focuses on the deliberate contamination of a drinking water supply at a point after the treatment plant or well house.

Engineering Requirements

EPA regulations require each state drinking water program to establish some method of assuring that the design and construction of new or substantially modified public water system facilities will be capable of compliance with the state primary drinking water regulations. The IEPA has adopted regulations that require the construction of drinking water systems to comply with the AWWA Standards and the Recommended Standards. Most states have adopted one or both of these sets of standards.

These construction standards require certain protection against contingencies: (1) physical protection of source areas, reservoirs and equipment against contamination or tampering and (2) in some cases, provision against possible loss of power or malfunction of treatment equipment. Well sites must be protected from potential sources of contamination through ownership, zoning, easements, leasing or, if required by the IEPA, fencing. Pumping stations must also be access-protected to prevent vandalism and entrance by animals or unauthorized persons. In addition, storage structures for treated water must be protected with fencing, locks, and other necessary precautions to prevent trespassing, vandalism, and sabotage. The Standards also require that any test wells or other drinking water wells not in use must be sealed to prevent contamination of the water source, preferably by filling the well with cement grout.

There must be provision for standby or backup power in facilities where power outages may result in pressure loss in the distribution system. Standby chlorination equipment of sufficient capacity to replace the largest unit in the supplier's treatment system is also required. In addition, where automated equipment is to be used, supporting data that (1) justify the equipment, (2) demonstrate proper maintenance of the equipment is within the capability of the plant manager, and (3) assure the existence of manual override and manual restart options must be supplied.

Applicants for a permit to build or substantially modify a water facility must submit design information sufficient to show that the finished facility will meet the AWWA Standards and the Recommended Standards. Older facilities are not subject to these requirements, unless or until a substantial overhaul is made.

Maintenance of Plans and Diagrams

The ICC requires each utility to maintain reasonably accurate, up-to-date supply diagrams and distribution maps showing equipment, capacities, and flow directions, including the locations of all mains, hydrants, and valves. The ICC reasons that these maps are necessary to enable the utility to promptly and accurately advise interested individuals on the location of mains and service lines. In the City of Chicago, applications for the installation of piping within buildings and for connection to the potable water system must include plans that clearly show the complete water supply system from the City service to the plumbing fixtures and other appliances, including fire protection equipment. This permitting requirement provides the City with detailed plans of internal plumbing.

Having complete, centralized information of this sort can affect infrastructure vulnerability in two conflicting ways. The general availability of this information may allow terrorists to easily obtain a working knowledge of the water supply distribution system and thereby determine the optimal point for introducing a contaminant or sabotaging critical treatment equipment. However, the information may also be essential for performing quick hazard analyses, repairs, isolation, or other actions in response to an act or threat.

Powers of Inspection

The IEPA and ICC have authority to enter and inspect any regulated drinking water supplier. In addition, the City of Chicago has authority to enter upon the premises and inspect the interior plumbing and any water connections. This authority to inspect might be valuable in preventing the intentional introduction of contaminants, if authorities have information on a possible threat. However, criminal investigation procedures would probably take precedence over civil inspection procedures in any such instance.

Cross Connections

Cross-connections are any physical connections or arrangements between two otherwise separate piping systems—one that contains potable water, and the other that contains water of unknown or questionable safety, steam, gases, or chemicals—whereby there may be a flow from one system to the other. Direct cross-connections are formed when a water system is physically joined to a source of unknown or unsafe substances. Indirect cross-connections are formed when an unknown substance can be forced, drawn by vacuum, or otherwise introduced into a safe water system.

The Board has issued regulations prohibiting such cross-connections unless there is a physical barrier preventing contamination of the potable water supply. These regulations were promulgated to protect the water supply from the unintentional introduction of industrial chemicals, pesticides, and nonpotable water at specific sites assessed by the IEPA for vulnerability (chemical processing tanks, fire hydrants, laboratory sinks, etc.).

Generally, where such connections exist, a backflow prevention device is installed. It provides a mechanical or air-gap separation between the two systems that prevents the back siphoning of untreated or contaminated water into the water supply system. Each supplier must conduct a cross-connection control survey of the distribution system every two years and must have an ordinance, tariff, or required condition for service that includes a plumbing code at least as stringent as the Illinois Plumbing Code (77 IAC 890). The City of Chicago, as a regulated water supply system, has issued plumbing ordinances to enforce this prohibition against cross-connections.

The impact of these regulations on water system security is that they prevent a situation in which water supplies can be easily contaminated, for example, by turning some valve in an industrial facility.

Operator Certification

EPA and IEPA regulations currently require water plant operating personnel to be certified. A personal interview is conducted; however, the subject is limited to the applicant's familiarity with Illinois requirements and general operating knowledge. Certification does not require a background check or other security investigation. Currently, under the SDWA Amendments of 1996, the EPA is to work with states, public water systems, and the public to develop additional recommended operator certification requirements.

Water Sampling and Monitoring

Water system operators are required to test their water periodically for the presence of certain contaminants. The types of contaminants and frequency of testing may vary depending on the size of the water system and other local circumstances. The list of contaminants that must be monitored is limited and focuses on the most common sanitary, industrial, and agricultural pollutant hazards. It appears unlikely that current monitoring practices would detect the presence of a contaminant deliberately introduced for hostile reasons. However, the requirements do establish a "monitoring infrastructure" of trained personnel and procedures that might be adapted to perform other types of monitoring as needed.

Water Treatment

Drinking water is generally treated by chlorination and other means in order to eliminate contaminants. This treatment effectively reduces biohazards. Board and IEPA regulations require chlorination be maintained at levels sufficient to provide adequate protection of human health. In addition, the regulations require that water suppliers maintain an active residual disinfectant concentration (RDC).[94] This requirements to maintain a particular RDC throughout the system at all times will continue to protect the water supply from the introduction of chlorine-sensitive contaminants at any point after the treatment plant. However, the level of disinfectant is currently under review in the proposed disinfectant/disinfectant by-product regulation. As stated above, if the new rulemaking lowers the level of chlorine residual that must be maintained in the water distribution system, although the level may be considered safe for normal microbial risk, it may no longer protect against the introduction of a more potent microbial agent.

---

94 Residual disinfection concentration is the concentration of disinfectant (free or combined chlorine) measured in mg/L in a representative sample of water.

Emergency Measures

As required by EPA regulations, the Board has adopted regulations concerning emergency operations. Whenever contamination is determined to persist in a public water supply, the owners of the supply must notify all consumers to boil all water used for drinking or culinary purposes until it is demonstrated that the water is safe for domestic use or appropriate corrective action approved by the IEPA has been taken. If the owner fails to take such action, the IEPA may issue a boil order directly to the consumers affected. Any emergency that results in water pressures falling below 20 pounds per square inch (psi) on any portion of the distribution system shall also be reason for immediate issuance of a boil order. Whenever the safety of a supply is endangered for any reason, the IEPA must notify the owner, and the supply system officials must take appropriate action to protect the supply.

City of Chicago ordinances provide that whenever the commissioner of water finds a continuous supply of water is indispensable in any building, structure or premises, the owner or occupant shall provide for a tank or other receptacle of sufficient capacity or for installation of redundant plumbing with an emergency connection of a second service pipe from an independent source. This authority could be useful for ensuring that critical emergency service facilities (e.g., hospitals, police stations, and designated shelters) have a second source of drinking water during emergency situations.

Source Protection

The EPA's water source protection programs (i.e., underground injection control program, sole source aquifer protection program, wellhead protection program, and watershed demonstration program) are not designed to protect a water source from the intentional introduction of contaminants. They are designed to increase awareness of activities that can, through mismanagement, result in the contamination of an underground water source through migration of surface contamination or through direct well injection into the aquifer. As discussed, however, the intentional introduction of a contaminant into a raw water source is probably not a logical avenue for terrorist activity, since the contaminant would be subjected to dilution and chemical treatment.

8.3.2 Measures for Improvement

Even though current drinking water regulations were not designed to prevent the intentional sabotage of drinking water supplies, they are generally consistent with the protection of the public drinking water supply. The following suggested measures for improvement have taken into consideration the existing regulatory scheme, along with the physical characteristics and operating requirements of a drinking water supply system.

1. *Access restrictions.* It would be impossible to physically protect all the water mains that serve a large metropolitan water system. The City of Chicago Waterworks system has 4,234 miles of water main and serves the City and 118 suburban communities. Similarly, any large metropolitan water system serves millions of residential, commercial, and industrial buildings and has thousands of outside connections such as sprinkler systems, fire hydrants, and water fountains. It would be impossible to adequately secure every service connection to a water main to prevent the intentional introduction of a contaminant. Therefore, to attempt to introduce regulations requiring some kind of physical protection of the entire distribution system would be futile and financially prohibitive.

It may be worthwhile, however, to identify key points in the system that are likely potential targets and consider additional (or more uniform) requirements for security at those points. In most systems, much of the piping is buried deep underground and relatively difficult to reach. Key aboveground facilities could probably be fairly easily identified and prioritized for protection.

EPA regulations, as adopted by most states, do not contain any requirements for restricting access to water production or treatment facilities. A common theme in the drinking water regulatory scheme is the certification or permitting of each water supply system by the state agency and the requirement that the construction of water production and treatment facilities meet certain engineering standards. As discussed, most states have adopted some sort of standards that require the physical protection of water production facilities (locked well houses, fenced pumping stations, etc.) and water treatment facilities. However, there is no uniform regulatory scheme for ensuring physical protection of these critical facilities. It may be possible to work with the existing engineering requirements (e.g., the AWWA Standards or Recommended Standards) to introduce additional security features in construction permits that would provide a measure of protection without imposing financial or operational hardships. Another area to consider for action would be setting minimum security requirements for retrofitting older facilities. Many older facilities already have security provisions such as locks and fences around critical facilities to ensure public safety and to prevent vandalism.

2. *Changes to monitoring and treatment standards.* Monitoring drinking water for contaminants is the primary focus of current drinking water regulations. Currently, all water companies must monitor for contaminants for which MCLs have been established as well as for a list of

"unregulated" contaminants. These monitoring requirements are very specific. Monitoring must be conducted by trained individuals, and samples must be analyzed by certificated laboratory personnel. This is an expensive process. Any additional contaminant testing might place a substantial financial burden on water suppliers. In addition, publication of these additional requirements would notify any hostile elments about which contaminants would most likely be detected. Nonetheless, it might be worthwhile to conduct research to identify test methods for detecting contaminants considered likely candidates for terrorist use. Even if not done routinely, such tests could be used following a threat or during periods of heightened security.

It also might be worthwhile to conduct research on the vulnerability of potential contaminants to existing water treatment technologies. Modest changes in treatment might provide additional protection against such contaminants. Again, techniques might be developed for use on a contingency basis if there is reason to suspect a potential for terrorist activity.

Another issue in this vein is the effect of the proposed changes in residual disinfectant requirements. It may be advisable for the EPA to consider intentional contamination scenarios in developing and implementing the disinfectant and disinfectant by-product rulemaking.

3. *Information restrictions.* Current regulations require that drinking water system information, including maps, diagrams, and construction details, be available to the public. Such information might be valuable to those planning to attack the system, and so restricting this information might provide a measure of protection. However, there are strong practical reasons why the information should not be restricted. Contractors of all types (e.g., road and building construction, utility construction, and plumbing installation) need this information to safely conduct their business without harming the water supply system. Instituting secrecy as to the configuration of the system might result in more damage and interruptions of service than terrorists are likely to cause.

4. *Personnel screening.* Requirements for certification of water system operators are currently under review by the EPA. It might be worthwhile to consider introducing requirements for background checks into this process.

5. *Emergency planning.* Per SDWA regulations, water suppliers are required to perform emergency planning to cover instances of system disruption. However, there appears to be a need for guidance as to how, in particular, to plan for sabotage-related scenarios. Development of guidance for emergency planners might be worthwhile. Such guidance could include appropriate responses to a threat of hostile action, such as:

- Increased monitoring and sampling;

- Increased facility security;
- Procedures for immediate public notification through the Emergency Alert System;
- Procedures for preventing ingestion of pollutants in the system (e.g., raising the residual chlorine levels in the water will give it an unappetizing taste and odor, indicating that it should not be drunk);
- Alternatives to damaged system facilities (e.g., the availability of portable or backup emergency treatment technologies from other systems in the area); and
- Alternative water supply strategies.

6. *Mechanisms for regulatory change.* The regulatory scheme for drinking water is relatively conducive to making regulatory amendments from the top down. Any regulation adopted by the EPA must be adopted by all states that have primary enforcement responsibility within two years of the date of promulgation of the federal regulation. The state regulations become applicable to all regulated water systems within the state, and if correctly worded, can be required to be codified in governmental water supply systems' ordinances, rules, or regulations. The water supply industry also has an appropriate forum for regulatory change; AWWA is a nationwide association that provides an information network that reaches most water supply companies, participates in EPA and state rulemakings on a regular basis, and publishes water supply system standards that are often incorporated into state construction permits or regulations.

9 EMERGENCY SERVICES INFRASTRUCTURE

9.1 General Description of Regulation

9.1.1 The Current Regulatory Environment

Overview of Emergency Services Organizational Structure

Emergency services consist of local, state, federal, and private response organizations. Local emergency services consist primarily of a very widely distributed network of local police, fire, emergency management, and emergency medical personnel. The wide distribution of responders allows for quick response. In virtually any emergency situation, the first responders at the scene will be local personnel, and they will continue to provide the main part of the response for at least the first several hours. As time progresses, depending on the nature of the emergency, local and state officials may call upon neighboring jurisdictions, the federal government, and private response organizations for assistance. If criminal activity is suspected, the U.S. Department of Justice (DOJ) may take over the primary investigatory role.

Local Response

The key players in local emergency response are generally fire and law enforcement departments. Emergency medical response in the field is provided by local fire departments (Emergency Medical System [EMS] units) and other public and private ambulance companies, in cooperation with local hospitals. Most large municipalities, countries, and states also have an emergency management agency (EMA) that performs disaster planning and provides operational oversight and coordination of response efforts in large-scale emergencies.

Local governments also have a host of local support organizations that they can call upon for assistance. Resources may be requested from other government agencies, such as public works despartments, and from volunteer organizations such as local American Red Cross (ARC) chapters, Radio Amateur Civil Emergency Services (RACES), and other community service and charitable organizations. The volunteer organizations typically have both local chapters and a national umbrella organization.

State Response

Each state has an emergency management statute that lays out state-level policy and authority for emergency response. This statute lays out procedures by which the governor is authorized to declare a state of emergency, thereby acquiring special emergency powers. During the state of emergency, the governor may direct action by all state personnel, commandeer property, set curfews, order evacuations, and otherwise take all steps necessary to respond to the event without going through the usual channels and procedures. Implementation of state-level emergency services is generally coordinated by the state EMA. This agency is usually tasked by statute with preparation and implementation of state-wide emergency planning, including coordination with local emergency services, neighboring states, and the federal government. Other major participants in emergency response at the state level generally include the state police and the state environmental protection agency, with support from various other state agencies such as public health and transportation. Most states have, in one department or another, units that can provide specialized response services such as monitoring for hazardous or radioactive materials.

Federal Response

At the national level, the Federal Emergency Management Agency (FEMA) is the primary Federal agency responsible for disaster response. FEMA maintains the Federal Response Plan (FRP) and the Federal Radiological Emergency Response Plan (FRERP) and coordinates federal assistance in the event of a Presidential declaration of emergency. FEMA also promotes emergency planning, training and preparedness on the part of state and local governments through a variety of guidance, training programs and funding mechanisms. Other agencies play significant roles too. The Environmental Protection Agency (EPA) and US Coast Guard, as co-chairs of the National Response Team, administer the National Contingency Plan (NCP),[95] which provides for response to releases of oil and hazardous substances nationwide. DOJ conducts investigations into disasters when criminal activity is suspected as the cause. Numerous other agencies, including the DoD, provide assistance with disaster response in accordance with the above-mentioned plans.

Table 9.1 briefly outlines the structure of emergency service organizations from the local to the national level.

Table 9.1: Outline of Emergency Response Organizations

Type of Organization	Local	State	Nationals
Law enforcement	Municipal Police Department County Sheriff Department	State Police Department	U.S. DOJ -FBI -ATF
Fire fighting	Local Fire Depts. (volunteer or professional) may include: -Emergency Medical Service - Hazmat Response Unit	State Fire Marshal (arson investigations and oversight)	National Fire Academy National Fire Protection Association (NFPA)
Medical	Local EMS (Fire Dept. unit) Private Ambulance Co.s Hospitals (Public/ Private)	State Public Health Department	U.S. Dept. of Health and Human Services—Centers for Disease Control
Emergency Management	Municipal and County Emergency Management Departments	State EM Department	FEMA
Technical Support	Municipal/County Agencies with support roles, e.g. Public Works, Social Services Local chapters of ARC, RACES and other volunteer org'ns.	State Agencies with support roles—e.g. Public Health Dept, Highway Dept. State chapters of volunteer organizations.	Federal Agencies cooperating under the NCP, FRP, and FRERP

95 40 CFR 300 *et seq.*

Physical infrastructure of Emergency Services

The physical infrastructure for emergency services includes the facilities and equipment used by the agencies that provide emergency services, and the facilities and equipment of public communications systems (e.g., radio broadcast stations) that are used to notify and instruct the public about the emergency situation. More concretely, the infrastructure includes police and fire stations, emergency operations centers, hospitals, telecommunications and data centers, radio and telephone systems used for communications among emergency workers, vehicles used by emergency workers, and radio and telephone station studios and transmission equipment.

Regulation of Emergency Services

Regulation of emergency services differs significantly from regulation of other infrastructures, since emergency services are primarily provided by governmental entities. Regulation of electric and gas utilities, telecommunications, banking and transportation has traditionally sprung from a perceived need to balance the providers' profit motive against public goals such as safety, universal service and reasonable pricing. Emergency services, on the other hand, are mainly provided by local governments and so are presumed to operate in the public interest; there is not the same kind of tension between public and private interests of the sort that has led to extensive regulation in other areas. The conduct of emergency services is controlled through political channels in the same manner as other governmental services.

Although not as all-encompassing as those controlling other types of infrastructure, there are regulations that affect how emergency services are provided. Some of these regulations set standards for certain technical aspects of response, such as medical treatment. Others represent procedures promulgated by response agencies, detailing how they will carry out their response actions. The following areas capture the most significant emergency services-related regulations:

- Implementation of federal response. Regulations and plans issued by FEMA and the EPA determine in detail how federal disaster-response efforts are conducted.
- Implementation of state response. State EMAs issue their own regulations setting out roles, authorities, and policies for state and local response.
- Hazardous material and radiological response. Federal regulations play a significant role in preparedness for specialized response capabilities such as hazardous material (hazmat) and radiological response.

- Communications. Federal Communications Commission (FCC) regulations significantly affect the way in which public safety communications are conducted.
- Medical response. Provision of emergency medical services, like all medical services, is closely regulated at the state level.

Details of these regulatory roles are discussed in Section 9.2.2 below.

Several nongovernmental organizations publish guidance on practice and performance standards related to the infrastructure for emergency services. One example is the American Society for Testing and Materials which has published several standard guides addressing emergency medical services (communications, vehicles, dispatch, and 911 enhanced telephone systems). Another example is the National Fire Protection Association, which publishes guides related to fire protection. Some of its guides, such as *Installation, Maintenance, and Use of Public Fire Services Communications Systems*, are relevant to emergency services infrastructure.

9.1.2 Current Trends in Regulation

The major recent trend in regulation of emergency services has been the growth of specialized programs aimed at technological hazards. Following the Three Mile Island accident in 1979, FEMA and the Nuclear Regulatory Commission (NRC) established regulatory standards for emergency preparedness at nuclear power plant sites. (10 CFR 50, App. E and 44 CFR 350 *et seq.*) Those regulations apply to both nuclear utilities and the communities surrounding the plants. Shortly thereafter, the Comprehensive Environmental Response, Compensation and Liability Act of 1980 (CERCLA) established a federal response program for oil and hazardous materials spills. Following the chemical-release disasters in Bhopal, India, and Institute, West Virginia, this program was supplemented by the Emergency Planning and Community Right-To-Know Act of 1986 (EPCRA), which established requirements for local and state level hazmat preparedness programs.

9.2 Description of Selected Regulatory Agencies

9.2.1 Selection Method

FEMA and the EPA are discussed below because of the roles they play in federal planning and in promoting preparedness on the part of state and local government and industry. The FCC is discussed because of its oversight of the Emergency Alert System and its role in assigning frequencies to various kinds of public safety and other emergency radio networks. At the state level, Illinois was selected as an example of a state with a well-developed system of regulations governing emergency operations and medical response. Regulations of the Illinois Emergency Management Agency (IEMA) and the

Illinois Department of Public Health (IDPH) are profiled. Lastly, the Association of Public-Safety Communications Officials (APCO) is profiled because of its role in setting standards for emergency communications.

9.2.2 Agency Descriptions

9.2.2.1 Federal Emergency Management Agency

FEMA is the federal agency with jurisdiction over the distribution of disaster relief and over emergency planning and preparedness. FEMA has carried out these responsibilities through publication of regulations and guidance. Some of the most significant are:

Stafford Act planning and regulations. The Robert T. Stafford Disaster Assistance and Emergency Relief Act (42 U.S.C.A. § 5121 *et seq.*) authorizes federal assistance in the event of a disaster or emergency to save lives and protect public health, safety and property. The Federal Response Plan (FRP), developed and published by FEMA in coordination with 27 other agencies and departments, lays out policies and procedures for provision of this assistance. FEMA is the primary agency for receiving disaster notification and initiating response measures. The FRP includes function-specific annexes addressing specific types of assistance, including among others transportation, communications, public works and engineering, firefighting, mass care, health and medical services, urban search and rescue, and hazardous materials response. A Terrorism Incident Annex was added in February 1997, describing coordinated arrangements among FEMA, DOJ, U.S. Department of Defense (DOD), U.S. Department of Energy (DOE), EPA, and U.S. Department of Human Health and Services (DHHS).

- Also under the Stafford Act, FEMA has published detailed regulations (44 CFR 206 *et seq.*) describing the process by which governors may apply for a Presidential emergency or disaster declaration. Such declarations authorize use of federal resources for response and financial assistance for reimbursement of state and local response expenses, repair of infrastructure, and aid to individuals affected by the disaster.
- Radiological emergency preparedness. FEMA reviews and approves state and local emergency plans and preparedness for addressing the off-site effects of a radiological emergency at a commercial nuclear power plant, including evaluation of biennial exercises. Standards for this activity are published at 44 CFR 350 *et seq.* FEMA and the NRC have also issued joint guidance on this topic: NUREG-0654/FEMA-REP-1, Rev. 1, *Criteria for Preparation and Evaluation of Radiological Emergency Response Plans in Support of Nuclear Power Plants.*
- FEMA's responsibilities also include providing instructions and technical guidance on emergency preparedness to states and their political subdivisions. To that end, FEMA has issued numerous guidance

documents, memoranda and letters providing specific advice and assistance in dealing with various kinds of emergencies.

9.2.2.2 Environmental Protection Agency

The EPA has had a significant role in developing and promoting preparedness against hazardous material releases. EPA regulations have driven development of hazmat response capabilities among a broad spectrum of organizations, as detailed below.

- The National Contingency Plan (NCP) (40 CFR 300) implements CERCLA, laying out procedures for federal response to spills of oil or hazardous substances. Many parts of the NCP are concerned with remediation and cleanup of sites that were contaminated long ago; however, there are also sections on immediate response to current spills. Under the NCP, certain types of spills must be promptly reported to the National Response Center, the single point of contact for the national response team (NRT). The Center immediately relays the notification to the appropriate predesignated federal on-scene coordinator (OSC),[96] who coordinates, directs and reviews the response effort in compliance with the NCP and the regional response plan for that area. The response may include assistance from other federal agencies according to their expertise. The NRT has representatives from almost every federal agency.[97] The 10 RRTs also have representatives from governor-designated state emergency response agencies.
- EPCRA regulations require preparedness against releases of "extremely hazardous substances" (EHSs) that pose particular danger to surrounding communities (40 CFR 355). The regulations direct states to set up state-level and local-level planning bodies to prepare against this hazard. Firms that use EHSs must offer assistance with this planning, and must immediately notify local authorities in the event of a release. The EPCRA regulations also require routine reporting of information such as stockpiles of EHSs and normal (non-emergency) releases (40 CFR 372).
- The EPA also has incorporated emergency preparedness provisions into several environmental regulations affecting industries that use

---

96 The OSC is a federal official who oversees the response. The OSC will be from the Coast Guard for all oil discharges to waters, or the EPA for all discharges or releases threatening inland areas, or from DOD or DOE for releases from facilities within their jurisdiction, custody or control.

97 In addition to the EPA and Coast Guard, the NRT has representatives from FEMA, DOJ, DOD, U.S. Department of Interior (DOI), U.S. Department of Commerce (DOC), U.S. Department of Agriculture (USDA), Department of Labor (DOL), DOE, U.S. Department of State (DOS), Research and Special Projects Administration (RSPA), NRC and DHHS.

hazardous materials. Regulations issued under the Resource Conservation and Recovery Act (RCRA) (applying to facilities that store or process hazardous waste [40 CFR 265.56]), Clean Air Act (CAA) (release prevention and response for hazardous air pollutants [40 CFR 68; 61 FR 31668, June 20, 1996; 61 FR 16958; and 61 FR 31730]), and Clean Water Act (CWA) (oil spill prevention and response [40 CFR 112]) all contain provisions addressing emergency preparedness. Under these regulations, facilities with permits for handling hazardous materials are required to have contingency plans in place; train employees on emergency response procedures; and have systems for providing warnings to employees and local authorities in the event of a spill.

9.2.2.3 Federal Communications Commission

The FCC was established by the Communications Act of 1934.[98] It has regulatory authority over many aspects of communications systems that serve the public, such as radio broadcast stations, television broadcast systems, and cable systems. The significance of the FCC for emergency services resides in (1) its oversight of the Emergency Alert System (EAS) for notifying and instructing the public; (2) its authority to assign frequency bands for public safety radio services; and (3) its authority to license special emergency radio services. "Public safety radio" includes the activities of local government radio, police radio, fire radio, highway maintenance, forestry-conservation radio, and emergency medical radio. "Special emergency radio services" includes the activities of medical services, rescue organizations, disaster relief organizations, school buses, and beach patrols. The FCC has established a new National Advisory Committee to provide coordination and direction in implementating the new rules and regulations for the EAS.

- Regulations pertaining to public safety radio services are found at 47 CFR 90. They lay out allowable frequencies, power output, and other technical specifications for radio equipment used by local and state public safety services.
- Requirements for the EAS are found in 47 CFR Parts 0, 11, 73 and 76. Essentially all radio and TV broadcast station and cable providers are required to participate in the national system. Participation in state and local systems (i.e., agreeing to transmit emergency messages originating at the state or local level) is voluntary. A redundant network is set up such that the failure of any one station will not cause other area stations to miss the message. Codes and protocols for activation are provided for.

---

98 48 Stat. 1064.

The new regulations require participating stations to phase in upgrades to their equipment. The new equipment will allow improvements in the system; for example, it will be possible to incorporate equipment in ordinary television sets so that they turn on automatically when there is an alert signal, similar to the way National Oceanographic and Atmospheric Administration (NOAA) weather radios work now.

9.2.2.4 Illinois Emergency Management Agency

State-level emergency management in Illinois is coordinated by the Illinois Emergency Management Agency (IEMA), pursuant to the Illinois Emergency Services and Disaster Agency Act of 1975. IEMA regulations appear in Title 29 of the Illinois Administrative Code. Those regulations include:

- Radiological Protection (Part 320). This part includes policy for the development and maintenance of a radiological protection program, specifications and procedures for testing and maintenance of radiological monitoring equipment, and standards and procedures for radiological training and recordkeeping.
- Individual and Family Grant Program (Part 410) and Public Disaster Assistance Program (Part 420). These parts lay out procedures for administering disaster-relief programs, in coordination with FEMA, to provide financial assistance to individuals, families, and local governments affected by disasters.
- Emergency and Written Notification of an Incident or Accident Involving a Reportable Hazardous Substance (Part 430). This part requires notification of IEMA whenever there is a reportable release of an extremely hazardous substance (as defined in EPCRA). IEMA serves as the point of contact for the State and in turn notifies other Illinois agencies as appropriate.
- Development, Annual Review, Coordination of Chemical Safety Contingency Plans (Part 610) and Emergency Planning and Community Right-to-Know (Part 620). These parts establish requirements and procedures for coordination between businesses and local governments regarding planning for hazardous material response, inventory reporting, release reporting, and emergency notifications.
- Emergency Services and Disaster Agencies: Establishment, Accreditation, and Workers' Compensation (Part 1300). This part requires establishment of local emergency services agencies by each county and by each municipality with more than 500,000 residents. It also sets standards for their accreditation and provides workers' compensation coverage for workers and volunteers engaging in emergency response activities, drills or exercises.

- Emergency Management Assistance Program (Part 1310) This part establishes administrative procedures and requirements for local emergency service agencies to receive financial assistance (grants) distributed by FEMA through the State of Illinois.

9.2.2.5 Illinois Department of Public Health

The Illinois Department of Public Health (IDPH) licenses hospitals and emergency medical transport providers (ambulances) within the State of Illinois. Some of the major components of the IDPH regulations relating to emergency medical services are:

- Hospital emergency room requirements. General-service hospitals are required to provide at least a minimum level of emergency care. (Some specialized hospitals, such as psychiatric hospitals, are not.) In addition, the regulations provide that facilities meeting certain additional requirements may be designated as primary and secondary Trauma Centers. The requirements at each level are extensive and include stipulations as to equipment, supplies, staffing, 24-hour availability, and other factors.
- Emergency medical transport requirements. Each service providing emergency medical transport, including fire department emergency medical services, ambulances, watercraft, and helicopters must be licensed by IDPH. Each such service must meet extensive requirements including training and certification of personnel, availability of medical equipment and supplies for enroute care, communications capabilities, and other factors.
- Emergency medical systems. The IDPH regulations establish emergency medical service regions and provide for a hierarchy of coordinated services within each region. Within each region, the primary resource hospital oversees and directs the delivery of emergency medical services performed by the transport providers and other participating hospitals.

9.2.2.6 Association of Public-Safety Communications Officials

The objectives of the Association of Public-Safety Communications Officials (APCO) are to foster the development and progress of the art of public safety communications; ensure greater cooperation in the correlation of the work and activities of town, county, state and federal agencies; promote cooperation between these agencies and the FCC; and conduct surveys, management and training seminars. APCO reviews and approves proposed practice and performance standards developed by organizations such as the American Society for Testing and Materials. It has recently participated in developing APCO-25, a suite of standards on digital radio communications that is intended to improve the effectiveness and reliability of interagency and interagency public safety radio communications.

9.3 Regulations and Critical Infrastructure Protection

9.3.1 Impact of the Regulations on Protection of Critical Infrastructure

The security of the emergency services infrastructure appears to be little affected by regulation. As noted earlier, regulation of emergency services is not pervasive; and among these regulations, few are concerned with the security of associated infrastructure. It should be noted, however, that the infrastructure of emergency services is decentralized and inherently resilient against attack. To keep response times short, police and fire services are provided out of numerous local stations. In a large city, there will be dozens or hundreds of stations. An attack on any one (or even several) of these locations cannot disable the entire response system. An attack on a police station will not even completely disable that station's personnel, since at any given time a large fraction of police officers are out on patrol. In addition, police and fire personnel are somewhat less vulnerable than most in that they have protective equipment on hand, and in the case of police at least can defend themselves against armed attack.

Emergency medical services and hospitals are also decentralized. Urban areas typically have several hospitals with emergency care capability within close range, as well as numerous transport services.

The strongest connection between regulation and emergency service infrastructure assurance appears to be in the area of communication to the public. The FCC's EAS system regulations establish network redundancies, passwords, testing mechanisms, and other measures directly intended to make the EAS more robust and less subject to single-point failure.

The regulation of emergency services *has* affected the ability of those services to respond to the types of emergencies that the Commission is concerned with. This has occurred through two mechanisms: one, coordinated planning for federal response, and two, the establishment of a network of public and private capabilities for response to emergencies involving hazardous or radioactive materials.

Coordinated Planning for Federal Response.

The development of emergency planning at the federal level, including the FEMA Stafford Act regulations, NCP, FRP, and FRERP, has strengthened response capabilities for major disasters of all types. These plans have been extensively verified through exercises and lessons learned from actual responses. Implementation of these plans has enhanced federal response capabilities. Also as a result of these plans, state and local governments are now in the habit of promptly notifying federal officials when major emergencies occur, and they routinely request assistance when it appears that local resources will be exhausted. This means that in the event of a significant terrorist event, local response officials could be expected to provide notice to key federal agencies within a matter of minutes. The resultant federal response will be well coordinated and will proceed according to established protocols for assistance to local officials.

Network of Hazmat/Rad Response

Regulations adopted since 1980 have created a network of public and private capabilities for response to radiological and hazardous material emergencies. EPA environmental regulations implementing EPCRA, RCRA, and CAA require firms that handle or store hazardous materials to have emergency preparations in place. Such firms are required to have in-house emergency response organizations, emergency plans, training programs, procedures for giving prompt warning to employees in the event of an emergency, and, if necessary, a method of notifying off-site authorities of the need to evacuate nearby areas. They are also required to promptly notify local and federal officials when there is a release of hazardous materials.

The EPCRA regulations set up a network of local and state emergency response organizations specifically to plan for and implement response to hazardous material emergencies. Each state has a state emergency response commission (SERC) to implement preparedness statewide, and local emergency planning committees (LEPCs), usually at the county or municipal level, to perform local planning functions. The LEPCs generally consist of officials from local response agencies and may also include industry representatives. Firms that store or use regulated hazardous materials are required to offer assistance and information to their local LEPCs. Thus, as a result of the EPCRA regulations, there is now a national network of state and local hazardous materials response agencies. These agencies in turn participate as state and local representatives on the RRTs under the NCP.

Radiological response capabilities are also relatively widespread, partly as a result of the regulation of nuclear power plants. Regulations adopted by the NRC and FEMA require radiological emergency response capability on the part of nuclear utilities and communities where nuclear power plants are located. Over 40 states, as well as numerous localities, participate in this program. They are required to have radiological emergency response plans, including the ability to assess radiological hazards and implement appropriate protective measures for the public.

One other agency also bears mention in this respect. The Occupational Safety and Health Administration (OSHA) administers regulations on workplace safety. Pursuant to EPCRA, in 1987 OSHA adopted regulations governing the safety of emergency response personnel who respond to hazardous material incidents (29 CFR § 1910.120(q)). Partly as a result of these regulations, hazmat response teams across the country are trained and equipped to operate in hazardous environments with appropriate personal protective gear.

9.3.2 Measures for Improvement

Review of the regulations described above, along with conversations with emergency response officials, indicates that the primary opportunity for improvement of response capabilities lies in enhancement of emergency communications. Two significant measures are already being undertaken. First, the FCC has instituted improvements in the EAS to increase speed and reliability of transmission of public instructional messages. The older Emergency Broadcast System (EBS) used a "daisy chain" of transmissions from a single primary entry station. If any link in the transmission was severed, the stations beyond the break would not receive the transmissions. Under EAS, a state is divided into regions, and each region has two primary entry stations that are continuously monitoring state emergency broadcast frequencies and each other. Other stations within the region monitor these stations. Thus EAS is a parallel system, with redundancies that make it more resilient against breakdown or attack.

The second measure that has been the adoption of technical standards for design of the next generation of communications equipment. Emergency service organizations and communications equipment manufacturers have cooperated to set standards to improve the effectiveness and reliability of public-safety communications equipment. They developed APCO 25, a suite of standards for digital radio communications for public safety applications that improves interoperability and talk-around capabilities (APCO-25 also refers to the process that developed these standards). Interoperability refers to the ability to use components made by different manufacturers in a single system. Talk-about refers to the capability to have portable-to-portable and/or mobile-to-mobile intercommunication without the use of a repeater. This is particularly valuable for tactical operations involving different agencies. Organizations representing governmental public safety radio users at each level of government (Association of Public Safety Communications Officials, National Association of State Telecommunications Directors, Department of Defense, National Telecommunications and Information Administration) and industry organizations (Telecommunications Industry Association, American National Standards Institute) participated in the APCO-25 process.

The action that remains to be taken is allocation of additional radio spectrum for use by emergency response agencies. The APCO-25 process will allow development of equipment that uses narrower radio bandwidths for communications. This will effectively double the number of channels available to public safety organizations. However, the scarcity of available channels remains an issue with emergency service organizations. Many such organizations are advocating reallocation of a portion of the radio spectrum that is currently reserved for television to public-safety radio communications. Such a reallocation would effectively double the available number of channels again. A bill to require the FCC to set aside an additional 24 MHz of radio spectrum for public safety was introduced in the Senate in February 1997 (S. 255, Law Enforcement and Public Safety Telecommunications Empowerment Act). This bill is currently inactive; however, similar language has been incorporated into the Senate version of the Budget Reconciliation Act (S. 947, A Bill to Provide for Reconciliation Pursuant to Section 104(a) of the Concurrent Resolution on the Budget for Fiscal Year 1998, June 23, 1997).

APPENDIX

ACRONYMS, AGENCIES AND ASSOCIATIONS, STATUTES AND REGULATIONS

Acronyms

ALM - automatic loan machines

ATM - automatic teller machine

BBS - bulletin board system

CFR - Code of Federal Regulations

CNG - compressed natural gas

EAS - Emergency Alert System

EDI - electronic data interchange

EFT - electronic funds transfers

EHSs - extremely hazardous substances

EOC - emergency operating center

EWGs - exempt wholesale generators

FR - Federal Register

FRERP - Federal Radiological Emergency Response Plan

FRP - Federal Response Plan

IAC - Illinois Administrative Code

IFR - instrumental flight rule

INFOSEC - information security

ISO - independent system operator

LDC - local distribution company

LEPC - Local Emergency Planning Committee

LNG - liquefied natural gas

LPG - liquefied petroleum gas

MCL - maximum contaminant level

NCP - National Contingency Plan

NCS - National Communications System

NRT - National Response Team

OASIS - Open Access Same-time Information System

OCS - outer continental shelf

OPSEC - operational security

OSC - on-scene coordinator

PSN - public switched network

PUC - public utility commission

QFs - qualifying facilities

RDC - residual disinfectant concentration

RSU - remote service units

SCADA - supervisory control and data acquisition

SERC - State Emergency Response Commission

SPR - Strategic Petroleum Reserve

SYSEC - system security

TAC - Texas Administrative Code

USC - United States Code

USCA - United States Code Annotated

VASI - visual approach slope indicator

VOR - very-high-frequency omni-range

Agencies and Associations

AEC - Atomic Energy Commission

AGA - American Gas Association

APCO - Associated Public Safety Communications Officers

API - American Petroleum Institute

AWWA - American Water Works Association

BLM - Bureau of Land Management

DOC - Department of Commerce

DOD - Department of Defense

DOI - Department of Interior

DOJ - Department of Justice

DOL - Department of Labor

DOS - Department of State

DOT - Department of Transportation

EPA - Environmental Protection Agency

ERDA - Energy Research & Development Administration

FAA - Federal Aviation Administration

FCC - Federal Communications Commission

FDIC - Federal Deposit Insurance Corporation

Fed - Federal Reserve System/Board

FERC - Federal Energy Regulatory Commission

FFIEC - Federal Financial Institutions Examination Council

FHFB - Federal Housing Finance Board

FHWA - Federal Highway Administration

FPC - Federal Power Commission

FRA - Federal Railroad Administration

FTA - Federal Transit Administration

GRI - Gas Research Institute

ICC - Illinois Commerce Commission

ICEC - Illinois Commission on Electronic Commerce

IDOT - Illinois Department of Transportation

IDPH - Illinois Department of Public Health

IEEE - Institute of Electronics and Electric Engineers

IEMA - Illinois Emergency Management Agency

IEPA - Illinois Environmental Protection Agency

INGAA - Interstate Natural Gas Association of America

IPCB - Illinois Pollution Control Board

MMS - Mineral Management Service

NAC - National Advisory Committee (FAA)

NASDAQ - National Association of Securities Dealers' Automated Quotation System

NERC - North American Electric Reliability Council

NFPA - National Fire Protection Association

NJBPU - New Jersey Board of Public Utilities

NOAA - National Oceanographic and Atmospheric Administration

NPC - National Petroleum Council

NRC - Nuclear Regulatory Commission

NRIC - Network Reliability and Interoperability Council

NTIA - National Telecommunications and Information Administration

OCC - Office of the Comptroller of the Currency

OSHA - Occupational Safety and Health Administration

OTS - Office of Thrift Supervision

PUCT - Public Utility Commission of Texas

RSPA - Research and Special Programs Administration

RTC - Resolution Trust Corporation

SEC - Security and Exchange Commission

STB - Surface Transportation Board

TRC - Texas Railroad Commission

Statutes and Regulations

APSPA - Accountable Pipelines Safety and Partnership Act of 1996 (Pub. L. 104-304)

CA of 1863 - Currency Act of 1863 (Chap. 106, 13 Stat. 99)

CAA - Clean Air Act (42 U.S.C.A. §§ 7401-7671q)

CPUC - California Public Utilities Code

CWA - Clean Water Act (33 U.S.C.A. §§ 1251-1387)

DIDMCA - Depository Institutions Deregulations and Monetary Control Act of 1980

ECPA - Energy Conservation and Production Act of 1976 (Pub. L. 94-385)

EPAA - Emergency Petroleum Allocation Act of 1973 (Pub. L. 93-159, 87 Stat. 627)

EPACT - Energy Policy Act of 1992 (Pub. L. 102-486, 106 Stat. 2776)

EPCA - Energy Policy and Conservation Act (Pub. L. 94-163, 89 Stat. 871)

EPCRA - Emergency Planning and Community Right-To-Know Act of 1986 (42 U.S.C.A. § 11001 - 11050)

FFIRREA - Federal Financial Institutions Reform, Recovery, and Enforcement Act of 1989 (Pub. L. 101-73, 103 Stat. 183)

FIRIRCA - Financial Institutions Regulatory and Interest Rate Control Act of 1978 (Pub. L. 95-630, 92 Stat. 3641)

FPA - Federal Power Act (16 U.S.C.A. § 791 *et seq.*)

FRA - Federal Reserve Act of 1913 (Pub. L. 63-43, 38 Stat. 251, 12 U.S.C.A. § 221)

Glass-Steagall - Glass-Steagall Act of 1933 (48 Stat. 162)

ILCS - Illinois Consolidated Statutes

IEPA - Illinois Environmental Protection Act (415 ILCS 5/1 *et seq.*)

IESA - Illinois Electric Supplier Act (220 ILCS 30/1 *et seq.*)

IGPSA - Illinois Gas Pipeline Safety Act (220 ILCS 20/1 *et seq.*)

IPUA - Illinois Public Utilities Act (220 ILCS 5/1-101 *et seq.*)

ISTEA - Intermodal Surface Transportation Efficiency Act of 1991 (Pub. L. 102-240, 105 Stat. 1914) (Amending 23 U.S.C.A. § 101 *et seq.*)

McFadden Act - McFadden Act of 1927 (Pub. L. 69-639, 44 Stat. 1224)

NBA - National Banking Act of 1864 (13 Stat. 99, 12 U.S.C. § 35 *et seq.*)

NGA - Natural Gas Act of 1938 (15 U.S.C.A. § 717-717z)

NGPA - Natural Gas Policy Act of 1978 (Pub. L. 95-621 (codified in various sections of 15 U.S.C., see especially 15 §§ 3301 *et seq.*)

NGWDA - Natural Gas Wellhead Decontrol Act (Pub. L. 95-617, 92 Stat. 3117)

PSA - Pipeline Safety Act (49 U.S.C.A. §§ 60101 *et seq.*)

PUHCA - Public Utility Holding Company Act of 1935 (15 U.S.C.A. §§ 79-79z)

PURPA - Public Utility Regulatory Policies Act of 1978

PURA95 - Public Utility Regulatory Act of Texas

Riegle-Neal - Riegle-Neal Interstate Banking and Branching Efficiency Act of 1994 (Pub. L. 103-328, 108 Stat. 2338)

SDWA - Safe Drinking Water Act (Public Health Service Act, 42 U.S.C.A. §§ 300f to 300j-26)

SA of 1933 - Securities Act of 1933 (48 Stat. 74)

SE of 1934 - Security Exchange Act of 1934 (48 Stat. 881)

Stafford Act - Robert T. Stafford Disaster Assistance and Emergency Relief Act (42 U.S.C.A. § 5121 *et seq.*)

CHAPTER 6
LEGAL AUTHORITIES DATABASE

A "LEGAL FOUNDATIONS" STUDY

Report to the
President's Commission
on Critical Infrastructure Protection
1997

This report was submitted to the President's Commission on Critical Infrastructure Protection, and informed its deliberations and recommendations. This report represents the opinions and conclusions solely of its developers.

LEGAL AUTHORITIES DATABASE

In the fall of 1996, the President's Commission on Critical Infrastructure Protection (PCCIP) commissioned the creation of a legal database. The PCCIP Legal Authorities Database (Database) now houses over 14,000 legal authorities bearing on infrastructure assurance, including Federal, state, local, international, and even private sector authorities such as standards and guidelines.

The Database grew out of the PCCIP's various legal "landscaping" projects. It was originally intended to serve as a study tool—a collection point for legal authorities that appeared relevant to infrastructure assurance efforts, but that defied ready categorization. As it turns out, many of the authorities collected do in fact defy ready categorization, which makes the Database all the more useful as a tool for researching avenues by which to implement infrastructure assurance recommendations.

The Database content specifically reflects the results of a *de novo* review of pertinent legal sources, and input from PCCIP Commissioners and staff from each of the five Sector Working Groups. Laws and regulations were included based on sector reports, briefings and additional authorities that came to the PCCIP's attention. The content also embodies contractor submissions, suggestions from numerous public officials, and reports from the private sector. The Database features separate fields for (among other things) the title and citation of the authority, a summary of the authority, relevant text, related citations, and a "current as of" date.

Uses for the Legal Authorities Database

The Database as a Research Tool

The Database was designed to assist interested persons in a number of key ways. First, by placing a substantial number of pertinent resources in one clearinghouse, the Database saves researchers considerable time and energy. Critical infrastructure issues are difficult to research. Numerous sources are required to address most critical infrastructure issues—Presidential materials, Congressional statutes, administrative rulings, unpublished guidelines, international laws, and private sector materials; these materials can now be accessed in one place.

Second, the Database provides a clearer context in which to analyze and understand critical infrastructure assurance issues. Many of the materials in the Database were originally entered around key words and concepts relevant to assurance efforts. These included topics such as indications and warnings, information sharing, education and awareness, information security, emergency response and the like. Although the Database has grown beyond these topics, they continue to serve as useful search terms to guide research efforts. Most interested persons will have limited experience with critical infrastructure assurance issues and the Database offers a compass for navigating through rugged terrain.

The Database as an Implementation Tool

Given the difficulties inherent in categorizing many of the Database entries, it would likely prove productive to enter search terms keyed to various recommendations; results of such a search are likely to produce an array of authorities—Federal laws and regulations, international agreements, and even private sector standards.

For example, a sample search for laws addressing cyber threats to the Nation's securities and commodities exchanges direct the user to pertinent portions of the Federal criminal code, Securities and Exchange Commission regulations and SEC Rules for emergencies, state computer crime laws, and NYSE, AMEX and NASDAQ floor rules, among others. The PCCIP Legal Authorities Database can assist implementation efforts by Federal and state agencies, lawmakers, and even private practitioners.

A Closer Look at the Legal Authorities Database

Appendix A contains a manual which accompanies the Database CD-ROM disk itself. The graphics illustrate the legal categories and critical infrastructure topic areas represented in the Database. The Legal Authorities Database includes relevant legal materials for each of the critical infrastructures. In addition, the Database includes up-to-date legal source materials on relevant "cross-cutting" issues. Cross-cutting topics include state computer crimes legislation, U.S. Sentencing Guidelines, the Federal Response Plan, reporting requirements for infrastructure owners and operators, information security and similar laws at the state level.

Appendix A also suggests the ease with which the Database may be used to locate relevant legal authorities. Instead of categorizing relevant laws in terms of issuing agency or general area of law (e.g., section of C.F.R. or U.S. Code topics), the Database is based on critical infrastructure issue areas (e.g., Energy, Telecommunications and Information, Vital Human Services, Banking & Finance, and Physical Distribution). Few commercial legal database sources include all of these materials in one searchable database file.

The Database manual contains graphic representations of the contents of the database for each of the infrastructure sectors and select cross-cutting issues. It is meant to give the reader a flavor of the scope and depth of the Database, and should not be taken as a comprehensive rendering of all authorities in each area.

APPENDIX A

LEGAL AUTHORITIES DATABASE MANUAL

The Legal Authorities Database will be forthcoming in the future.

CHAPTER 7

INFRASTRUCTURE PROTECTION SOLUTIONS CATALOG

A "LEGAL FOUNDATIONS" STUDY

Report to the
President's Commission
on Critical Infrastructure Protection
1997

This report was submitted to the President's Commission on Critical Infrastructure Protection, and informed its deliberations and recommendations. This report represents the opinions and conclusions solely of its developers.

CONTENTS

INTRODUCTION

"Never hype a problem unless you have a solution." (*statement in a meeting with Commissioners and staff of the President's Commission on Critical Infrastructure Commission*)

> **Michael R. Nelson, Director, Technology Policy**
> **Office of Plans and Policy**
> **Federal Communications Commission**

In early 1997, the requirement arose in the President's Commission for Critical Infrastructure Protection (PCCIP) for maintaining a running catalog of ideas relating to potential solutions for infrastructure protection. In assembling this catalog, we relied heavily on the work of other bodies that preceded the President's Commission. The *Infrastructure Protection Solutions Catalog* embodied those preliminary efforts and was maintained through March 1997, the time at which the Commission began to formulate its own potential recommendations. It is a compendium of prior recommendations made by informed individuals, study commissions, and other groups who have weighed in on issues related to infrastructure assurance.

Possible solutions are arranged within broad solution categories. Within each category is a general description of its scope and an "external considerations" section. External considerations are specific solution-related ideas, proposals or recommendations that came to our attention by way of the papers and briefings referenced in the Bibliography to this document.

These efforts led to the development of another "Solutions Catalog" used during the U.S. Infrastructure Assurance Prosperity Game and Planning Event Number 2, held on March 23-25, 1997 in Chantilly, Virginia. The event was sponsored by the President's Commission on Critical Infrastructure Protection, the National Communications System and the Department of Energy, and was produced by Sandia National Laboratories and the Prosperity Institute. This catalog is also included here in Appendix A, as it appeared in gaming materials, to demonstrate the full range of solution ideas that were brought to the attention of the PCCIP to inform its deliberations

I. Solutions Relating to the Creation of a New Federal Policy Entity

Coordinate Public/Private Assurance Efforts

This category encompasses solutions aimed at creating an entity to carry on the work of the PCCIP. This entity could take the form of an executive branch agency, a Federal spokesperson or "czar" model, an independent regulatory commission (created by Congress), or a standing executive branch advisory committee. The scope of the entity's authority could include the full range of critical infrastructures, as well as the ability to leverage Cabinet-level agency resources. Private sector participation, substructures within the entity for each critical infrastructure, and the appropriate level of staffing are all issues that will need to be considered.

External Considerations

Establish a joint office for system, network and infrastructure design to promote utility, resiliency, reparability and security in infrastructures.	(Defense Science Board Task Force on Information Warfare-Defense, November 1996).
Reduce the expectation of an all-encompassing response from the Department of Defense.	(Logan, Michael; 5th International Conference on Information Warfare, 1996).
Need effective collaboration between the federal national security community and private sector industries because of the degree of dependence of the government on privately provided transportation, telecommunications and energy.	(Volpe Center).
Establish an Information Security Foundation.	(National Research Council, "Computers at Risk," 1991).
Stand up an Information Assurance Committee (IAC) under the PDD-29 Security Policy Board structure to create information assurance policy for Government systems processing classified and national security information. The IAC would also propose legislation, policy, and regulations for the Executive branch and influence private and non-government entities important to national security.[1]	(U.S. Security Policy Board, "White Paper on Information Infrastructure Assurance," 1996).
Broaden NSTAC's charter to reflect the full scale of NII issues beyond telecommunications, security and national security.	(U.S. Security Policy Board, "White Paper on Information Infrastructure Assurance," 1996).
Establish an Information Assurance focus within the National Security Council.	(U.S. Security Policy Board, "White Paper on Information Infrastructure Assurance," 1996).

Assign a focal point for Federal government leadership in support of a coordinated U.S. response to the strategic IW threat. Also conduct risk assessments, discuss proper government role, address preparedness.	(RAND, "Strategic Information Warfare: A New Face of War," 1996).
Create an NSTAC-like advisory committee for the electric power industry.	(NSTAC Information Assurance Task Force, "Electric Power Information Assurance Risk Assessment," 1996).
Establish individual infrastructure sector advisory committees, based on the NSTAC model, to facilitate communication and information sharing and generate policy recommendations.	(Center for Strategic International Studies (CSIS) Information Assurance Working Group, internal PCCIP memo by Commissioner Brenton Greene dated October 2, 1996).
Advances need to be made in the following policy areas: -achieving consensus on problems and solutions -enhancing government/industry cooperation for identifying and characterizing reliability challenges -enhancing Federal and state interaction to ensure consistent and appropriate attention is placed on infrastructure reliability.	(OSTP, "Cybernation," Draft of January 17, 1997).
Devise an integrated program combining security technology, managerial practices, and US government policy into a mutually supporting coherent program for the protection of U.S. financial institutions.	(Nickson, Mark; "Vulnerability of US Financial Markets to an Information Warfare Attack," Draft of March 1997).

May require replacing or modifying PL 100-235, NSD-42, OMB Circular A-130, Paperwork Reduction Act, and other subordinate directives.

Identify and Leverage Sources of Funding for Public/Private Assurance Efforts

This subsection will be used to catalog potential sources of funding that can be used to build and maintain the entities and capabilities described in items I-III. The range of possibilities for funding may include earmarking a certain percentage of available funds expressly for assurance efforts or recommending that state and local governments identify and develop similar funding sources, "matching" or incentive programs to further assure infrastructures deemed critical to their own sustained operations.

II. Solutions Relating to the Creation of a Federal Coordinating Entity Including Operational Responsibilities (e.g. Funding Authority, Rulemaking Authority)

"Manager" of Interagency Indications and Warning Capability

Solutions under this category address creation of a capability for coordinating and maintaining a nationwide indications and warnings system. Such a capability should allow for the exchange of appropriate information with the private sector and federal, state and local governments, including the intelligence and law enforcement communities. It may also include an appropriate international liaison capability. It could also serve as a repository for the operational resources (personnel, equipment, etc.) that would comprise the Federal government's emergency response and response management capabilities described in the next section. This "manager" may also be able to assume many of the responsibilities currently delegated to the Infrastructure Protection Task Force (IPTF) by E.O. 13010.

External Considerations

Create a center for intelligence indications and warnings, current intelligence and threat assessments within NSA.	(Defense Science Board Task Force on IW-D, 1996).
Create a center for information warfare defense operations within DISA, to have links with any successor of the IPTF.	(Defense Science Board Task Force on IW-D, 1996).
Create a National Information Infrastructure Threat Center with representatives from law enforcement, intelligence and the Defense communities to perform liaison with the private sector, act as a clearinghouse for intrusion reports, and have "real time" 24-hour operational capabilities.	(U.S. Senate Permanent Subcommittee on Investigations, Staff Statement: Hearings on Security in Cyberspace, 1996).
Promote regular vulnerability assessments, or "red teaming," of government agencies, especially those outside of the Department of Defense, to operate in a manner similar to that of DISA in assessing the armed forces.	(U.S. Senate Permanent Subcommittee on Investigations, Staff Statement: Hearings on Security in Cyberspace, 1996).
Establish an early warning capability that uses internet messages to help identify developing situations overseas that could lead to security threats.	(Swett, Charles, OSD; "Strategic Assessment: The Internet," 1995).
Use a flexible integrated response to counter information terrorists, employing information warfare tactics tailored to countering gray-area phenomena, but also reserving the use of conventional counter-terrorism operations.	(Devost, Houghton, & Pollard, "Information Terrorism: Can You Trust Your Toaster?," 1996).

Response structure should incorporate assets from both the military and law enforcement. The military should take an advisory role in domestic incidents, and law enforcement assets a similar role in overseas incidents.	(Devost, Houghton, & Pollard, "Information Terrorism: Can You Trust Your Toaster?," 1996).
Establish mechanisms to support focused information exchange, provide information on threats, vulnerabilities and mitigation measures, and to facilitate identification of recommended actions by industry and government to ensure the viability and security of the National Information Infrastructure (NII).	(Information Infrastructure Task Force, "NII Risk Assessment: A Nation's Information at Risk," 1996).
Facilitate the inclusion of NII risk data in other relevant data collection activities, such as those conducted by the Census Bureau or the National Economic Council.	(Information Infrastructure Task Force, "NII Risk Assessment: A Nation's Information at Risk," 1996).
Establish a red team for independent assessments of vulnerabilities.	(Defense Science Board Task Force on IW-D, 1996).
Establish a method to facilitate the flow of intrusion and threat data between and among government agencies, especially law enforcement and intelligence communities, and the private sector. Mechanism should include a process to provide effective threat warnings and pattern analysis to aid in the protection and assurance of information and information-based control systems.	(CSIS Information Assurance Working Group, PCCIP Commissioner Greene memo dated October 2, 1996).
Establish a new agency in the Executive Office of the President to address Information Assurance. Functions of the agency would include review, coordinate, propose, and, where appropriate, direct Executive Branch NII protection-related policy, resource allocation, education, training and awareness, legislation, technology development and application, international liaison, and threat/vulnerability assessment.	(U.S. Security Policy Board, "White Paper on Information Infrastructure Assurance," 1996; SPB Briefing to Commission, September 23, 1996).
Establish a NII Threat Center with 24 hour, 7 days a week operations.	(Nickson, Mark; "Vulnerability of US Financial Markets to an Information Warfare Attack," Draft of March 1997).

Create a national process for strategic information warfare threats Indications and Warnings to include the following features: -dual track watch centers to monitor IW indicators, infrastructure status and intrusion patterns -reporting channels to communication information between two tracks centralized warning function -reporting channels to communication warnings to key decision makers IW threat management process -standing National Security and Emergency Coordination (NS/EP) programs for strategic IW threat -mechanisms to integrate offensive skills into defense programs -national level test and exercise regimes.	(NSR, "Indications and Warnings of Strategic Information Warfare," October 1996).

III. Solutions Relating to the Need for a Permanent "IPTF" Capability

Solutions in this category address whether certain functions are set out in Executive Order 13010 that are not captured in the operational entity discussed in the category above. This category contemplates creation of an ongoing IPTF (Infrastructure Protection Task Force) capability to fill these needs.

Issue Threat and Warning Notices

Solutions relating to a threat and warning capability are closely related to the solutions relating to the creation of an indications and warnings "manager." Solutions falling under this subsection relate specifically to placing that capability under the purview of a permanent IPTF capability. These solutions will include a broad range of ideas that vary according to the amount of private sector and law enforcement involvement.

External Considerations

Create a center for intelligence indications and warnings, current intelligence and threat assessments within NSA.	(Defense Science Board Task Force on IW-D, 1996).
Create a National Information Infrastructure Threat Center with representatives from law enforcement, intelligence and defense communities to perform liaison with the private sector, act as a clearinghouse for intrusion reports, and have "real time" 24-hour operational capabilities.	(U.S. Senate Permanent Subcommittee on Investigations, Staff Statement: Hearings on Security in Cyberspace, 1996). See also, Nickson, Mark, "Vulnerability of US Financial Markets to an Information Warfare Attack," Draft of March, 1997. Software Publishers Association (SPA) also supports creation of a clearinghouse.
Promote creation of an international computer crime bureau with emergency response capability.	(U.S. Senate Permanent Subcommittee on Investigations, Staff Statement: Hearings on Security in Cyberspace, 1996).

Secretary of Energy should establish within the Department of Energy a focal point for energy security matters and make it responsible for collecting relevant intelligence information from cognizant government agencies and providing the petroleum industry with advance warnings of potential dangers.	(Comptroller General Report to Congress, 1979).
Establish and staff a national-level NII operations center within the Executive Office of the President (or National Security Council (NSC), or Federal Emergency Management Agency (FEMA)) to execute national 24/7 operational indications and warnings, detection/assessment, and reaction functions.	(SPB Briefing to the PCCIP, September 23, 1996).
Create new legislation to provide for monitoring critical parts of the NII for National Infrastructure Assurance purposes with built in safeguards to prevent government abuse.	(SPB Briefing to the PCCIP, September 23, 1996).
Share threat information available to the government with the private sector.	(CERT Coordination Center, Report to PCCIP, January 1997).
Support the growth and use of global detection mechanisms by using incident response teams to identify new threats and vulnerabilities.	(CERT Coordination Center, Report to PCCIP, January 1997).
Create an equitable, institutional means, with clear statutory boundaries, for the timely two-way flow of relevant intelligence information and incident data between government and private utilities, which protects business-sensitive data as well as sources and methods. This institution, which would bring government and industry together, would provide early warning of hostile activities.	(OSTP, "Cybernation," draft of January 17, 1997).
Establish threat response teams of technical experts, system operators, and law enforcement officials to respond to requests from infrastructure operators. Teams may be ad hoc, but small permanent organization needs to refine response process, coordinate action and advise system operators.	(Lukasik, Steve; "Public and Private Roles in the Protection of Critical Infrastructure," March 1997).

Provide Training and Education on Reducing Vulnerabilities and Responding to Attacks

This category will capture solutions relating to creation of an ongoing training and education role for a follow-on IPTF capability.

External Considerations

Create an international computer crime bureau to provide education and awareness to foreign law enforcement agencies in order to promote the creation of dedicated computer crime units as well as uniform investigative and computer forensic practices.	(U.S. Senate Permanent Subcommittee on Investigations, Staff Statement: Hearings on Security in Cyberspace, 1996).
Promote regular vulnerability assessments, or "red teaming," of government agencies, especially those outside of the Department of Defense, to operate in a manner similar to that of DISA in assessing the armed forces.	(U.S. Senate Permanent Subcommittee on Investigations, Staff Statement: Hearings on Security in Cyberspace, 1996).
Federal government should work with public and private infrastructure operators in planning and executing simulations, tests, and exercises to help identify vulnerabilities, attack signatures and defensive concepts. This should be jointly funded.	(Lukasik, Steve; "Public and Private Roles in the Protection of Critical Infrastructure," March 1997).

Provide/Facilitate Provision of Expert Guidance to a Halt or Confine Attack and Restore Service

External Considerations

Create a center for information warfare defense operations within DISA, to have links with any successor of the IPTF.	(Defense Science Board Task Force on IW-D, 1996).
Establish a national coordinating and response center, and individual response centers for each of the infrastructure sectors, to facilitate consequence management and capability restoration.	(CSIS Information Assurance Working Group, Commissioner Greene memo dated October 2, 1996).

IV. Solutions Relating to Enhancement of Protection, Mitigation, Recovery and Emergency Response Capabilities, Including Development and Refinement of Coordinated Emergency Response Plans

Federal

Solutions in this category propose means of enhancing the nation's emergency response mechanisms including protection, mitigation and recovery efforts. Solutions could include changes to the Federal Response Plan such as changing definitions to address cyber issues, greater use of mitigation tactics, and appropriate funding. All solutions will need to adequately consider the need for prioritization in the event of an emergency, as well as clear delineation of responsibilities among interested parties.

External Considerations

Expand the definition of disaster to include information warfare and expand recovery mechanisms to prepare for and recover from information disasters.	(Logan, Michael; 5th International Conference on Information Warfare, 1996).
Need for clearer policy assigning responsibilities for coordinating emergency actions to deal with threats to water quality.	(U.S. Congress, "Water Quality: A Catalog of Related Federal Programs.").
Recommendations regarding development of emergency plans for telecommunications: -establish a Nationwide Emergency Telecommunications Service -deploy NCS and FCC priority service through industry for use in declared emergencies -add route diversity and node diversity -increase radio access capabilities through either terrestrial or satellite radio transmission -use simulated disaster and recovery scenarios to develop strategies for network use during emergencies -establish software security measures to protect public network from penetration by hostile users -exploit value-added networks -promote inter-network gateways.	(National Research Council, "Growing Vulnerability of the Public Switched Networks," 1989).
Assure adequate emergency response capability on the National Information Infrastructure.	(Information Infrastructure Task Force Security Issues Forum, "NII Security: The Federal Role," 1996).
Improve national security/emergency preparedness capabilities.	(Information Infrastructure Task Force Security Issues Forum, "NII Security: The Federal Role," 1996).
Flexible, automated, prioritized responses to disruption should be implemented: -design the Defense Information Infrastructure for automated detection, differentiation, warning, response, and recovery from disruptions -design data centers, network components, and network control centers for ease of repair.	(DISA, "Planning Considerations for Defensive Information Warfare-Information Assurance," 1993).
Increase robustness of U.S. infrastructure systems.	(RAND, "Risks to the U.S. Infrastructure from Cyberspace," 1996).
Army should review legal constraints on military participation in civil disaster relief.	(Schrader, K.J.; "The Army's Role in Domestic Disaster Support: An Assessment of Policy Choices," 1993).
Army should support formal acceptance of civil disaster response as a mission for both active and reserve forces.	(Schrader, K.J.; "The Army's Role in Domestic Disaster Support: An Assessment of Policy Choices," 1993).

The Secretary of Energy should establish contingency plans for minimizing supply shortages which could result from prolonged disruptions in the flow of petroleum through important pipeline systems.	(Comptroller General Report to Congress, 1979).
Create a stockpile of transformers beyond the level of normal spares or require utilities to back up each important transformer.	(U.S. Congress, Office of Technology Assessment, "Physical Vulnerability of Electric Systems to Natural Disasters and Sabotage," 1990).
Establish and advertise national policies for protection of vital infrastructures and promulgate strategies for deterring structured attacks.	(CSIS Information Assurance Working Group, Commissioner Greene memo dated October 2, 1996).
Assess current federal emergency preparedness for inter-jurisdictional gaps and overlaps; adequacy to deal with network risks and organization of FEMA through creation of a Congressional Joint Committee.	(CSIS, "America's Hidden Vulnerabilities: Crisis Management in a Society of Networks," 1984).
Congress should fund an inter-industry Emergency Preparedness Council of industry and academic representatives to provide an industry-government interface and serve as a pilot instrument for private sector cooperation in emergency planning and the setting of standards.	(CSIS, "America's Hidden Vulnerabilities: Crisis Management in a Society of Networks," 1984).
Create a set of emergency restart sites for re-establishing networks after an Internet failure.	(Hughes Electronics Corporation, Report to PCCIP, March 1997).

State

Solutions relating to state enhancement of protection, mitigation, recovery and response to emergency situations could cover a broad range of activities based on availability of resources and state cooperation. A model disaster recovery statute is one solution with a modest amount of federal resources and involvement required; however additional solutions, such as creating incentives for enhancing existing state plans, may also be considered.

V. Solutions Relating to the Identification, Creation and Maintenance of A "Minimal Essential Infrastructure," Including Critical Node Identification

External Considerations

Identify the elements and processes of national information, both public and privately owned which collectively comprise the "vital national information interests." Such interests should include both physical infrastructure components and virtual processes, capabilities, and some forms of information.	(CSIS Information Assurance Working Group, Commissioner Greene memo dated October 2, 1996).
Create a minimum essential information infrastructure for use in restoring services and adapting to wide-scale outages.	(Defense Science Board Task Force on IW-D, 1996).
Make investments to protect key electrical system facilities, particularly substations. Base levels of protection on importance, physical characteristics, location and the nature of the threat.	(U.S. Congress, Office of Technology Assessment, "Physical Vulnerability of Electric Systems to Natural Disasters and Sabotage," 1990).
Establish and advertise national policies for protection of vital infrastructures and promulgate strategies for deterring structured attacks.	(CSIS Information Assurance Working Group, Commissioner Greene memo dated October 2, 1996).
Revisit FEMA's proposed National Asset Program.	(U.S. Senate Subcommittee on Governmental Affairs, "Draft Committee Report on Network Vulnerabilities to Terrorist Attack, 1989).
Consider advisability and feasibility of creating a minimum essential information infrastructure.	(RAND, "Risks to the U.S. Infrastructure from Cyberspace," 1996).

VI. Solutions Relating To Creation of a Centralized Data Collection and Analysis Capability

Government (Federal)

External Considerations

Create an information clearing house and analytical activity with authority to task all Executive Branch elements to provide it with information; to accept information from other government agencies and to develop retrievable system of records; to deal with and protect information at all classification levels (to receive proprietary information from private sector without being subject to FOIA); to create a staff of analytical and substantive experts from all elements of government on a reimbursable basis; to disseminate its analytical products.	(Keyes, David; "Stove-Pipe City: It's Not a New Computer Game, It's Why Infrastructure Assurance Needs Centralized Analysis," October 17, 1996).
Create an interim clearing house capability using either the IPTF or the Joint Intelligence Community Law Enforcement process.	(Keyes, David; "Stove-Pipe City: It's Not a New Computer Game, It's Why Infrastructure Assurance Needs Centralized Analysis," October 17, 1996).
Establish a method to facilitate the flow of intrusion and threat data between and among government agencies, especially law enforcement and intelligence communities, and the private sector. Mechanism should include a process to provide effective threat warnings and pattern analysis to aid in the protection and assurance of information and information-based control systems.	(CSIS Information Assurance Working Group, Commissioner Greene memo dated October 2, 1996).
Congress should support the interface of selected civil and military agency resource databases and crisis control systems with White House systems for emergency coordination.	(CSIS, "America's Hidden Vulnerabilities: Crisis Management in a Society of Networks," 1984).
Refocus National Computer Emergency Response Teams and other incident collection centers to collect data not now collected on what information resources or services supporting critical infrastructures were impacted or affected. Need to guarantee absolute confidentiality to facilitate reporting of computer intrusions from the private sector.	(CSIS Information Assurance Working Group, Commissioner Greene memo dated October 2, 1996).

Designate a single, independent, trusted organization to be responsible for collecting, analyzing, and reporting incident data. Organization should be neither government nor commercial and should not have other responsibilities such as public policy, investigation, or enforcement.	(CERT Coordination Center, Report to PCCIP, January 1997).
Create an equitable, institutional means, with clear statutory boundaries, for the timely two-way flow of relevant intelligence information and incident data between government and private utilities, which protects business-sensitive data as well as sources and methods.	(OSTP, "Cybernation," draft of January 17, 1997).
Encourage private industry to create an anonymous clearinghouse to allow businesses to report attacks while maintaining privacy.	(Nickson, Mark; "Vulnerability of US Financial Markets to an Information Warfare Attack," Draft of March 1997).
Establish a point within the Federal government for system owners and operators to report threats to infrastructure. May be parallel structure in private industry. Reporting of threats to federal systems needs to be mandatory.	(Lukasik, Steve; "Public and Private Roles in the Protection of Critical Infrastructure," March 1997).

VII. Solutions Relating To The Federal Government's "Model" Performance (i.e., Influencing Private Sector Behavior through Unilateral Efforts At Information Sharing, Standardization and Procurement)

Solutions in this category suggest means by which the Federal government can promote infrastructure assurance objectives through unilateral changes to its own practices. Solutions relate to both role of the government as model user and specific security measures that should be adopted in that role.

External Considerations

Government has the responsibility to act as a model user by protecting information in its possession and its own security requirements through good management processes.	(Information Infrastructure Task Force Security Issues Forum, "NII Security: The Federal Role," 1996.)
The government should lead by example in achieving a robust and secure information environment by demonstrating a will to purchase and implement products and practices that promote system security.	(CSIS Information Assurance Working Group, Commissioner Greene memo dated October 2, 1996).

DoD should design a curriculum for the study of cyber security including system design, detection of intrusion, intrusion response, firewall design and implementation, and so on. Curriculum could form a new category of engineering in the military academics of an Associates Degree from the Industrial College of the Armed Forces.	(Commissioner Keyes memo, November 6, 1996).

VIII. Solutions Relating to Existing Federal Department and Agency Jurisdiction and Authority

This category collects solutions suggesting that clearer boundary lines, or new boundary lines, be drawn to delineate jurisdiction and responsibilities between existing federal government entities, including the law enforcement, intelligence, and defense communities.

External Considerations

Roles and missions among those responsible for security must be clarified and coordinated.	(RAND, "Risks to the U.S. Infrastructure from Cyberspace," 1996).
Congressional oversight is needed to ensure top-management involvement in security, refocusing federal policy activity on technology neutral policy, and determining where federal authority for safeguarding unclassified information in the civil agencies should reside.	(U.S. Congress, Office of Technology Assessment, "Issue Update on Information Security and Privacy in Network Environments," 1995).
Explore the manner in which the responsibilities levied by E.O. 12656 will be implemented since federal agencies do not have the authority to require cooperation of the owner of a private asset designated as a key facility.	(U.S. Subcommittee on Governmental Affairs, Draft Committee Report on Network Vulnerabilities to Terrorist Attack, 1989).
Clarification of the missions, responsibilities, and authorities of national defense, intelligence, and law enforcement are needed for dealing with cyberspace threats. This will involve all three branches of government.	(OSTP, "Cybernation," Draft of January 17, 1997).
Federal government should address its own complex organization and settle jurisdictional issues for responding to infrastructure threats.	(Lukasik, Steve; "Public and Private Roles in the Protection of Critical Infrastructure," March 1997).

Law Enforcement Community

External Considerations

Examine the proper role of law enforcement in policing the online world.	(German, Jerry; Testimony on the right to free speech, free association, and privacy on the Internet in the aftermath of Oklahoma City, Subcommittee on Terrorism, Technology, and Government Information, 1995).
Maintain cautious interpretation of Attorney General's guidelines for online investigations.	(German, Jerry; Testimony on the right to free speech, free association, and privacy on the Internet in the aftermath of Oklahoma City, Subcommittee on Terrorism, Technology, and Government Information, 1995).
Treat proposals to expand surveillance authority with great caution so that protection efforts do not infringe upon civil liberties.	(German, Jerry; Testimony on the right to free speech, free association, and privacy on the Internet in the aftermath of Oklahoma City, Subcommittee on Terrorism, Technology, and Government Information, 1995).

IX. Solutions Relating to Administrative and Regulatory Requirements for Government

Solutions in this category will address whether government-wide administrative and regulatory requirements, such as mandatory reporting of significant intrusion incidents, would contribute substantially toward achieving an effective indications and warnings system, or whether an expanded program of voluntary compliance would provide sufficient data. Solutions will also address the appropriate vehicles for promulgation and enforcement of such administrative or regulatory requirements including whether agencies should promulgate and enforce their own regulations to comport with government standards, or whether such regulations should be promulgated and enforced by either of the entities proposed above. Solutions could take the form of recommendations regarding mandatory encryption of Supervisory Control And Data Acquisition system (SCADA), information, use of certified intrusion detection systems, or specified reserve capacities or redundancy margins.

External Considerations

Create Government Computer Security Specialist and Computer Systems Administrator Career Fields with potential for career progression and incorporate specialized computer security training.	(U.S. Senate Permanent Subcommittee on Investigations, Staff Statement: Hearings on Security in Cyberspace, 1996). Concur. (Defense Science Board Task Force on IW-D, 1996).

Mandate the reporting of intrusions and attempted intrusions in government and government interest systems. Federal agencies should develop protocols and procedures for reporting computer intrusions and for referral to appropriate criminal and other agencies.,	(U.S. Senate Permanent Subcommittee on Investigations, Staff Statement: Hearings on Security in Cyberspace, 1996).
Make logon warning banners mandatory for all government and government interest systems.	(U.S. Senate Permanent Subcommittee on Investigations, Staff Statement: Hearings on Security in Cyberspace, 1996).
Protect system from information warfare (IW) attacks using firewalls, enhanced domain compartmentalization, controlled transfer between domains; with monitoring and probing tools, deploying network administration/management, real-time monitoring, and reactive capability; and embedded encryption.	(Naval Research Advisory Council (NRAC) Study, 1996).
Address the vulnerabilities of military use of commercial SATCOM systems that provide geolocation and develop technology and operating procedures to minimize the impact of denial of service and avoid geolocation.	(Naval Research Advisory Council (NRAC) Study, 1996).
Make IW-D a higher priority by including it as part of exercises, doing operational analysis, and designating IW-D as an acquisition reform model to reduce time to achieve operational capability.	(Naval Research Advisory Council (NRAC) Study, 1996).
Create a capability for detection and early warning of cyber attacks on systems. Consider using a screening process similar to that used for airline passenger for the cyber arena; a screening process to filter out malicious code as it passes through telephone switches; adapting current tools for indications and warnings purposes.	(Commissioner Keyes memo, December 16, 1996).
Legislate for mandatory reporting of Information Warfare attacks in the public sector.	(Nickson, Mark; "Vulnerability of US Financial Markets to an Information Warfare Attack," Draft of March 1997).

Federal (Defense Community)

External Considerations

Create a Government Computer Crime Investigators Career Field that includes potential for career progression and specialized computer crime training.	(U.S. Senate Permanent Subcommittee on Investigations, Staff Statement: Hearings Security in Cyberspace, 1996).

Consider the proper role of the federal government in guaranteeing the reliability of interstate energy distribution and telecommunications networks, to the extent it is beyond the capability of individual states.	(U.S. Senate Subcommittee on Governmental Affairs, Draft Committee Report on Network Vulnerabilities to Terrorist Attack, 1989).

X. Solutions Relating to Joint Public-Private and Government-Assisted Research, Development and Technology

Research and development solutions in this category take three basic forms. Solutions may suggest topics in need of research and development; forms of government, government-private cooperative, or purely private research and development initiatives; or incentives or mechanisms to encourage research and development within the recommended areas. Options may include the funding and establishment of a government subsidized industrial institute; Congressionally approved Cooperative Research and Development Authorities (CRADAs); or expansion of the existing capabilities of the National Labs to provide partially subsidized research services for qualified assurance-related projects.

Areas of Study

External Considerations

Fund and pursue needed research on computer and communications security.	(National Research Council, "Computers at Risk," 1991).
Federal Government should promote research and development in critical and high-risk areas.	(Information Infrastructure Task Force Security Issues Forum, "NII Security: The Federal Role," 1996.)
Real-time control mechanisms to enhance information assurance should be developed.	(DISA, "Planning Considerations for Defensive Information Warfare-Information Assurance," 1993).
Focus research and development on robust survivable system architectures; techniques and tools for modeling, monitoring, and management of large-scale distributed/network systems; tools and techniques for automated detection and analysis of localized or coordinated large-scale attacks; and tools and techniques for automated detection and analysis of localized or coordinated large scale attacks.	(Defense Science Board Task Force on IW-D, 1996).
Set a national agenda and prioritize research and development in information technology security, computer security and computer investigative fields to safeguard national infrastructures.	(CSIS Information Assurance Working Group, Commissioner Greene memo dated October 2, 1996).

Government should create a capability for detection and early warning of cyber attacks on systems. Consider using a screening process similar to that used for airline passengers in the cyber arena; a screening process to filter out malicious code as it passes through telephone switches; adapting current tools for indications and warnings purposes.	(Commissioner Keyes memo, December 16, 1996).
Create testing programs to enhance information assurance.	(DISA, "Planning Considerations for Defensive Information Warfare-Information Assurance," 1993).
Government should fund a demonstration project to examine the costs, benefits and pitfalls of attempting to use encryption and authentication to prevent unauthorized access to SCADA control systems.	(Commissioner Keyes memo, December 16, 1996).
Fund research and development in areas of security and survivability of unbounded systems' architectures with distributed control.	(CERT Coordination Center, Report to PCCIP, January 1997).
Encourage the development of comprehensive system/security administrators' toolkits.	(CERT Coordination Center, Report to PCCIP, January 1997).
Support the development of techniques for comprehensive, continuous risk identification and mitigation programs.	(CERT Coordination Center, Report to PCCIP, January 1997).
Develop and strengthen tools and procedures for detecting, reporting and reacting to network problems.	(OSTP, "Cybernation," Draft of January 17, 1997).
Construct profiles of attack sources to develop a better understanding of attacker motivations and goals in order to develop better search parameters for them.	(Lukasik, Steve; "Public and Private Roles in the Protection of Critical Infrastructure," March 1997).
Develop technology to identify and collect information about a potential attacker while they are probing the system.	(Lukasik, Steve; "Public and Private Roles in the Protection of Critical Infrastructure," March 1997).
Public infrastructure operators should take the responsibility for developing protection for those systems and to stimulate the transfer of that technology to the private sector.	(Lukasik, Steve; "Public and Private Roles in the Protection of Critical Infrastructure," March 1997).
Government infrastructure operators should test commercial audit system products and publish the results. Government would act as a test bed for new technology. A joint industry-government working group could be established to set requirements for system audits and provide information on the state of the technology.	(Lukasik, Steve; "Public and Private Roles in the Protection of Critical Infrastructure," March 1997).

Bodies to Engage in Study (Government)

External Considerations

Federal government must support the needs of industry by supporting research and development.	(NSTAC, "An Assessment of the Risk to the Security of the Public Network," 1995).
Government should closely coordinate research and development within the government to avoid duplication of efforts and ensure that all critical areas are being addressed.	(Commissioner Keyes memo, December 16, 1996).

Bodies to Engage in Study (Joint Government-Private)

External Considerations

Fund national laboratories and government agencies and encourage academic and commercial participation.	(CSIS Information Assurance Working Group, Commissioner Greene memo dated October 2, 1996).
Work with National Science Foundation to develop research in U.S. computer science and computer engineering programs and educational programs for curriculum development at the undergraduate and graduate levels in resilient system design practices.	(Defense Science Board Task Force on IW-D, 1996).
Government should support joint R&D with infrastructure operators and industry for information infrastructure defense.	(Lukasik, Steve; "Public and Private Roles in the Protection of Critical Infrastructure," March 1997).

Bodies to Engage in Study (Private)

External Considerations

Industry should significantly increase its involvement with the university community and support academic research with grants, joint R&D projects, equipment, and fellowships.	(Center for Strategic and International Studies, "R&D for National Strength," 1982).
Industry should utilize and encourage academic and governmental technology-transfer programs.	(Center for Strategic and International Studies, "R&D for National Strength," 1982).
Academe should provide support and encouragement for those who work in military R&D and inculcate attitudes of risk-taking, innovation, and creativity among engineering and science students.	(Center for Strategic and International Studies, "R&D for National Strength," 1982).
Academe should increase part-time academic positions for industrial, governmental and military R&D personnel.	(Center for Strategic and International Studies, "R&D for National Strength," 1982).

Incentives

External Considerations

Congress assess whether amendment or re-interpretation of anti-trust legislation would encourage the organization of private sector industrial combinations and consortia strong enough to address major research and develop issues with their own resources.	(Center for Strategic and International Studies, "R&D for National Strength," 1982).
Congress should use educational subsidies to encourage development of disciplines and research that contribute to military needs.	(Center for Strategic and International Studies, "R&D for National Strength," 1982).
Congress should create new tax incentives to encourage major stimulation of the U.S. R&D industrial base and long-term investment in science and technology.	(Center for Strategic and International Studies, "R&D for National Strength," 1982).
Terms of procurement contracts should be changed to allow commercial market rights to government contractors as an incentive to private developers and to facilitate technology transfer to the private sector.	(Lukasik, Steve; "Public and Private Roles in the Protection of Critical Infrastructure," March 1997).

XI. Solutions Relating To Certifications, Standards and Guidelines

Solutions under this category will relate to the broad range of options for government or private sector action in setting standards or creating certifications for products or services that enhance assurance or assurance-related capabilities. Factors to be considered in formulating individual solutions should be the use of existing standard setting vehicles in the federal government such as NIST or NSA, interaction with other areas of concern for the Commission such as insurance and liability issues, and funding issues, and the need for monitoring and sanctions to ensure continued compliance with standards.

External Considerations

Create a Government Computer Security Specialist and Computer Systems Administrator Career Fields with potential for career progression and incorporate specialized computer security training.	(U.S. Senate Permanent Subcommittee on investigations, Staff Statement: Hearings on Security in Cyberspace, 1996).
Promulgate comprehensive generally accepted system security principles, as well as methods, guidelines, and facilities for evaluating products for conformance.	(National Research Council, "Computers at Risk," 1991).

Manufacturers must ensure that the security capabilities of their goods and services adequately reflect the needs of the marketplace.	(NSTAC, "An Assessment of the Risk to the Security of the Public Network," 1995).
Users must subscribe to and pay for appropriate levels of privacy and security.	(NSTAC, "An Assessment of the Risk to the Security of the Public Network," 1995).
Government has the responsibility to protect information in its possession and protect its own security requirements through good management processes.	(Information Infrastructure Task Force Security Issues Forum, "NII Security: The Federal Role," 1996).
Telecommunications carriers should review their implementation of best practices to insure that alarms are in place and maintained, and that an effective alarm strategy is in place to promptly identify and address power problems.	(Alliance for Telecommunications Industry Solutions-Network Reliability Steering Committee (ATIS-NRSC), "Analysis of Power Related Network Outages," 1996).
Information assurance standards, technologies, tools, and guidelines should be developed.	(DISA, "Planning Considerations for Defensive Information Warfare-Information Assurance," 1993).
Design products flexible enough to serve a broad spectrum of security needs at the operating system level, the application level, the organizational level, and the site level.	(NIST, "Assessing Federal and Commercial Information Security Needs," 1992).
Vendors should consider new mechanisms that directly address discretionary and nondiscretionary controls, such as role-based access controls, separation of duties, separation of transactions, and user-oriented least privilege.	(NIST, "Assessing Federal and Commercial Information Security Needs," 1992).
Secretary of Energy should establish minimum security standards for critical pipelines and related facilities.	(Comptroller General Report to Congress, 1979).
Establish an industry-wide standard for reliability reporting, analogous to the financial accounting and reporting guidance of the Financial Accounting Standards Board, for infrastructure operators. Board would help put discipline into the process of capturing and applying lessons learned.	(OSTP, "Cybernation," Draft of January 17, 1997).
Establish legitimized, generally accepted, and broadly applicable standards for network design and installation for enhanced reliability. The National Fire Protection Agency's National Electrical Code can be used as a model.	(OSTP, "Cybernation," Draft of January 17, 1997).

Develop institutionalized product reliability testing and certifications, similar to Underwriters Laboratories, for network components. Such certifications or testing would be of "immense value" for comparing products. Certifications or standards would ideally be created by a joint government-private sector effort.	(OSTP, "Cybernation," Draft of January 17, 1997).
Federal government could license and certify routes for the Internet and collateral attributes such as topography and survivability.	(Hughes Electronics Corporation, Report to PCCIP, March 1997).
Federal government license and certify ISPs and related service providers based on quality, performance and security.	(Hughes Electronics Corporation, Report to PCCIP, March 1997).
All networks should authenticate routing.	(Hughes Electronics Corporation, Report to PCCIP, March 1997).
Private sector should develop voluntary, industry-wide, accepted best practices for information security. Government should support industry in this effort as requested, but especially in gaining support in the international arena for such practices.	(Copeland, Guy; "Information Assurance Insurance," February 1997).
Government set standards need to be closely aligned with commercial and consumer interests to be effective.	(Neumann, Peter; "Security Risks in the Emerging Infrastructure," Senate Testimony June 1996).
Regulators of infrastructure systems should take primary responsibility for encouraging their industries to take cooperative action through existing administrative procedures for the setting of standards.	(Lukasik, Steve; "Public and Private Roles in the Protection of Critical Infrastructure," March 1997).
Government infrastructure operators should test commercial audit system products and publish the results. Government would act a test bed for new technology. A joint industry-government working group could be established to set requirements for system audits and provide information on the state of the technology.	(Lukasik, Steve; "Public and Private Roles in the Protection of Critical Infrastructure," March 1997).

XII. Solutions Relating to Influencing Existing Liability Climates

External Considerations

Determine priorities in a natural emergency and relieve lending utilities of liability for power outages in their own areas.	(U.S. Congress, Office of Technology Assessment, "Physical Vulnerability of Electric Systems to Natural Disasters and Sabotage," 1990).

Convene a panel with representatives from insurance, information technology, and government sectors to examine and propose a plan for government underwritten insurance for information driven losses. Long term goal is to start up this area of insurance and to provide economic motivation for enhancing protection against cyber attacks.	(Copeland, Guy; "Information Assurance Insurance," February 1997).
Quality of Service agreements, agreements between infrastructure service providers or state-federal regulation should be used to encourage reallocation of system load.	(Lukasik, Steve; "Public and Private Roles in the Protection of Critical Infrastructure," March 1997).
Government should act to require minimum levels of coverage (as was done with auto insurance) for high risk pools while being careful to maintain the financial health of insurers.	(Lukasik, Steve; "Public and Private Roles in the Protection of Critical Infrastructure," March 1997).

XIII. Solutions Relating to Administrative and Regulatory Requirements for the Private Sector

Solutions under this category will generally suggest appropriate requirements and vehicles for encouraging or requiring the private sector to act for greater infrastructure assurance. Included within this category will be solutions that suggest whether industry-wide administrative and regulatory requirements, such as mandatory reporting of significant intrusion incidents, would be necessary to maintain an effective indications and warnings system, or whether expanded voluntary compliance, or a system of incentive-based compliance, would provide adequate information when taken in conjunction with information that would be reported by government entities.

External Considerations

Encourage private industry and the private sector to report intrusions into private information systems through anonymous clearinghouses and similar methods.	(U.S. Senate Permanent Subcommittee on Investigations, Staff Statement: Hearings on Security in Cyberspace, 1996).
Establish independent system operators for the power grid, with access to utility sensitive network information to maintain reliability.	(Coy, Peter; "Who's Watching the Power Grid," Business Week, 1996).
Make investments in improving the technology of firewalls, testers and evaluation tools. Use simulation and training systems for operators which cover a wide range of natural outages as well as malicious attacks.	(DARPA, "Electronic Power Distribution Case Study, Defensive Information Warfare Study," 1995).

Ensure that utilities establish contact with the FBI and coordinate utility emergency plans.	(U.S. Congress, Office of Technology Assessment, "Physical Vulnerability of Electric Systems to Natural Disasters and Sabotage," 1990).
Strengthen end system robustness in the financial services industry with multi-function smart cards with cryptographic services, selective cryptographic functions, real-time fault tolerant operating systems, monitor for "anticipatory" system behavior, and use computational service "user agreements."	(DARPA, Defensive Information Warfare Summer Study Final Report, 1995).
Strengthen network robustness in the financial services industry by using high bandwidth circuits, use authenticated, encrypted virtual circuits, pursue development of bandwidth reservation, use network "user agreements," employ selective partitioning and routing, and employ a realm-based certification authority structure.	(DARPA, Defensive Information Warfare Summer Study Final Report, 1995).
Manufacturers must ensure that the security capabilities of their goods and services adequately reflect the needs of the market place. Service providers must ensure reliability and assurance of their network services.	(NSTAC, "An Assessment of the Risk to the Security of the Public Network," 1995).
Users must subscribe to and pay for appropriate levels of privacy and security.	(NSTAC, "An Assessment of the Risk to the Security of the Public Network," 1995).
Congress promote the convergence of government and private sector interest in cryptography through hearings, evaluation of the Administration's market study, and encouraging timely, open, and productive dialog.	(U.S. Congress, Office of Technology Assessment, "Issue Update on Information Security and Privacy in Network Environments," 1995).
Encryption and authentication should be applied to all communications sessions.	(DARPA, "Electronic Distribution Case Study, Defense Information Warfare Study," 1995).
Encourage Internet Service providers to develop security incident response teams and other security improvement services for their customers.	(CERT Coordination Center, Report to PCCIP, January 1997).
Government should facilitate the development and use of security mechanisms for information in cyberspace that do not undermine commerce or intrude on basic freedoms.	(CERT Coordination Center, Report to PCCIP, January 1997).
Develop, implement, and follow internal reporting procedures within companies that are infrastructure operators.	(OSTP, "Cybernation," Draft of January 17, 1997).

Develop means for private sector companies with a stake in infrastructure operations to share (within legal limits) information on vulnerabilities, incident data, technical solutions, and best practices, that will adequately protect proprietary or business-sensitive information. Use Network Security Information Exchange (NSIE), NSTAC, within telecommunications industry as a model.	(OSTP, "Cybernation," Draft of January 17, 1997).
Financial Institutions should implement a robust computer and information security program to include: -vulnerability and risk assessment analysis -separation of duties among software -development, test and installation personnel -stringent access controls to software and equipment -audit trails -written policies -written contingency plans -CERT capabilities -robust training programs -full-time security positions.	(Nickson, Mark; "Vulnerability of US Financial Markets to an Information Warfare Attack," Draft of March 1997).
"The Internet must build in its own protection mechanisms to ensure its survivability."	(Hughes Electronics Corporation, Report to PCCIP, March 1997).
"All telecommunications providers should initiate efforts to offer Quality of Service (QOS) commitments, as a contractual guarantee."	(Hughes Electronics Corporation, Report to PCCIP, March 1997).
Telecommunications providers should take measures to contain damage to the local level, making sure architectures are not susceptible to the domino effect.	(Hughes Electronics Corporation, Report to PCCIP, March 1997).
Create a "national incentive program to encourage geographic diversity of network routes," and robustness.	(Hughes Electronics Corporation, Report to PCCIP, March 1997).
Increase network survivability by: -creating a rich path fabric with automatic alternate routing -using geographic dispersal and redundant deployment of critical network facilities -using physical hardening of and water protection for critical network facilities -using distributed control -creating redundant database with automatic switchover -developing well-designed interconnection standards for use by multiple interfacing -rigorous interoperability testing prior to deployment of service.	(IDC/LINK, "The U.S. Electronic Distribution Infrastructure: Size, Ownership, Geography, and Vulnerabilities," March 1997).

Satellite, cable, PCS, and wireless telephone network providers should address the network reliability risk by putting in place the following: -firewalls -safeguards to prevent the exportation of problems to other networks -diversity/redundancy to avoid failures in interconnections to the PSTN/ cellular/PCS/cable telephone networks -practices to reduce likelihood of AIN related failures -improved standards-based network management systems -cooperation/participation policies in design and development of multi-network interconnection standards -compliance polices and practices for implementing multi-network interconnection interoperability standards and related conformance testing -provision for hardening and physically dispersing interconnection facilities.	(IDC/LINK, "The U.S. Electronic Distribution Infrastructure: Size, Ownership, Geography, and Vulnerabilities," March 1997).
Regulatory agencies should mandate the use of audit systems and establish standards for assuring the integrity of the infrastructure systems under their jurisdiction. Use of public input should be used to balance interests of shareholders, ratepayers, and national leaders.	(Lukasik, Steve; "Public and Private Roles in the Protection of Critical Infrastructure," March 1997).

XIV. Solutions Relating To Public-Private Investment Incentives

This category will contain solutions for the implementation of a public-private investment plan (or plans) with incentives to infrastructure owners to cause their voluntary investment in enhanced assurance measures. These measures might include, for example, providing tax credits or changing depreciation schedules for certain percentages of investment in infrastructure assurance. Options for implementation may include the determination of whether a particular protective measure qualifies for receipt of an incentive being tied to the government certifications and standards described previously.

External Considerations

Make Federal security products and techniques available for use in the NII.	(Information Infrastructure Task Force Security Issues Forum, "NII Security: The Federal Role," 1996).
Promote private sector development of high quality security products and services.	(Information Infrastructure Task Force Security Issues Forum, "NII Security: The Federal Role," 1996).

Consider incentives for industry to develop appropriate security features.	(NSTAC Information Assurance Task Force, "Electric Power Information Assurance Risk Assessment," 1996).
Draw on expertise of insurance industry to develop models and guidelines to better balance the risk equations for infrastructure systems by reflecting the differences in information protection practices.	(CSIS Information Assurance Working Group, Commissioner Greene memo dated October 2, 1996).
Encourage academic research into the economic value of information assurance and security to business, by focusing on identifying the value, attributes of information, and models to manipulate information value attributes.	(CSIS Information Assurance Working Group, Commissioner Greene memo dated October 2, 1996).
Use incentives, such as tax credits, and government indemnification to improve private sector communication with government.	(SPB Briefing to the Commission, September 23, 1996).

XV. Solutions Relating to the Development of New Risk Management Models

External Considerations

Emphasize individual, commercial, and economic needs in public policy, as well as government and military needs.	(CERT Coordination Center, Report to PCCIP, January 1997).

XVI. Solutions Relating to Public Awareness and Education

Awareness and education solutions might address not only the need for greater awareness of infrastructure vulnerabilities, but also the specific content of such education, the proper entity to conduct such efforts, as well as suggesting proper audiences and methodologies. The effectiveness of advocating or sponsoring a public awareness campaign aimed at increasing awareness of effective assurance practices should be considered. It has been suggested that a more limited, targeted awareness campaign directed at corporate security officers and personnel might be more appropriate to the subject matter at hand.

External Considerations

Stimulate dialogue and awareness of security risks, needs, and solutions.	(Information Infrastructure Task Force Security Issues Forum, "NII Security: The Federal Role," 1996).

Create an international computer crime bureau to provide education and awareness to foreign law enforcement agencies in order to promote the creation of dedicated computer crime units as well as uniform investigative and computer forensic practices.	(U.S. Senate Permanent Subcommittee on Investigations, Staff Statement: Hearings on Security in Cyberspace, 1996).
Assign an appropriate agency to develop and conduct an awareness program to increase awareness of threats, vulnerabilities and solutions within the electric power industry.	(NSTAC Information Assurance Task Force, "Electric Power Information Assurance Risk Assessment," 1996).
Conduct an awareness campaign to ensure senior-level government and industry representatives are aware of the vulnerabilities and appreciate the implications of Information Warfare. Consider large scale demonstrations, simulations, and experiments, as well as expanding outreach to the public, industry, Commanders-in-Chief (CINCs), services end Agencies.	(Defense Science Board Task Force on IW-D, 1996).
Charter an awareness plan to be promoted by the White House to stimulate increased awareness re-enforcing industry, academic, and government sector efforts. The plan should stress the interests of individual computer users and organizations.	(CSIS Information Assurance Working Group, Commissioner Greene memo dated October 2, 1996).
Create consumer protection scheme to alert consumers to security risks and protection controls available in computer and network products and services.	(CSIS Information Assurance Working Group, Commissioner Greene memo dated October 2, 1996).
Promote academic degree programs in information assurance.	(CSIS Information Assurance Working Group, Commissioner Greene memo dated October 2, 1996).
The Federal government should promote awareness among the private sector of potential vulnerabilities and costs to them of such incidents, as well as the advantage of taking precautions.	(Volpe Center, "Emerging Areas in Transportation Information Infrastructure Security," 1996).
Establish a mechanism for sanitizing and disseminating data on security problems, data that help the network community understand the scope and cost of the overall problem.	(CERT Coordination Center, Report to PCCIP, January 1997).
Government should support the development of educational material and programs about cyberspace for all users, both adults and children.	(CERT Coordination Center, Report to PCCIP, January 1997).
Government should work with industry to create quality user training and comprehensive independent accreditation services.	(Hughes Electronics Corporation, Report to PCCIP, March 1997).

Regulatory proceeding can be used to raise public awareness of infrastructure security threats and requirements through notice of inquiry, notice of proposed rulemaking, or industry advisory panels. Executive branch commissions, task forces and advisory committees or Congressional-hearings can serve a similar function.	(Lukasik, Steve; "Public and Private Roles in the Protection of Critical Infrastructure," March 1997).
Federal government should use its role as national collector and distributor of national statistics to raise awareness of losses from infrastructure attacks.	(Lukasik, Steve; "Public and Private Roles in the Protection of Critical Infrastructure," March 1997).

Federal Role (DoD)

External Considerations

Design a curriculum for the study of cyber security including system design, detection of intrusion, intrusion response, firewall design and implementation, and so on. Curriculum could form a new category of engineering in the military academics of an Associates Degree from the Industrial College of the Armed Forces.	(Commissioner Keyes memo, November 6, 1996).
Stress need for proper security training for system administrators, network managers, and chief information officers with long-term goal of promoting undergraduate and master's level training in network and information security.	(CERT Coordination Center, Report to PCCIP, January 1997).

Private Role

External Considerations

Academe should encourage training in the areas of advanced technology and science, including emphasis on primary and secondary school curricula.	(Center for Strategic and International Studies, "R&D for National Strength," 1982).

Curriculum (Ethics)

External Considerations

Provide early training in security practices and ethics.	(National Research Council, "Computers at Risk," 1991).

Create a grammar school and adult education curricula or class modules to acquaint children and adults (especially parents) with proper ethics and protection pitfalls for interaction in Cyberspace (e.g. property rights, abuse of resources, credit cards, and access codes).	(CSIS Information Assurance Working Group, Commissioner Greene memo dated October 2, 1996).
Government should support programs that provide early training in security practices and appropriate use that is integrated into general education about computing.	(CERT Coordination Center, Report to PCCIP, January 1997).

XVII. Solutions Relating to Professional Training

This category is devoted to solutions pertaining to enhancing training and professional licensing or certifications for those who work most closely with critical infrastructures.

External Considerations

Government should support programs that provide early training in security practices and appropriate use that is integrated into general education about computing.	(CERT Coordination Center, Report to PCCIP, January 1997).
Design a curriculum for the study of cyber security including system design, detection of intrusion, intrusion response, firewall design and implementation, and so on. Curriculum could form a new category of engineering in the military academics of an Associates Degree from the Industrial College of the Armed Forces.	(Commissioner Keyes memo, November 6, 1996).
Stress need for proper security training for system administrators, network managers, and chief information officers with long term goal of promoting undergraduate and master's level training in network and information security.	(CERT Coordination Center, Report to PCCIP, January 1997).
Create a Government Computer Security Specialist and Computer Systems Administrator Career Fields with potential for career progression and incorporate specialized computer security training.	(U.S. Senate Permanent Subcommittee on Investigations, Staff Statement: Hearings on Security in Cyberspace, 1996).
Create a Government Computer Crime Investigators Career Field that includes potential for career progression and specialized computer crime training.	(U.S. Senate Permanent Subcommittee on Investigations, Staff Statement: Hearings on Security in Cyberspace, 1996).
Develop measures to increase the reliability of network operators.	(OSTP, "Cybernation," Draft of January 17, 1997).

Separate the duties of development, test, and installation personnel to minimize success of disgruntled employees and potential effects of innocent errors.	(Nickson, Mark; "Vulnerability of US Financial Markets to an Information Warfare Attack," Draft of March 1997).
Computer system and software professionals should be encouraged to perform in ways similar to those in the engineering fields through substantial enforcement of licensing, accreditation, responsibility, ethical behavior, legal liability, and incentives for risk management.	(Neumann, Peter; "Security Risks in the Emerging Infrastructure," Senate Testimony, June 1996).
Government should provide long-term support to existing centers of excellence in information and network security to stabilize their funding and ensure their continued existence.	(Spafford, Eugene; COAST/Purdue Univ.; Information Security Education).
Private industry should become more involved with network and information security education and research through funding, personnel and sharing of expertise.	(Spafford, Eugene; COAST/Purdue Univ.; Information Security Education).
Establish programs of scholarships or forgiveable loans to students majoring in information security at the graduate level. Make these programs also available to retain appropriately personnel already in the computing profession.	(Spafford, Eugene; COAST/Purdue Univ.; Information Security Education).

XVIII. Solutions Relating to Enhancing International Cooperation and Participation in Assurance Practices

External Considerations

The Federal Government should promote international cooperation for the protection of the National Information Infrastructure.	(Information Infrastructure Task Force Security Issues Forum, "NII Security: The Federal Role," 1996).
Create an international computer crime bureau to provide education and awareness to foreign law enforcement agencies in order to promote the creation of dedicated computer crime units as well as uniform investigative and computer forensic practices.	(U.S. Senate Permanent Subcommittee on Investigations, Staff Statement: Hearings on Security in Cyberspace, 1996).
Industry should take the lead within existing international industry standards groups to establish international recognition and adaptation of effective security technical and management standards.	(CSIS Information Assurance Working Group, Commissioner Greene memo dated October 2, 1996).
Make national policy and operations decisions with the awareness that cyber security issues are international in scope and require international cooperation.	(CERT Coordination Center, Report to PCCIP, January 1997).

Federal government has an obligation to work with other countries to develop compatible cyberspace legal structures and to foster worldwide cooperation among law enforcement agencies.	(OSTP, "Cybernation," Draft of January 17, 1997).
National Security and law enforcement branches of the federal government should facilitate cooperation in the international arena to identify infrastructure threats.	(Lukasik, Steve; "Public and Private Roles in the Protection of Critical Infrastructure," March 1997).
Federal government should attempt to further international cooperation in the identification of infrastructure threats by reviewing laws and considering use of formal defense alliance model, such as NATO.	(Lukasik, Steve; "Public and Private Roles in the Protection of Critical Infrastructure," March 1997).
Government should work toward engineering of international interfaces between national systems. The International Telecommunications Union could provide a valuable model.	(Lukasik, Steve; "Public and Private Roles in the Protection of Critical Infrastructure," March 1997).

XIX. Solutions Relating to Export and Trade Policy

External Considerations

Clarify export criteria and set up a forum for arbitration for computer and computer security technology.	(National Research Council, "Computers at Risk," 1991).
Launch a dedicated program to establish bilateral and multilateral agreements to adopt compatible legal standards for the use and protection of information systems and products as well as appropriate international criminal codes and definitions of jurisdiction.	(CSIS Information Assurance Working Group, Commissioner Greene memo dated October 2, 1996).
U.S. Government should encourage the use of strong encryption technologies in the private sector, especially the financial sector.	(Nickson, Mark; "Vulnerability of US Financial Markets to an Information Warfare Attack," Draft of March 1997).
U.S. Government should eliminate regulatory control of development or use of strong crypto, key escrow and key recovery.	(Hughes Electronics Corporation, Report to PCCIP, March 1997).
U.S. Government must take a strong position relating to the protection of personal and corporate privacy including use of nontrivial individual authentication and encryption.	(Neumann, Peter; "Security Risks in the Emerging Infrastructure," Senate Testimony June 1996).

XX. Solutions Relating to Enhancing Deterrence

Solutions relating to a statement of Administration or Department of Defense policy delineating the degree of proof required to trigger a presumption that a given deleterious act against a critical infrastructure resulted from the concerted activities of a foreign nation (and are thereby akin to an act of war) belong within this category. This category will also include solutions regarding possible statements of U.S. retaliatory intentions, to serve as an effective deterrent and discourage foreign government sponsorship of threatening activities, and enhanced criminal provisions.

International (DoD)

External Considerations

Establish and advertise national policies for protection of vital infrastructures and promulgate strategies for deterring structured attacks.	(CSIS Information Assurance Working Group, Commissioner Greene memo dated October 2, 1996).
Establish a theory of deterrence for large scale infrastructure attacks analogous to a theory of nuclear deterrence.	(Lukasik, Steve; "Public and Private Roles in the Protection of Critical Infrastructure," March 1997).

XXI. Solutions Relating to period of Transition to War

External Considerations

Information workers should begin to train as defensive information warriors.	(DISA, "Planning Considerations for Defensive Information Warfare-Information Assurance," 1993).
Readiness exercises and war games for defensive information warfare should begin.	(DISA, "Planning Considerations for Defensive Information Warfare-Information Assurance," 1993).
Using an anonymous response, the U.S. government could strike at information terrorists without large display or legitimizing the terrorists, both of which occur with a physical response.	(Devost, Houghton, & Pollard, "Information Terrorism: Can You Trust Your Toaster?," 1996).

XXII. Solutions Relating To Government and Civilian Responsibilities in Time of War (Declared War)

(none identified)

XXIII. Solutions Relating to Existing Legislation/Regulations

Solutions based on amending existing legislation or regulations generally fit into this category. The solutions in this category are aimed at providing infrastructure assurance through modifications of existing legislation, rather than the creation of wholly new laws.

External Considerations

The Federal Government must develop laws to enable prosecution of those who attack public networks.	(NSTAC, "An Assessment of the Risk to the Security of the Public Network," 1995).
The Federal Government needs to review criminal law to protect the public interest in the National Information Infrastructure.	(Information Infrastructure Task Force Security Issues Forum, "NII Security: The Federal Role," 1996.)
Maintain cautious interpretation of the Attorney General's guidelines for online investigations.	(German, Jerry; Testimony on the right to free speech, free association, and privacy on the Internet in the aftermath of Oklahoma City, Senate Subcommittee on Terrorism, Technology, and Government Information, 1995).
Redefine the federal role and responsibilities for assuring the safety of intrastate pipelines, including the hazardous liquid pipelines.	(Comptroller General, "Need to Assess Federal Role in Regulating and Enforcing Pipeline Safety," 1984).
Determine whether there are sufficient hazards involving personal injury or environmental damage to warrant regulation of certain gas and liquid pipeline facilities or commodities not presently covered by federal regulations.	(Comptroller General, "Need to Assess Federal Role in Regulating and Enforcing Pipeline Safety," 1984).
Consider legislation to give the Department of Energy clear authority to make on-site visits to pipeline facilities necessary for identifying and analyzing critical pipelines and related facilities; to develop minimum physical security standards and establish penalties for non-compliance and administrative procedures for appeal; to conduct periodic inspections for determining compliance and for reassessing physical security requirements.	(Comptroller General Report to Congress, 1979).

Create mandatory reporting requirements to support a National Coordinating and Response Center and National deterrence policy for information assurance.	(CSIS Information Assurance Working Group, Commissioner Greene memo dated October 2, 1996).
Create "technology-smart" legislation for cyber-related crimes.	(Hughes Electronics Corporation, Report to PCCIP, March 1997).
Allow users to make copies of copyrighted information and licensed software to protect against loss.	(Hughes Electronics Corporation, Report to PCCIP, March 1997).

XXIV. Solutions Relating to Short-Term Assurance Measures (Low-Hanging Fruit)

The solutions in this category have been grouped together because they represent small steps that the Federal government could take, independent of larger endeavors, to shore up the nation's critical infrastructures. Though these are small steps, they represent important first steps for jump-starting infrastructure assurance efforts.

External Considerations

Actions suggested include: -develop security policies -form computer emergency response teams -use sound methodology and modern technology to develop high quality software -security standards and participate actively in their design -use technical aids to foster secure operations.	(National Research Council, "Computers at Risk," 1991).
Adapt current oversight processes to meet the challenges of the NII.	(Information Infrastructure Task Force Security Issues Forum, "NII Security: The Federal Role," 1996.)
Create Government Computer Security Specialist and Computer Systems Administrator Career Fields with potential for career progression and incorporate specialized computer security training.	(U.S. Senate Permanent Subcommittee on Investigations, Staff Statement: Hearings on Security in Cyberspace, 1996).
Create a Government Computer Crime Investigators Career Field that includes potential for career progression and specialized computer crime training.	(U.S. Senate Permanent Subcommittee on Investigations, Staff Statement: Hearings on Security in Cyberspace, 1996).
Make logon warning banners mandatory for all government and government interest systems.	(U.S. Senate Permanent Subcommittee on Investigations, Staff Statement: Hearings on Security in Cyberspace, 1996).
DoD intelligence agencies should routinely monitor Internet traffic (that is readily accessible to the general public) related to their responsibilities.	(Swett, Charles, OSD; "Strategic Assessment: The Internet," 1995).

Raise the bar with high payoff, low-cost items including training and awareness programs; improving the security of unclassified computers by eliminating fixed passwords, improving identification and authentication; and promoting use of government approved commercial security technologies.	(Defense Science Board Task Force on IW-D, 1996).
DoD should design a curriculum for the study of cyber security including system design, detection of intrusion, intrusion response, firewall design and implementation, and so on. Curriculum could form a new category of engineering in the military academies of an Associates Degree from the Industrial College of the Armed Forces.	(Commissioner Keyes memo, November 6, 1996).

BIBLIOGRAPHY

ATIS, Network Reliability Steering Committee, "Analysis of Power Related Network Outages," 1996

Center for Strategic and International Studies, "R&D for National Strength," 1982

CERT Coordination Center, Report to PCCIP, January 1997

Comptroller General Report to Congress, 1979

Comptroller General, "Need to Assess Federal Role in Regulating and Enforcing Pipeline Safety," 1984

Copeland, Guy; "Information Assurance Insurance," February 1997

Coy, Peter; "Who's Watching the Power Grid" in Business Week, 1996

CSIS Information Assurance Working Group, Greene memo dated October 2, 1996

CSIS, "America's Hidden Vulnerabilities: Crisis Management in a Society of Networks," 1984

DARPA, "Electronic Power Distribution Case Study, Defensive Information Warfare Study," 1995

DARPA, Defensive Information Warfare Summer Study Final Report, 1995

Defense Science Board Task Force on Information Warfare-Defense, November 1996

Devost, Houghton, & Pollard, "Information Terrorism: Can You Trust Your Toaster?," 1996

DISA, "Planning Considerations for Defensive Information Warfare- Information Assurance," 1993

German, Jerry; Testimony on the right to free speech, free association, and privacy on the Internet in the aftermath of Oklahoma City, Subcommittee on Terrorism, Technology, and Government Information, 1995

Hughes Electronics Corporation, Report to PCCIP, March 1997

IDC/LINK, "The U.S. Electronic Distribution Infrastructure: Size, Ownership, Geography, and Vulnerabilities," March 1997

Information Infrastructure Task Force Security Issues Forum, "NII Security: The Federal Role," 1996

Information Infrastructure Task Force, "NII Risk Assessment: A Nation's Information at Risk," 1996

K. J. Schrader, "The Army's Role in Domestic Disaster Support: An Assessment of Policy Choices," 1993

Keyes memo, December 16, 1996

Keyes memo, November 6, 1996

Keyes, David; "Stove-Pipe City: It's Not a New Computer Game, It's Why Infrastructure Assurance Needs Centralized Analysis," October 17, 1996, 15

Logan, Michael; 5th International Conference on Information Warfare, 1996

Lukasik, Steve; "Public and Private Roles in the Protection of Critical Infrastructure," March 1997

National Research Council, "Computers at Risk," 1991

National Research Council, "Growing Vulnerability of the Public Switched Networks," 1989

Naval Research Advisory Council (NRAC) Study, 1996

Neumann, Peter; "Security Risks in the Emerging Infrastructure," Senate Testimony June 1996

Nickson, Mark; "Vulnerability of US Financial Markets to an Information Warfare Attack," Draft of March 1997

NIST, "Assessing Federal and Commercial Information Security Needs," 1992

NSR, "Indications and Warnings of Strategic Information Warfare," October 1996

NSTAC Information Assurance Task Force, "Electric Power Information Assurance Risk Assessment," 1996

NSTAC, "An Assessment of the Risk to the Security of the Public Network," 1995

OSTP, "Cybernation," Draft of January 17, 1997

RAND, "Risks to the U.S. Infrastructure from Cyberspace," 1996

RAND, "Strategic Information Warfare: A New Face of War," 1996

Spafford, Eugene; COAST/Purdue Univ.; Information Security Education

SPB Briefing to the Commission, September 23, 1996

Swett, Charles, OSD; "Strategic Assessment: The Internet," 1995

U.S. Congress, "Water Quality: A Catalog of Related Federal Programs

U.S. Congress, Office of Technology Assessment, "Issue Update on Information Security and Privacy in Network Environments," 1995

U.S. Congress, Office of Technology Assessment, "Physical Vulnerability of Electric Systems to Natural Disasters and Sabotage," 1990

U.S. Security Policy Board, "White Paper on Information Infrastructure Assurance Infrastructure Assurance," 1996

U.S. Senate Permanent Subcommittee on Investigations, Staff Statement: Hearings on Security in Cyberspace, 1996

U.S. Senate Subcommittee on Governmental Affairs, "Draft Committee Report on Network Vulnerabilities to Terrorist Attack, 1989

Volpe Center, "Emerging Areas in Transportation Information Infrastructure Security," 1996

APPENDIX A

U.S. INFRASTRUCTURE ASSURANCE
Prosperity Game and Planning Event
March 23-25, 1997
Chantilly, Virginia

Solutions Catalog

Introduction

The Solutions Catalog consists of 22 broad categories, each comprised of a number of options. Categories and options are presented as "starters" to focus interaction and speed play. They are not intended to discourage players from devising novel solutions, or from modifying those presented. A number of the categories and options are derived from specific observations, proposals and recommendations of prior bodies that have addressed infrastructure assurance issues. Some of the categories and options are included because they reflect solutions invoked during play of the predecessor Prosperity Game™ and Planning Event. Still others appear in the catalog to stir provocative discussion. Players should not take the inclusion (or omission) of any particular category or option to reflect the views or preferences of the sponsors or process managers of this event.

To promote ease of use (and encourage at least a little pre-preparation), the catalog is organized into a workbook format. You are encouraged to review and consider the solution categories (summarized, for convenience, in the table of contents on the following page) prior to the start of the game. You may wish to register your rough sense of preference for a particular solution category in the large check-boxes provided. (Preferences may change through play, so use a pencil!) Players may also wish to register their pre-game preferences for some of the more specific options that are presented within a particular solution category. Smaller check-boxes have been provided for this purpose.

Note: Although some categories contain options that appear to be mutually exclusive, other categories allow or even encourage the selection of multiple options.

Please feel free to use these workbooks to capture notes to aid in play of the game. The books will *not* be collected, reviewed, or evaluated by the game staff. However, you will be asked to identify your top six solution category preferences at dinner on the first evening of the event.

Table of Solution Categories

	Vigilance
I.	Increase public awareness and education
II.	Enrich training programs for cyber-security professionals, consider licensing or certification
III.	Mandate administrative and regulatory requirements for government and/or the private sector to promote information system security and early warning of threatening cyber attacks
IV.	Develop coordinated national infrastructure assurance policies between government and the private sector
V.	Encourage coordination of national infrastructure assurance policies within the government
VI.	Support creation of a permanent infrastructure protection capability
VII.	Encourage reconsideration of existing Federal department and agency jurisdiction and authority
VIII.	Enhance protection, mitigation, recovery, and emergency response capabilities through development and refinement of coordinated emergency response plans
IX.	Enhance deterrence domestically and internationally
	Network Management
X.	Develop security *standards* for software, hardware, and network design
XI.	Develop security *certifications* for software, hardware, and network design
XII.	Promote international cooperation and participation in assurance practices
XIII.	Improve the Federal government's "model" performance ability (i.e.
XIV.	Support creation of a centralized data collection and analysis capability to further development of an effective indications and warnings system
XV.	Accelerate reform of the liability climate
XVI.	Support adoption of public-private investment plans and incentives
XVII.	Encourage government efforts to identify, create, and maintain a "minimal essential infrastructure"
	Technology
	Advance specific technology needs and requirements (cyber)
XIX.	Advance specific technology needs and requirements (physical)
XX.	Promote public-private and government-assisted research and development of specific technology needs and requirements
XXI.	Support mandated adoption of specific technology needs and requirements
XXII.	Develop new risk management tools, models, and techniques
XXIII.	Write in Other Solution Categories:

SOLUTION CATEGORIES AND SUPPORTING OPTIONS

Vigilance

☐ **I Increase Public awareness and education**

☐ A. Support additional government funding of Federal, state, and local educational programs to expand public awareness of physical and cyber-security issues.

☐ B. Support creation of grammar, middle, high school, and adult education curricula to acquaint children and adults with rules and norms for interactions in cyber space (including ethics, property rights, abuses of resources, theft of services, etc.).

☐ C. Charge government entities such as the National Science Foundation to develop educational programs in resilient system design and practices for implementation at undergraduate and graduate levels.

☐ D. Support a tax credit for private insurance companies to conduct educational programs to include infrastructure assurance measures in business practices and reduce the likelihood of debilitating loss through insurance claims.

☐ E. Organize resources to participate in joint government-private sector programs to increase threat and vulnerability awareness of senior-level government and industry policy makers, to include, for example, demonstrations and simulations of key vulnerabilities.

☐ F. Same as E., above, but targeted at middle managers.

☐ G. Other (describe)____

____

____

☐ **II. Enrich training programs for cyber-security professionals, consider licensing or certification**

☐ A. Support development and adoption of degree programs in cyber security (including system design, intrusion detection and response, firewall design, etc.) for inclusion in military college curricula.

☐ B. Support addition of security training modules to existing computer science curricula.

☐ C. Support development and operation of a government licensing system for certain categories of computer professionals (e.g., "driver's licenses" for particular programmer and system administrator positions) with criteria for technical and ethical excellence. Licensing would be required for those holding certain sensitive positions with government and infrastructure service providers.

☐ D. Support development and operation of a private sector certification body for certain categories of computer professionals. Certification, though not required, would likely carry insurance and liability-related benefits for private-sector participants.

☐ E. Support and fund the development of a government-private-sector, performance-based certification standard for computer professionals, to be used by government and the private sector to guide personnel decisions.

☐ F. Other (describe)____

____

____

☐ III. Mandate administrative and regulatory requirements for government and/or the private sector to promote information system security and early warning of threatening cyber attacks

A. Regulations are to be binding on (check all that apply):

☐ 1. the Federal government, including military installations

☐ 2. state and local governments

☐ 3. schools and universities (as prerequisite to receiving Federal funds)

☐ 4. all infrastructure service providers

☐ 5. all businesses having more than 50 employees with remote access capabilities

☐ 6. Other(describe)____

____

____

B. Regulations are to provide for (check all that apply):

☐ 1. mandatory reporting of significant infrastructure service outages

☐ 2. mandatory reporting of significant cyber-intrusion incidents

☐ 3. mandatory use of authentication controls such as one-time password generators

☐ 4. mandatory use of strong access controls (e.g., firewalls, strong authentication) to protect Supervisory Control and Data Acquisition (SCADA) information

☐ 5. mandatory encryption of SCADA information

☐ 6. mandatory reserve capacities and redundancy margins

☐ 7. Other (describe)____

____

____

☐ IV. Develop coordinated national infrastructure assurance policies between government and the private sector

☐ A. Support tasking of an existing Federal cabinet-level agency to assume policy-level infrastructure assurance responsibilities, including coordinating public-private infrastructure assurance activities and devising appropriate funding vehicles for these activities.

☐ B. Advocate creation of a new cabinet-level agency to accomplish the above (e.g., an "Infrastructure Protection Agency").

☐ C. Establish an information security/infrastructure assurance focus within an existing White House policy entity, such as the National Security Council or National Economic Council, to assume policy-level infrastructure assurance responsibilities, as in IV.A. above.

☐ D. Support legislation or other legal measures to create a new Federal policy entity, within the White House, to assume infrastructure assurance responsibilities as in IV.A. above.

☐ E. Support legislation or other legal measures creating a governmental public policy entity, such as the Federal Reserve Board, to assume policy-level infrastructure assurance responsibilities as in IV.A. above. (The board would be comprised of Presidential appointees, confirmed by the Senate, who are appointed to terms as full-time government employees).

☐ F. Support legislation or other legal measures creating individual infrastructure advisory committees, based on a National Communications System - National Security Telecommunications Advisory Committee (NCS-NSTAC) model, to assume policy-level responsibilities as in IV.A. above and coordinate pre-competitive approaches to infrastructure assurance. (NSTAC is a Presidential Advisory Committee of 23 corporations representing the telecommunications infrastructure. The NCS represents the Federal agencies responsible for telecommunications in the U.S. and operates in close coordination with the NSTAC).

☐ G. Accomplish policy level input to the government through an exclusively private-sector consortium of corporate officers assembled according to their ability to represent diverse infrastructure interests.

☐ H. Other (describe)____

____

____

☐ V. Encourage coordination of national infrastructure assurance policies within the government

☐ A. Support tasking of an existing Federal cabinet-level agency to coordinate infrastructure assurance efforts of the Federal government.

☐ B. Establish an information security/infrastructure assurance focus within an existing White House policy entity, such as the National Security Council or National Economic Council, to assume policy-level infrastructure assurance responsibilities, to include organization and funding of public-private assurance efforts.

☐ C. Support appointment of a full-time government board, made up of qualified Federal-government employees, to facilitate implementation of national infrastructure assurance policies throughout government, and to coordinate government activities to improve efficiency and avoid redundancy.

☐ D. Support appointment of a full-time, public-private-sector board to facilitate implementation of national infrastructure assurance policies throughout the government, with enhanced ability to track developments in the private sector.

☐ E. Fund and support a full-time private-sector entity, to work in parallel with government efforts and consult with the government as to how best to facilitate implementation of national infrastructure assurance policies.

☐ F. Other (describe)____

____

____

☐ VI. Support creation of a permanent infrastructure protection capability

☐ A. Support the development, within the Federal government, for the exclusive benefit of Federal, state, and local governments, of a capability to provide or facilitate provision of expert guidance to halt or confine attack and restore service.

☐ B. Support the development, within the Federal government, for the benefit of governments *and* the private sector, of a capability to provide or facilitate provision of expert guidance to halt or confine attack and restore service.

☐ C. Support the development, within the Federal government, of a capability to receive specific threat information, and issue threat and warning notices to government and the private sector based on such reports and information collected and analyzed by the mechanism described above.

☐ D. Support the development, within the academic community and private sector, of a capability to receive specific threat information, and issue threat and warning notices based on such reports (e.g., an expanded Carnegie-Mellon Computer Emergency Response Team (CERT) capability).

☐ E. Support the development, within the Federal government, of an independent administrative body, such as the National Transportation Safety Board, to hire qualified staff to perform investigations of infrastructure-threatening incidents, then issue recommendations for further action.

☐ F. Other (describe)____

____

____

☐ VII. Encourage reconsideration of existing Federal department and agency jurisdiction and authority

☐ A. Ensure appropriate Federal agencies have the authority, mission, and responsibility to assist other Federal agencies in information infrastructure protection efforts, including review of proposed or planned system architectures and sharing of information on countermeasures.

☐ B. Ensure appropriate Federal agencies have the authority, mission, and responsibility to assist other Federal agencies, state and local governments, and industry, in information infrastructure protection efforts, including review of proposed or planned system architectures and sharing of information on countermeasures.

☐ C. Expand existing defense community jurisdiction to address threats to the infrastructure (by, for example, providing technical support to law enforcement in criminal investigations and disaster recovery services to the private sector).

☐ D. Expand existing intelligence community jurisdiction to address threats to the infrastructure (by, for example, allowing limited collections relating to vital infrastructure protection functions, whether such needs arise in the U.S. or abroad).

☐ E. Expand existing law enforcement community jurisdiction to address threats to the infrastructure (by, for example, expanding its ability to compel assistance from the defense and intelligence communities to fulfill its investigative or protective functions).

☐ F. Expand jurisdiction of other regulatory and administrative bodies and/or Federal agencies to address threats to the infrastructure.

☐ G. Other (describe)____

____

____

☐ VIII. Enhance protection, mitigation, recovery and emergency response capabilities through development and refinement of coordinated emergency response plans

☐ A. Task military branches to accept and adopt civil disaster response as a mission for active and reserve forces.

☐ B. Amend the Federal Response Plan and associated authorities to expressly include ability to address critical cyber-incidents, greater use of mitigation tactics, and funding authority to more effectively coordinate services following an infrastructure attack.

☐ C. Provide incentives for proactive state-government disaster mitigation efforts (e.g., improved building codes, emergency response, training for disaster response personnel), and other measures to reduce damage from a major, infrastructure threatening occurrence.

☐ D. Support formation of a public-private "emergency corps" to assist in repair and recovery following infrastructure threatening occurrences.

☐ E. Other (describe)____

____

____

☐ IX. Enhance deterrence domestically and internationally

☐ A. Adopt or amend legislation to more clearly criminalize intentional or willful attacks on, damage to, or destruction of critical infrastructures.

☐ B. Enhance deterrence domestically through aggressive public awareness campaigns equating damage to government systems with sabotage, and intrusions into private systems with other, analogous crimes.

☐ C. Encourage the U.S. Government to make known, through "back door" diplomatic channels, its ability to anonymously strike back at information terrorist without the display of force or publicity that would otherwise legitimize the terrorist activity.

☐ D. Encourage the U.S. Government to make widely known the degree of proof that would be required to trigger a presumption that certain infrastructure failures resulted from concerted activities of a foreign nation, thereby constituting an act of war subject to retaliatory action through cyber means or deployment of conventional forces.

☐ E. Other (describe)____

____

____

Network Management

☐ X. Develop security *standards* for software, hardware and network design

☐ A. Support efforts by government standard-setting bodies, such as the National Institute of Standards and Technology (NIST), to develop security standards for software, hardware, and network configuration for Federal government systems. (Although not binding on the private sector, compliance with standards would likely carry insurance and liability-related benefits).

☐ B. Support efforts by an appropriate policy authority (such as described in Category IV., above) to develop security standards for providers of infrastructure services, compliance to be enforced through administrative sanctions.

☐ C. Support creation of exclusively private-sector bodies to develop appropriate security standards to apply to providers of infrastructure services, compliance to be enforced through administrative sanctions. (This may require that certain legal restrictions be relaxed for this limited purpose).

☐ D. Support creation of exclusively private-sector bodies to develop appropriate security standards for providers of infrastructure services, compliance with which is voluntary but will likely carry insurance and liability-related benefits.

☐ E. Other (describe)____

____

____

☐ XI. Develop security *certifications* for software, hardware, and network design

☐ A. Encourage Federal standard-setting agencies, such as the National Institute of Standards and Technology (NIST), to develop a system of *government* certifications for software, hardware and network configuration. (Although the certifications would not be available to the private sector, compliance with underlying requirements would likely carry insurance and liability-related benefits).

☐ B. Support tasking of an appropriate policy authority (see Category IV., above) to develop thresholds and procedures for issuance of certifications to qualified government and private-sector applicants (i.e., the rough equivalent of an Underwriter's Laboratory for cyber certifications).

☐ C. Fund and support existing private-sector standards bodies to develop thresholds and procedures for businesses to receive certifications as above.

☐ D. Encourage creation and funding of new private-sector standards bodies to develop thresholds and procedures for businesses to receive certifications as above.

☐ E. Other (describe)____

____

____

☐ XII. Promote international cooperation and participation in assurance practices

☐ A. Support creation of an international computer crime bureau to provide education, awareness, and investigative assistance to foreign law-enforcement agencies.

☐ B. Encourage Federal standard-setting agencies, such as the National Institute of Standards and Technology (NIST), to participate in the development of voluntary international standards or certifications for software, hardware, network configuration, and trusted network personnel.

☐ C. Support the tasking of an appropriate policy authority (see Category IV., above) to participate in the development of voluntary international standards or certifications as above.

☐ D. Support and fund exclusively private-sector bodies to develop voluntary international standards or certifications as above.

☐ E. Other (describe)____

____

____

☐ XIII. Improve the Federal government's "model" performance ability (i.e., its ability to influence private-sector action through unilateral efforts at standardization and improved procurement practices)

☐ A. Mandate that all new legislation considered by Congress and all deregulation actions considered by Federal agencies that affect critical infrastructures include an "Infrastructure Assurance Impact Statement" that fully explores expected positive and negative effects of these measures on infrastructure assurance objectives.

☐ B. Support legislative initiatives that require Federal, state, and local governments to unify and improve government procurement standards and practices to reflect preferences for purchasing and implementing products that promote system security.

☐ C. Allocate resources to work with Federal, state, and local governments, through a policy authority (such as in Category IV., above) to accomplish as stated in B., above.

☐ D. Support and fund exclusively-private-sector bodies to recommend to Federal, state, and local governments appropriate procurement standards and practices.

☐ E. Other (describe)____

____

____

☐ XIV. Support creation of a centralized data collection and analysis capability to further development of an effective indications and warnings system

☐ A. Support legislation requiring the government to share sufficiently reliable threat information with the private sector.

☐ B. Support tasking of existing government resources within the defense, intelligence, and law-enforcement communities (and state and local governments) to develop centralized data collection and analysis at existing levels of funding, and share threat information with the private sector.

☐ C. Support measures that dramatically increase funding for the development of government data collection and analysis capabilities with threat information to be shared with the private sector, *without* having to rely on specific threat or vulnerability input from the private sector.

☐ D. Support measures that increase contributions by the private sector to fund additional government data collection and analysis capabilities, to then share threat information, *and* contribute proprietary threat or vulnerability information under conditions that such information would remain confidential.

☐ E. Same as above, but only under the condition that specific threat and vulnerability information from the private sector be contributed anonymously.

☐ F. Designate/create a single, independent, trusted organization to be responsible for collecting, analyzing, and reporting incident data, with no public policy, investigation, or enforcement authority.

☐ G. Other (describe)____

____

____

☐ **XV. Accelerate reform of the liability climate**

☐ A. Create a new Financial Accounting Standards Board (FASB) standard relating to information security so that companies regulated by the Securities and Exchange Commission (SEC) would be audited against it and, if their efforts are not adequate, will have a contingent liability noted on their financial statements.

☐ B. Extend the Federal Deposit Insurance Corporation (FDIC) concept to insure against catastrophic infrastructure loss by creating an entity to insure deposits against losses that are not currently insurable, such as losses from electronic commerce and harm from acts of terrorism. Funding would come from a tax on transactions in each respective infrastructure.

☐ C. Support legislation creating liability for providers of infrastructure services who fail to meet standards of due care—which failure then results in damaging loss of service to customers.

☐ D. Support legislation capping liability for providers of infrastructure services who meet certain standards or certifications.

☐ E. Encourage providers of infrastructure services to form coalitions to develop guidelines for reliability, reserve capacity, etc., to inform and influence standards of due care within infrastructures.

☐ F.Other(describe)____

____

____

☐ **XVI. Support adoption of public-private investment plans and incentives**

☐ A. Create an organizational entity within the Executive Branch, similar to the "superfund" structure, financed by tax dollars, to assess and ameliorate the most critical vulnerabilities to infrastructure assurance when responsibility for the deficiencies cannot be assigned to a responsible and financially viable legal entity.

☐ B. Tie receipt of targeted financial incentives to compliance with standards or certifications such as those described in Categories X. and XI., above.

☐ C. Support permanent (not year-to-year) tax credits for investment in research, development, or deployment of measures specifically devoted to increasing the assurance of the infrastructure.

☐ D. Support an accelerated depreciation schedule for investments in information related hardware to coincide with the actual lifetime of the equipment so that companies will be motivated to more aggressively upgrade their level of service with accompanying improvements in security, reliability, and interoperability.

☐ E. Change Federal income tax rules, which currently treat all funds retained at the end of the tax year as taxable profit, to encourage private insurers to retain adequate funds in reserve for emergencies.

☐ F. Other (describe)____

____

____

☐ XVII. Encourage government efforts to identify, create and maintain a "minimal essential infrastructure"

☐ A. Support Federal government programs to identify critical nodes within the government, assess their vulnerabilities and interdependencies, and make recommendations to funding authorities for cost-effective protections and improvements.

☐ B. Support Federal efforts to revitalize critical asset assurance and protection programs to include involvement by Federal, state, and local governments.

☐ C. Support government efforts to identify a minimal essential information infrastructure for use in restoring services and adapting to wide-scale outages.

☐ D. Support Federal-government programs to identify elements of public and privately-owned information which collectively comprises "vital national information interests," including physical infrastructure components, virtual processes, and some forms of information.

☐ E. Support government efforts to subsidize additional reserve capacities for assets comprising the minimal essential infrastructure.

☐ F. Work with private sector consortia to develop infrastructure-wide agreements establishing service priorities in the event of major infrastructure outages.

☐ G. Other (describe)_______________________________________

____

____

Technology

☐ XVIII. Advance specific technology needs and requirements (cyber)

☐ A. Develop robust, survivable system architectures that include advanced firewall technology.

☐ B. Develop techniques and tools for modeling, monitoring, and managing large-scale distributed/network systems (including practical techniques and tools for system administrators to determine the security condition of their network including the strength of the various security mechanisms and the system's vulnerability to threats).

☐ C. Develop techniques and tools for auditing, detecting, and responding to intrusions into large-scale distributed/network systems.

☐ D. Develop techniques and tools for using encryption technologies to address surety and security deficiencies.

☐ E. Develop administrator-friendly strong access controls (e.g., superior firewalls, stronger authentication, effective intrusion countermeasures, and isolation techniques that do not directly depend on encryption) to protect Supervisory Control and Data Acquisition (SCADA) from cyber attack.

☐ F. Research and develop truly adaptive software that scans a control system's sensor suites and SCADA programs for evidence of unusual activity and provides warnings and indicators of impending problems from critical nodes, intrusions, unexpected interdependencies, or outdated hardware or software.

☐ G. Research and develop new breakthrough architectures, comparable to the invention of packet switching in the 1960's, that are self-healing independent of traffic rate to improve robustness and survivability under cyber attack and under abnormal loading.

☐ H. Develop means for dynamically reallocating communication resources to give priority to critical messages such as emergency preparedness, national security, or system security communications even if the network is overloaded with normal traffic.

☐ I. Other (describe)____________________________________

____

____

☐ XIX. Advance specific technology needs and requirements (physical)

☐ A. Develop and deploy a sensor system (such as a UV laser-induced fluorescence system for non-intrusive screening or neutron-induced activation with x-ray emission) that remotely interrogates vehicles for explosives.

☐ B. Develop cost-effective chemical/biological detectors (including procedural guidelines governing their use) for placement in facilities such as buildings, subways, and other places where large numbers of people congregate.

☐ C. Other (describe)____________________________________

____

____

☐ XX. Promote public-private and government-assisted research and development of specific technology needs and requirements

☐ A. Support tasking existing government resources to coordinate government research and development efforts and expand government R & D on cyber-resilient technologies, such as firewalls, auditing, monitoring, and intrusion detection technologies, and encryption. Expansion of these efforts will occur at the expense of current government R & D priorities.

☐ B. Support enhanced funding for existing government mechanisms to coordinate and expand government research and development efforts on cyber-resilient technologies, and to procure targeted research and development from the private sector.

☐ C. Support grants for research and development through Cooperative Research and Development Agreements (CRADAs) specifically targeting infrastructure assurance objectives, and provide Federal agencies, the national labs, and industry consortia with incentives to develop and deploy new technologies.

☐ D. Create a process whereby the government directly develops technology and transfers it to the private sector through an "Infrastructure Investment Bank" for targeted use in infrastructure strengthening.

☐ E. Support the creation of a government corporation (such as the U.S. Synthetic Fuels Corporation) to guide sizable government investment in private sector research and development on cyber-resilient technologies.

☐ F. Support the creation of a joint government-industry research corporation (such as Sematech) to employ government, academia, and private sector experts for limited terms to focus on research and development of cyber-resilient technologies, and to provide support to other emerging centers of excellence.

☐ G. Support and fund coordinated programs to facilitate technology transfer from government to the private sector with respect to new cyber-resilient technologies.

☐ H. Support coordinated and focused private-sector research and development efforts through the creation of exclusively private sector alliances and through the suspension of existing impediments to enhanced cooperation.

☐ I. Other (describe)____________________________________

____

____

☐ XXI. Support mandated adoption of specific technology needs and requirements

☐ A. Support legislation mandating the adoption, when ready for deployment, of selected technology needs and requirements, such as those identified in Categories XVII. and XIX., above, by Federal, state, and local governments.

☐ B. Support legislation mandating the adoption, when ready for deployment, of the selected technology needs and requirements, such as those identified in Categories XVII. and XIX. above, by Federal, state, and local governments and infrastructure service providers.

☐ C. Other (describe)______________________________

____

____

☐ XXII. Develop new risk management tools, models and techniques

☐ A. Facilitate inclusion of risk data in ongoing government data collection activities, such as those conducted by the Census Bureau or the National Economic Council.

☐ B. Establish a separate risk category for information system security for regulators to use in their risk-based oversight of critical infrastructure industries.

☐ C. To reduce risk of harmful insider threats, develop and deploy new personnel and management approaches that keep employees positively engaged.

☐ D. Develop practical metrics for determining the security condition of a network, including the strengths of various security mechanisms, current known and potential threats and vulnerabilities, and potential impacts in a graded methodology to assist system administrators.

☐ E. Develop tools, models, and methods that help system designers make decisions affecting the system-level robustness in their economic selection studies.

☐ F. Support establishment of a joint government/private sector institute to study and develop new risk management models, and to provide education to senior-level government and industry policy makers.

☐ G. Same as above, but limited to a private-sector consortium.

☐ H. Other (describe)______________________________

____

____

☐ **XXIII. Other**
(describe)____

CHAPTER 8

MAJOR FEDERAL LEGISLATION

A "LEGAL FOUNDATIONS" STUDY

Report to the
President's Commission
on Critical Infrastructure Protection
1997

This report was submitted to the President's Commission on Critical Infrastructure Protection, and informed its deliberations and recommendations. This report represents the opinions and conclusions solely of its developers.

CONTENTS

PART ONE
INTRODUCTION

Some of the most sweeping Federal legislation relevant to efforts to protect the critical infrastructures was originally conceived, passed into law and implemented long before the proliferation of computer and computer networks, and before the emergence of serious threats to the infrastructures. While the long-standing divisions of authority created by such legislation and the mechanisms that flourished thereunder still appear to be fundamentally sound, some of this legislation may now require modernization so that it may continue to serve its originally intended purpose.

Several pieces of legislation that appear relevant to infrastructure assurance objectives were written before the emergence of a recognizable cyber threat. It is not clear whether such authorities would apply, and should apply, to a cyber-related event. Until the dynamics of such cyber event are better understood, including the necessary response vehicles, sweeping legislative changes would be premature. However, it is nonetheless possible to identify key issues and to make general recommendations to begin the process of incorporating the full range of infrastructure assurance issues within the legislative framework. It is also possible to identify the additional pieces of legislation relevant to achieving infrastructure assurance objectives and consider whether those acts should be amended or revised, or should act as models to guide the implementation of other specific infrastructure assurance objectives.

Research Issues

The questions at hand include:

- Do major areas of Federal legislation, such as the Defense Production Act of 1950, the Stafford Act/Federal Response Plan, the Computer Security Act of 1987, the War Powers Resolution, and Nunn-Lugar-Domenici legislation adequately take into account threats to, and incidents arising from attacks on critical infrastructures?
- Should these laws be modernized to better account for and address physical or cyber threats, and incidents arising from physical or cyber attacks?
- Can modernization of these laws also accomplish other related infrastructure assurance objectives?
- Alternatively, can infrastructure assurance objectives be more effectively implemented through new and different legislative schemes?

Defense Production Act of 1950

The Defense Production Act serves as an important legal mechanism for security of the Nation's industrial and technological base. Its authorities, including priorities in contracts, financial incentives, and voluntary agreements, may serve as important vehicles to assist in the reconstitution or recovery of a critical infrastructure necessitated by a physical or cyber event. It may prove productive to review the DPA authorities and triggering mechanisms, and current efforts underway at modernization, to ensure that these powers would available and adequate in light of emerging threats, vulnerabilities, and related challenges.

Background

The Defense Production Act of 1950 (DPA)[1] was enacted to ensure that national security and defense interests would be supported by a strong and dependable industrial and technological base. The DPA vests a great many powers in the President during both times of peace and of national emergency to garner support for national defense from private industry. It is a tool for response *and* prevention. Among the provisions of the DPA are priorities for performance of contracts, requirements for stockpiling, availability of loans and a DPA fund, and recognition of the need to plan and train for emergencies. The DPA, as amended, may also be invoked during civil disasters covered by the Stafford Act, whether catastrophic earthquakes or terrorist events.

Uses of the DPA

The most recent uses of the DPA have centered on the use of priority contract performance to provide military operations (*e.g.* Desert Shield/Storm) needed computer and communications equipment, global positioning systems, chemical warfare protective clothing and equipment, and medical supplies.[2] The DPA contains a specific priority contract provision that relates to "domestic energy supplies" which was used to ensure the timely completion of the Trans-Alaska pipeline and to support accelerated development of oil and gas reserves from the Alaska North Slope.[3] Funds available under the DPA have also been used for innovation in defense technology, such as the $7.6 million spent to qualify active matrix liquid crystal displays for use on Apache Longbow helicopters.[4] The focus

---

1 Codified at 50 App. U.S.C. § 2061 *et seq.*

2 *See* Interagency Working Group on Modernization of the Defense Production Act, *Final Draft Report on Modernization of the DPA* (December 24, 1997).

3 *DOE Offers U.S. Refineries Assistance Under Defense Production Act,* Inside Energy with Federal Lands, Sept. 17, 1990, available in DIALOG, File No. 624.

4 *Pentagon Proposes $50 Million Commercial Insertion Program,* Vol. 15 Defense Week No.

on research and development has been to provide incentives for private industry to develop technologies specifically needed by the defense community.[5]

The DPA is a powerful law that, if ever needed, could do much to further infrastructure assurance objectives. However, it was written during an era when national defense was more dependent on such things as energy and weaponry produced by the private sector, than the networks and communications that are currently the focus of information warfare concerns. The powers contained in the DPA may be suitable for use in response to infrastructure-related emergencies, but the law, and its triggering mechanisms, should be reviewed to ensure its availability to respond to serious cyber as well as physical incidents.

Application To Cyber-Incidents

Under the DPA. the President is given broad powers to invoke certain measures after making particularized findings. Though the Act is clearly available during times of national emergency, including civil defense emergencies, as declared by the President, most of its provisions are available prior to such an occurrence in order to prevent a national defense or national security-related crisis. Thus for individual provisions of the Act, priorities in contracts or various financial incentives may be available if an item, resource, or specific industry is essential to the national security strategy and is potentially scarce.

A cyber-related situation could satisfy the requirements to invoke many of the provisions of the DPA. For example, a minor disruption in the energy industry could trigger the authority to set priorities in contracts without regard to the cause. For other industries, however, the bar is set much higher. Thus, it may be necessary to review the individual triggering mechanisms as well as the policy statement for the statute as whole to ensure its availability to prevent crises that may arise out of cyber-related events. A statement specifically establishing the nexus between the critical infrastructures, including the information and communications infrastructure, and national security may be sufficient to update the Act for infrastructure assurance generally, as well as for addressing emerging cyber threats and vulnerabilities to infrastructures.

Current Efforts To Amend The DPA

In the 1995 Amendments to the DPA, an inter-agency study group was given one year to prepare a report on the need to modernize the statute. That report is expected to be transmitted to Congress in September, 1997.

---

15 (IAC April 8, 1996).

5 Under an agreement between DOD and OMB, the only DPA financial incentives currently in use are purchases and purchase commitments.

The modernization effort focuses on addressing new threats in the post Cold War era, enhancing financial incentives for the availability of materials, services and technologies, and revitalizing the DPA fund. The legislative changes suggested may enhance the availability of DPA powers for infrastructure contingencies, but additional modifications may be in order.

Specifically, the Modernization Report is expected to recommend that the Act be broadened to include "acts of terrorism." Depending on how terrorism is defined, it may or may not be broad enough to encompass cyber threats to critical infrastructures. In addition, expanding the Act to be responsive to "acts of terrorism" essentially only addressed those portions of the act concerning incident response N not prevention. Many of the provisions of the DPA that could be identified as potentially useful for use in the protection of the critical infrastructures could be, and perhaps should be, available *prior to* an incident. These preventive-type measures have been used in the energy industry, for example, to facilitate constructing and repairing pipelines.

Additional Considerations Relating to Infrastructure Assurance

It may be advisable to consider additional modifications to the DPA in order to implement infrastructure assurance measures or to serve infrastructure assurance objectives. While additional study may be needed before recommendations regarding the DPA may be put in specific terms and implemented, there are a few areas which seem to merit attention. First, DPA funds may be available for infrastructure assurance-related research and development or expansion of production capacity of critical infrastructures. Second, DPA priorities may be available to assist in reconstituting critical infrastructures (*e.g.*, by expediting delivery of parts). Third, stockpiling provisions could be considered as possible avenues for ensuring the availability of rare and crucial parts for critical infrastructures.[6] Finally, voluntary agreements may provide a planning mechanism for preparing for infrastructure incidents.

The authority to accomplish many of these uses of the DPA for infrastructure assurance objectives is in many instances already in place. Promoting such uses is a matter of education and awareness of both government and industry of the availability of such resources especially in response to new types of incidents. However, the DPA, even in light of current modernization efforts, may not provide such resources to all of the necessary critical infrastructures under all the necessary circumstances. The statute addresses energy resources specifically, but not communications. While other avenues may be available for maintaining telecommunications in the

---

6 Other independent authorities exist for stockpiling which could be pursued for infrastructure assurance objectives. *See* 50 U.S.C. § 98; 44 C.F.R. Part 328.

event of emergency, it may be important to give consideration to the need for and applicability of DPA provisions to all of the critical infrastructures to ensure adequate coverage in the event of a major infrastructure disruption. For example, the priorities in contracts authorities may be invoked when necessary for "national defense" for all of the critical infrastructures, but in the energy industry, specific priorities in performance of contracts may be granted when materials, services or facilities are "scarce, critical and essential" to maintain or expand exploration, production, refining or transportation; to conserve energy supplies; or to construct or maintain energy facilities. Thus, the authorities may be available for cyber incidents in the energy industry, but depending on the prevailing definition of "national defense," not for the telecommunications industry or another critical infrastructure.

Actions for Consideration

Maintenance of the Status Quo

One possibility is based on recent Congressional and Administration attention, that the DPA will be appropriately modified to address emerging threats, such as those from cyber attacks. The most recent amendment to the DPA came in 1995 when Congress asked the President to report on the need to modernize the DPA. This report is due September, 1997. In the 1995 amendments, Congress also expanded "national defense" and "defense" to include emergency preparedness functions described in the Stafford Act. As of this writing, the President's report to Congress is currently in draft and while it does not specifically address infrastructure assurance, it does propose to expand the scope of the DPA to include a wider range of national security and national defense issues including globalization. It also proposes increased funding for the DPA fund. In light of these trends toward expanding the DPA, it may be reasonable to conclude that the DPA will be modernized appropriately under current Congressional thinking.

There are pros and cons to this approach. The DPA is a complex piece of legislation with a significant history, and thus, infrastructure assurance objectives may be more easily achieved apart from DPA structure. It may also be prudent to avoid creating threat of government authority encroaching on private industry resources and functions. However, the Administration might not want to ignore the DPA when designing implementation strategy, and it may make more sense to try to work with what is already there since the President's report on DPA modernization may not adequately address threats to critical infrastructures.

Incorporating Critical Infrastructures Specifically Within the Defense Production Act

Though current, and even proposed expansions of policy address national security, national defense, and civil emergencies, there may nonetheless be value in specifically including critical infrastructures in the statement of policy. Many of the provisions of the DPA could be valuable for infrastructure assurance objectives (*e.g.*, DPA funds for research and development, priorities in contract performance, etc.). However, without specific inclusion of critical infrastructures into the policy of the DPA, such authorities may be presumed unavailable for infrastructure assurance.

The advantages of such an approach would allow critical infrastructure to be included within the DPA's coverage without a series of technical amendments. It would also leave the basic authorities of the DPA intact and unchanged, and technically will only expand the areas to which they may apply. It would also be necessary to ensure that this approach does not become a quick patch which fails to adequately account for lack of coverage for specific critical infrastructures, or that it does not read as overly broad.

In addition there are precedents in priorities in contracts related to energy which can be studied as a potential model for the other critical infrastructures in order to ensure continued operation and clear authority for use in reconstituting such infrastructures after an event. This is because the DPA designates energy as a strategic and critical resource to receive special treatment within the priorities and allocations section of the statute. This treatment allows the priorities in contracts provisions to be used for a wide variety of purposes including supporting energy construction and repair projects. Such uses do not require as stringent requirements in terms of Presidential findings as do other industries and resources covered generically within the DPA due to energy's special status. This provision of the DPA is potentially useful for avoiding serious disruptions of critical infrastructures, regardless of their cause, or reconstituting the infrastructures once an incident has occurred. It may therefore be prescient for future needs for the Administration and Congress to consider: (1) whether the energy provision is an appropriate model for other critical infrastructures; and (2) whether critical infrastructures would benefit from such additional authorities. This will require consultation with Federal and private bodies and research into analogous authorities contained in other bodies of law, for example in the telecommunications industry.

DPA Fund and Financial Incentives

Additionally, there is the use of DPA funds for research and development to support the resiliency and security of critical infrastructures. DPA funds are currently used to support a wide range of activities including development of new technologies to support national defense. At current rates of spending the DPA fund will soon be depleted. In addition, an agreement between the Office of Management and Budget (OMB) and the Department of Defense (DOD) limits the types of financial incentives that DOD can make available to industry including loan guarantees and loans. The draft of the President's report on modernization of the DPA recommends new funding for the DPA.

Additional funding of the DPA may create a pool of resources that can be used to further infrastructure assurance objectives without needing to specifically make the case for spending on infrastructure assurance to Congress. However, there is no guarantee that funds will be allocated for infrastructure assurance objectives even if put toward the DPA fund.

Incorporation of Provisions by Agencies with Existing DPA Authorities

The Department of Energy recently sent a letter to refinery companies it works with to explain the availability of a certain provision of the DPA to expedite delivery of needed parts during an emergency or other contingency.[7] While no one has yet taken DOE up on the offer, the letter did increase awareness both within the agency and the private sector of the availability of certain authorities in times of necessity. It may be possible to further expand the level of awareness of the DPA and its provisions by other agencies with connections to critical infrastructures undertake a study of the DPA and determine which provisions may be used during an infrastructure incident. These agencies and other partners may opt to incorporate the provisions into their planning for infrastructure emergencies.

Such actions would allow an agency-by-agency/infrastructure-by-infrastructure review of the statute, a process which may also identify any needed legislative changes. It would also allow close coordination with the private sector on how the DPA can be used.

The Stafford Act/Federal Response Plan

The Stafford Act and Federal Response Plan set out the parameters of the Federal response to major disasters as declared by the President. The Federal Emergency Management Agency's (FEMA) authority to prevent, mitigate, and respond to incidents affecting the operation of the critical infrastructures may be unclear under the triggering mechanism currently contained in the statute.

---

7 *See* Inside Energy, *supra* note 3.

Background

The Stafford Act, and the management structure of the Federal Response Plan (FRP) are the primary means by which the Federal government responds to domestic disasters. The purpose of the Stafford Act is to provide assistance to state and local governments to enhance response capabilities. This assistance may take the form of materials and resources, including personnel and expertise, or Federal funds. However, before such assistance to be available, the statute and thus the FRP management structures must be triggered.

Stafford Act/FRP Triggering Mechanisms

The triggering event for the Stafford Act is a "major disaster." This includes natural catastrophes (*e.g.*, flood, earthquake, tornado) and fire, flood or explosions regardless of cause.[8] Once a major disaster has occurred, the governor of the affected state must implement the state emergency plan before asking the President to declare a "major disaster." The declaration of a disaster allows the powers of the Stafford Act and the response mechanism set out in the FRP to be activated. Among the assistance the Stafford Act provides are agency resources (personnel, equipment, supplies, facilities, technical and advisory services)[9]; technical and advisory assistance for performance of essential community services; issuance of warnings of risks and hazards, public health and safety information, provision of health and safety measures; and management control and reduction of immediate threats to public health and safety[10]; DOD resources for clearance and removal of debris and wreckage and temporary restoration of essential public facilities and services;[11]and emergency telecommunications[12]and public transportation services.[13]Authorities also exist to take post-disaster steps toward mitigation of future damage.[14]

The potential for infrastructure-based emergencies and even cross-infrastructure catastrophes may one day necessitate that the Federal government mobilize for a coordinated response. The Stafford Act and Federal Response Plan already provide a framework for mobilizing such a response. However, it is not clear that the trigger mechanisms, as they are

---

8 *See* 42 U.S.C. § 5122.

9 42 U.S.C. § 5170a(1).

10 42 U.S.C. § 5170a(2).

11 42 U.S.C. § 5170b.

12 42 U.S.C. § 5185.

13 42 U.S.C. § 5186.

14 *See e.g.*, 42 U.S.C. § 5170c(a) (Federal funds available for 75 percent of costs of future hazard mitigation efforts); 42 U.S.C. § 5176 (minimum standards for public and private structures in conjunction with disaster loans or grants).

currently defined, would allow FEMA to become involved in prevention, mitigation and recovery actions for cyber-based infrastructure events (or, for example, to proactively pre-position resources when faced with a sufficiently credible threat). FEMA (and the FBI) would clearly be within statutory bounds to respond to a physical attack on a critical infrastructure involving a bomb or other explosive device. In fact, the Terrorism Annex has recently been added to the FRP to address such incidents and those involving Weapons of Mass Destruction. However, if the disruption to the power supply to a largo a geographic area were caused by a cyber-based attack, it is not clear that FEMA would have the necessary authority to act. Such an event is neither a "natural catastrophe" nor need it necessarily cause a "fire, flood, or explosion."

Preparation and Mitigation Authorities

The Stafford Act also has many provisions which could be revised, or merely promoted, to enable actions to protect or mitigate against infrastructure damage. Provisions related to insurance requirements, reimbursement of costs by those liable for damage, hazard mitigation measures and minimum rebuilding standards in conjunction with disaster loans could all be leveraged to further infrastructure assurance objectives. Most of the mitigation-related provisions apply to areas affected by major disasters and receiving Federal assistance. The provisions often do not reach private businesses and do not apply at all to areas that have not yet been struck by such a disaster.

FEMA's authorities also extend to planning. Under the Stafford Act, the President does have authority to establish, *"a program of disaster preparedness that utilizes services of all appropriate agencies and includes:*

> *(1) preparation of disaster preparedness plans for mitigation, warning, emergency operations, rehabilitation, and recovery;*
>
> *(2) training and exercises;*
>
> *(3) post-disaster critiques and evaluations;*
>
> *(4) annual review of programs;*
>
> *(5) coordination of Federal, State, and local preparedness programs:*
>
> *(6) application of science and technology;*
>
> *(7) research."*[15]

---

15 42 U.S.C. § 513 (a).

Other Preparation and Mitigation Authorities

One program, outside of the Stafford Act/FRP framework, where substantial emphasis has been placed on mitigation is the National Earthquake Hazards Reduction Assistance Program.[16] This program was established by a separate piece of legislation, the Earthquake Hazards Reduction Act, in 1977.[17]The program may provide a model for the types of cost-sharing, mitigation-oriented programs be appropriate for FEMA to pursue in the area of infrastructure assurance. Under the program, areas that are at risk apply to FEMA for funds, which it must match, for use in planning for and mitigating the damage that results from earthquakes. Such legislation contrasts with implementation of the Stafford Act which has not to date focused on mitigation and preparation activities to the extent necessary to satisfy some infrastructure assurance objectives.

Scope of Authority

An additional area of concern is the availability of Federal assistance to private industry, including infrastructure owners and operators, in the event of a major infrastructure-disrupting event. The Stafford Act does not provide assistance for private industry. Instead, the Act is designed to create a conduit for assistance from the Federal government to state and local agencies. Then, through the state and local agencies, assistance may be available to private concerns. In terms of direct aid, the Stafford Act provides funding and other assistance authority only for public entities and facilities and for private nonprofit facilities (which may include some utilities, hospitals, and emergency services, but certainly does not include many of the critical infrastructures).[18] A broad reading of certain provisions of the Stafford Act may allow Federal resources to be used to restore essential services to affected communities, which arguably would include electric power or telecommunications. However, the statute's wording indicates a preference for Federal provision of essential services (*e.g.*, telecommunications) until private providers are restored rather than a Federal role in restoring the private service providers.[19] Other statutory mechanisms do provide some assistance to private industry in restoring operations after a disaster or other emergency. The DPA, for example, can be used to speed delivery of items or provide financial assistance (*i.e.* loans) to assist in recovery.

The Stafford Act/FRP scheme has been an effective mechanism for responding to natural disasters. There may be significant advantages to

---

16 *See* 44 C.F.R. Part 361 (1997).

17 42 U.S.C. § 7701 *et seq.* (1997).

18 *See, e.g.*, 42 U.S.C. § 5172 (repair, restoration, and replacement of damaged facilities).

19 *See* 42 U.S.C. §§ 5185-86.

considering the benefits of expanding this scheme to cover additional type of incidents. A shift in focus toward greater mitigation and preparedness activities may also serve infrastructure assurance objectives. This may include predisaster mitigation measures, conduct of training exercises, and greater integration of cyber elements in current planning.

Actions for Consideration

Maintenance of the Status Quo

In light of recent additions to the FRP (*e.g.* the Terrorism Annex) and growing responsibilities of FEMA for such things as consequence management with respect to Weapons of Mass Destruction attacks, FEMA may undertake to prepare its authorities as necessary to respond to a critical infrastructure-related incident. Congress may also take appropriate steps to ensure preparedness for infrastructure-related incidents either by tasking FEMA or through another analogous piece of legislation.

Conduct an Additional Study

It may be desirable for a specific agency to engage in a full-scale study of the Stafford Act/Federal Response Plan to determine:

- Its likely applicability to critical infrastructure emergencies, particularly those with a cyber dimension;
- Current mitigation efforts and the need for the further legislative or administrative action to supplement them;
- Applicability and desirability of Federal assistance for critical infrastructure owners and operators through the Stafford Act/FRP;
- Suitability of the current management structure of the FRP to respond to critical infrastructure emergencies (and whether there should be a specific annex to the FRP to address such incidents or even an analogous response structure based on the FRP).

While such a study would focus primarily on the Stafford Act and FRP, it could also incorporate other disaster preparedness and response statutes, regulations and programs at the Federal and state levels. If so, it would allow for detailed approach and careful consideration of possible effects of each amendment or revision, and also allow experts on emergency preparedness to drive process and tailor provisions to critical infrastructure issues.

Making Minimal Modifications to Key Definitions to Ensure Response To Serious Cyber Incidents

It may be that definitions of "emergency" and "major disaster," as used in the Stafford Act, ought to be revised to specifically include critical infrastructure incidents and cyber attacks. Amending the triggering mechanism for the Stafford Act and Federal Response Plan would ensure that the response mechanism they create will be available if needed to respond to a critical infrastructure crisis. In addition, amending the circumstances which trigger the Stafford Act also allows the DPA to apply to those incidents.

Such a streamlined approach to amending legislation is likely to provide coverage of cyber-incidents without potentially controversial expansion of scope to private sector, and may be more than adequate to cover infrastructure assurance concerns. However, this strategy could be overly simplistic, may create powers in the President and his delegates that are more broad than is either necessary or prudent, or, may neglect to create specific requirements, uses of programs or funds, and authority that are necessary for infrastructure assurance objectives. It may also stretch the FRP in directions which it was never intended to go.

Enacting Substantive Amendments

There may be a need for more comprehensive amendments to the Stafford Act to address specific cyber-related issues, including definitional clarifications (see above) and others. These may include revisions to the triggering mechanism, mitigation programs, and funding requirements. Significant study would need to be given before going forward with a specific recommendation for amendment.

New Legislation

New legislation to specifically address similar issues as Stafford Act for the cyber arena may be appropriate. It could include a management structure for responding to emergencies analogous to the FRP.

Information Security In The Federal Government: The Computer Security Act Of 1987 & OMB Circular No. A-130

Background

The primary legal instruments providing for government information security are the Computer Security Act of 1987 (CSA) and OMB Circular A-130 (A-130). The CSA tasks the National Institute of Standards and Technology (NIST) with developing information security standards for the Federal government and A-130 sets out Federal agency requirements for adhering to the NIST standards. While this seems to be a fairly simple paradigm, a number of factors have prevented it from being implemented with maximum success.

NIST Responsibilities Under The CSA

The CSA defines NIST's mission as "developing standards, guidelines, and associated methods and techniques for computer systems."[20] These standards are to be uniform for unclassified systems within the Federal Government.[21] The primary purpose of these standards and guidelines is to protect the integrity and privacy of the information in such systems and to prevent fraud and misuse.[22] The Act requires that NIST submit their recommendations of standards to the Secretary of Commerce for review and promulgation.[23] NIST's other responsibilities under the Act include: setting guidelines for training employees in security awareness and procedures, and developing validation procedures for, and evaluating the effectiveness of, the standards and guidelines they develop. This validation process is to be coordinated with other government and private agencies.[24]

To accomplish the mission and responsibilities of the Computer Security Act, NIST is given the authority to engage in research and studies to determine vulnerabilities and to design security measures for Federal systems; to coordinate with other Federal agencies including, but not limited to, DOD, DOE, NSA, GAO, OTA, and OMB; to work with the Office of Personnel Management (OPM) in developing training regulations for computer security; and to provide technical assistance to Federal agencies for implementation of NIST standards and guidelines.[25] The Act also authorized NIST to "assist the private sector, upon request, in using and applying the results of the programs and activities" developed under the Act.[26]

New NIST/NSA Partnership

Recently, NIST and NSA have teamed up to address information security issues. The primary focus on their effort will on creating testing standards for information systems. Through this partnership, they will then train and certify private sector testers.

Many of the difficulties that appear to arise from the Computer Security Act do not spring directly from a change in technology, but rather from the framework for creating standards and guidelines for security of Federal government systems set out in the CSA and as it has been implemented. One of the primary and most persistent problems has been the lack of funding for NIST to carry out its responsibilities under the Computer Security Act.[27] A new authorization bill, sponsored by Representative Connie Morella (R-MD), should go to the floor of the House of Representatives

---

20 15 U.S.C. § 278g-3(a)(1) (1997).

21 15 U.S.C. § 278g 3(a)(2).

22 *Id.*

23 15 U.S.C. § 278g-3(a)(4).

24 15 U.S.C. 278g-3(a)(6).

25 15 U.S.C. 278g-3(b)(2)-(6).

26 15 U.S.C. 278g-3(b)(1).

27 See Gov't Secrecy 104; Computers at Risk 197.

in September that would increase funding for Computer Security, NIST and the Federal Aviation Administration (FAA) by $14 million. Funding for computer security in the era of the balanced budget has also been a problem for individual agencies. While it has been suggested that information system security be made a budget line item, others have noted that this is usually one of the first areas cut in an agency budget crunch.

OMB and Circular A-130

OMB's role in implementing and enforcing NIST standards and guidelines may also require examination. Circular A-130 sets forth an ambitious agenda. But the Report of the Commission on Protecting and Reducing Government Secrecy states that OMB has put inadequate resources and attention toward computer security.[28] Specifically, the Report states that OMB has only two people assigned to oversee the Federal government's information systems.[29] As it currently stands, there is no oversight of research and development to reduce redundancies in efforts within the Federal government. Even Congressional oversight of computer security issues is dispersed across the jurisdiction of several committees and subcommittees, including House and Senate Committees on Appropriations, Commerce, Government Reform and Oversight, National Security, and Science, to name just a few.[30]

While OMB may need to put more resources toward overseeing the implementation of A-130, the current division of responsibility in A-130 need not be altered. A-130 is specifically designed to put responsibility for information security with the individual agency Chief Information Officers (CIOs). Legislation such as the Information Technology Management Reform Act (ITMRA) requires CIOs to establish performance measures for information technology systems and report on them as part of the OMB budget review. Similarly, the enforcement mechanism for A-130, as recently revised upon the suggestion of the National Performance Review, requires the reporting of "material weaknesses" in information security also through a budgetary review process (*see* Federal Managers Financial Integrity Act). The strength of this enforcement mechanism is dependent on the weakness being found and reported. Without a mandated audit or additional oversight procedure, this is unlikely to occur.

Additional oversight, under the auspices of OMB, but with greater technological capability and resources, may be a welcome addition to the current framework. Such an entity could have operation responsibility for studying and leading the reform of information security for the government.

---

28 Gov't Secrecy 102.

29 Id. at 107.

30 Id. at 103.

This would require extensive consultation with the private sector as well as links to the technological resources of NIST and NSA. Ultimately, the process would involve audits of information security systems at each of the individual agencies. Such an oversight body would require substantial study before launching and may require revisions to core Federal information security authorities (such as the CSA, A-130, ITMRA).

Despite volumes of literature criticizing the CSA, its purpose and intent remain fundamentally sound. The Federal government does need minimum information security standards for its unclassified systems. Likewise, the mechanism for enforcing adherence to those standards as set out in A-130 may also become adequate with time and increased attention and awareness. However, in order for the government to have a truly successful information security program these two primary elements must be fully realized. This may require additional resources and concrete direction to agencies already bearing the responsibility, and additional commitments from individual agencies to give information security the attention it deserves.

Actions for Consideration

Maintenance of the Status Quo

Efforts by Representative Morella and others to re-authorize NIST and provide adequate funding may result in satisfactory improvements to the existing legislative scheme.

Need for Task Force to Study of Methods for Enhancing Compliance with Standards and Requirements for Information Security in OMB Circular A-130

As the Federal government moves toward management methods more closely resembling those used in the private sector, it may be appropriate to begin to explore the use of internal auditors for such areas as information security. An auditing scheme would fit squarely within the enforcement framework as currently designed and would not require any major modifications to existing legal authorities. However, those studying the issue would likely consider the need for additional legislation or revision of key authorities such as A-130 as part of their study. The study may also consider whether the Inspector General offices within agencies would be an appropriate place to house such a function and what would be necessary to make the expertise needed to perform Information Technology (IT) audits available to IG offices. The utility of such a study would be dependent on the development of a set of standards against which agencies could be audited.

Joint NIST-NSA Partnership Study and Report on an Effective Enforcement Vehicle for Computer Security Within the Federal Government

Rather than creating a task force to study the need for greater enforcement of computer security standards within the Federal government, this approach would utilize the existing partnership and expertise of NIST and NSA. The scope of the study would be essentially the same as in the previous suggested action, but would allow for greater coordination with the standards development process.

Existing resources can be used to undertake the study, which allows considerable knowledge and expertise in computer security issues to look at a difficult problem. However, computer security expertise may not translate into expertise in management of the Federal government. The study group must be sufficiently diverse enough o encourage new thinking in the area.

The War Powers Resolution

'When natural resources were the dominant factor of production, the conquest and control of territory seemed a reliable way to enhance national power. Today, conquest of territory is rarely worth the cost to the nation.' It is both much easier and more profitable to conduct information warfare against an adversary's knowledge resources than to conduct a conventional war against its armed forces.[31]

Background

The War Powers Resolution is one part of the legal and policy structure that enables the Federal government to defend the nation. The Resolution was written before the emergence of information warfare and cyber threats. As these threats evolve, the mechanisms that enable defense of our nation will need to be reviewed and updated to ensure they enable an adequate and effective response, and instill adequate deterrence where appropriate.

While information warfare is still emerging as a threat to national security, resources and attention are being devoted by the Department of Defense and others charged with protecting the nation. Thus far, the resources have been devoted largely to estimating the threat and developing our own capabilities. The focus has not yet come to rest on the rather important question of how information-based warfare fits within our current structure—legal, policy, and even practical— for detecting and responding to attacks on the nation.

---

31 Sean P. Kanuck, *Information Warfare: New Challenges for Public International Law*, 37 Harv. Int'l L.J. 272, 290 (1996) (quoting Albert J. Edmonds, Address at Seminar on Intelligence, Command, and Control, John F. Kennedy School of Government, Harvard University (Apr. 20, 1995)).

The current legal structure for responding to an "attack" has domestic and international components. As a fundamental matter, one must first determine what is an "attack" or act of war. Does it include mechanisms driven by a computer? If a bomb disables a major defense computer station or an important telephone switch, that is clearly an attack. Is it likewise an attack if accomplished with a computer? And what has to be accomplished to be an attack? Electrons crossing a national border is not enough under current law or practice to constitute a violation of a nation's sovereignty.

Depending on the result of an incident, how, considering the anonymity and ready availability at low cost associated with computer technology, can the victim country distinguish between an accident and an attack? There may also be challenges associated with convincing allies to join in retaliatory efforts based on the available evidence. Defining the terms that drive domestic and international warfare policies, such as "attack," "act of war," or "force," will undoubtedly provide an important first step in clarifying the application of such laws, rules and customs to computer-based acts of aggression. However, additional, important questions remain. For example, international law requires that the force of a retaliatory attack be proportionate to that of the aggressor's attack. If a country is subject to an information warfare attack, would it be appropriate for them to use conventional military forces to retaliate? If the country retaliates through electronic means, may it pass through the systems of a neutral country?[32]

As important as each of these questions are to understanding the dynamics of the problem and creating an international consensus on how to approach information-based attacks, an even more fundamental question exists: Are there benefits from this current state of confusion that outweigh the possible benefits of coming to consensus on this issue?

Under the current regimes, it would be difficult to show that any country, whether as an aggressor or in retaliation, acted inappropriately. This can benefit intelligence gathering, defensive maneuvering, and expansion of offensive capabilities. Clearly, when considering how the War Powers Resolution fits into the larger picture and the options that exist for changing the domestic and the international regime of which it is a part, such questions need to be serious considered. The options themselves demonstrate the complexity of the issue. Definitions are an appropriate beginning option, but would have to carefully crafted so as not to be excessively inclusive of seemingly benign applications of information technology. Early warning systems and limiting the information portion of weaponry could also achieve a deterrent effect. However, one of the most unlikely, but

32 Many of these questions are raised in Greenberg, Goodman & Soo Hoo, *Old Law for a New World: The Applicability of International Law to Information Warfare*, ISAC, Stanford University (February 1997).

perhaps effective forms of deterrence may be the growing transnational complexity and transparency of systems, so that an attacker may not be able to harm the systems of one nation or target without potentially harming his own.[33]

"Inforwar" Attacks Under The War Powers Resolution

One part of this structure is indicative of the problem that would confront the nation in the event of an "inforwar" attack. The War Powers Resolution, written in the post-Vietnam era, does not contemplate an information-based attack within its parameters. The statute, which sets out the authorities and responsibilities of the President with respect to introducing U.S. troops into hostilities, was written well before the advent of the computer-based threat. Its language clearly anticipates warfare as a physical attack.

In the event of a cyber attack, quick and decisive action may be necessary to launch an effective response. The speed of the nation's response may be significantly slowed if a Congressional declaration of war or a statute authorizing the introduction of forces into hostilities is required. The President is, however, only authorized to introduce forces into hostilities without Congressional approval in the limited circumstances of a national emergency created by an attack upon the U.S.[34] Whether the definitions of "hostilities" and "attack on the U.S." are broad enough to include a cyber attack is unclear.

A similar problem exists with respect to international law. Article 51 of the U.N. Charter uses "armed attack" as the basis for determining whether an armed response is appropriate. The U.N. has not yet addressed how Information Warfare would be treated under its law of use of force. Recommendations in the area include: specifically defining "armed attack" to either include or exclude information warfare; coordinating and harmonizing computer crime laws and prosecutions on an international scale; and developing a network of agreements, similar to those used for aviation, to protect critical systems on an international level.[35]

The question is, then, does the War Powers Resolution or its international analogues apply to an information-based attack or does the response structure fall outside of its bounds? Should there be a separate response structure for such "inforwar" attacks? While the President's ability to act outside the War Powers Resolution has been well documented,[36]a cohesive

---

33 For further discussion of these and other options for deterring information warfare, *see* Timothy Thomas, *Deterring Information Warfare: A New Strategic Challenge*, Parameters 81 (Winter 1996-1997).

34 *See* 50 U.S.C. § 5141 (1997).

35 *See Greenberg, Goodman, et. al, supra* note 32.

36 *See, e.g.*, Hansen & Banks, *From Vietnam to Desert Shield: The Commander in Chief's*

policy of deterrence and plan for defense to information-based attacks should clarify the manner in which cyber attack fits into the War Powers Resolution. This undertaking is likely broader than amending only the War Powers Resolution. It may require creating a structure capable of answering the question as to whether the U.S. is under cyber attack; setting out jurisdictional boundaries; assigning roles and missions; and perhaps most importantly, translating the national strategy into a comprehensive statement of deterrence. However, to jump-start such a process, it may be possible to draw attention to the need for a new paradigm by highlighting the inadequacies of the War Powers Resolution in the face of these new cyber threats.

The technology necessary to answer many of these policy questions may not yet be available. As these technologies and the analytic capability necessary to support them develop, many important groups will likely begin to address these concerns. It will be important for these groups, in the process of developing a national policy of deterrence, to give substantial thought and consideration to the legal regime that is currently in place and the potential impact of such a regime on national policy.

Actions for Consideration

Maintenance of the Status Quo

This resolution has historically reflected tension between the Executive and Legislative branches of government, and may not be a good lead vehicle for addressing a complex new issue such as infrastructure assurance.

Development of an "Inforwar" Strategy

The President may direct the National Security Council and Department of Defense develop a comprehensive strategy for responding to cyber attack, taking into account effects of the War Powers Resolution and international law. Such a strategy could clearly outline under what circumstances a cyber incident would be considered an act of war, thus acting as a bridge to existing authorities. The strategy may also provide an opportunity to clearly enunciate a policy of deterrence for information-based attacks on the U.S. and its critical infrastructures.

Articulating what will be considered an "attack" may alleviate some confusions regarding Federal jurisdiction for cyber incidents and provide an opportunity to make a clear statement to U.S. enemies regarding U.S. intentions in event of an information-based attack. Such a definition would also serve to clarify responsibilities under the War Powers Resolution and International law.

Spending Power, 81 Iowa L. Rev. 79 (1995).

On the other hand, it may be difficult and potentially undesirable for the U.S. to establish clear definitions, for purposes of deterrence and clarity of international obligations, of concepts that might in turn restrict U.S. activities or available options. There may be significant strategic benefits to operating in "murky waters" until a greater degree of international consensus develops.

Amendment of Key Definitions

Congress may wish to add a definitions section to the War Powers Resolution that specifically includes cyber attack and cyber responses within the framework set out by the Act. This enhancement would ensure that cyber attack is covered by the framework for responding to attacks on U.S., and would also serve as a de facto deterrence policy indicating U.S. readiness to consider certain serious cyber attacks as potential acts of war. However, for the reasons mentioned earlier, the controversial history of the Act may preclude quick action.

Nunn-Lugar-Domenici-Legislation

The Nunn-Lugar-Domenici legislation creates a partnership involving many Federal agencies and state and local governments in an effort to increase the capability to respond to weapon of mass destruction incidents across the United States. This legislation focuses on providing training, access to equipment, and information to local first responders. While these programs are in their initial phases, already state and local police, fire, and medical officials are requesting an expanded effort in this area. Infrastructure assurance concerns may require more resources for training and equipment and may also demand that the effort be expanded in scope to address infrastructure-threatening events.

Background

The Nunn-Lugar-Domenici legislation included in the Department of Defense Authorization for 1997 sets out a comprehensive plan for dealing with terrorist threats from "Weapons of Mass Destruction." The legislation defines a Weapon of Mass Destruction (WMD) as, "any weapon or device that is intended, or has the capability, to cause death or serious bodily injury to a significant number of people through the release, dissemination, or impact of . . . toxic or poisonous chemicals . . ., a disease organism; or radiation or radioactivity."[37] The legislation is specifically designed to "enhance the capability of Federal, state, and local emergency response agencies to prevent and respond to domestic terrorist incidents involving weapons of mass destruction."[38]

---

37 P.L. 104-724, § 1043 (1996).

The DOD role in assisting emergency response to WMD events has focused on providing assessments to major cities and conducting training for local first responders. In addition, DOD was tasked with developing a rapid response team and is creating a hotline and help line to provide technical advice during emergencies. A web site will provide information on WMD issues on a routine basis. To accomplish these goals, DOD is working with FEMA, DOE, HHS, EPA and the Department of Justice.

In the *President's Report on Government Capabilities to Respond to Terrorist Incidents Involving Weapons of Mass Destruction*, dated February 26, 1997, additional needs are highlighted. The report generally identifies the initiatives undertaken by the FBI, DOD and FEMA, including the training and support function of DOD in relation to state and local governments. Among the identified needs to continued progress in the fight against WMD, the President's report identifies: continued funding with State and local contributions, expanded training programs and updated materials, specialized equipment and protective gear for the FBI, annual updating of the Rapid Response Information System, additional exercises and legislation.

Such expanded efforts may be consistent and necessary to further objectives for infrastructure assurance as they relate to Chemical or Biological weapons. In addition, depending on the success of the Nunn-Lugar-Domenici legislation, it may be an appropriate vehicle for objectives not currently included within its scope or an appropriate model for a separate initiative.

Actions for Consideration

Maintenance of the State Quo

It may be preferable for Congress to make no changes regarding the scope or coverage of Nunn-Lugar-Domenici to infrastructure events, continue to support the program initiated by the original legislation and expand its coverage as necessary. It may also be deemed that other avenues would be more appropriate for achieving desired infrastructure assurance objectives.

Nunn-Lugar-Domenici is Expanded to Include Attacks on Critical Infrastructures

On the other hand through a simple legislative revision, coverage could be extended to critical infrastructures. Considerable thought and planning would be required to implement the Nunn-Lugar-Domenici program for infrastructure assurance efforts (*e.g.*, deciding what sort of training, equipment and other resources to provide), which may require further study by either Congress or the Administration.

---

38 DOD Background Briefing-Part I, 1997 WL 10369097 (May 16, 1997).

Nunn-Lugar-Domenici is Studied as a Potential Model for Infrastructure Assurance Legislation

Nunn-Lugar-Domenici legislation may be highlighted as an embodiment of useful principles, including partnership, the Federal government enabling State and local responders, and information sharing. It appears to accomplish many of the goals which would benefit infrastructure assurance, including training and education, information sharing, enhanced cooperation, sharing of resources, and funding.

A study group from the Administration or Congress perhaps should study the Nunn-Lugar-Domenici legislation as a model for an infrastructure assurance statute and determine.

- Whether the legislation is suitable to use as a model;
- Which infrastructure assurance objectives could be achieved through such legislation;
- What sort of revisions would be necessary;
- What additional programs would need to be designed and who should be responsible
- What level of funding would be necessary.

The group could then draft legislation for Congressional consideration.

PART TWO
CONCLUSIONS

Several pieces of legislation appear relevant to infrastructure assurance objectives, but were written before the emergence of a cognizable cyber threat. It is not clear whether such authorities would apply, and should apply, to a cyber-related event. Until the dynamics of such a cyber event are better understood, including the necessary response vehicles, sweeping legislative changes would be premature. Nevertheless, it is possible to identify key issues and to make general recommendations to begin the process of incorporating the full range of infrastructure assurance issues within a legislative framework. It is also possible to identify additional pieces of legislation relevant to achieving infrastructure assurance objectives, and to consider whether those acts should be amended or revised, or even act as models to guide the implementation of infrastructure assurance objectives.

Defense Production Act

The Defense Production Act serves as an important legal mechanism for security of the Nation's industrial and technological base. Its authorities, including priorities in contracts, financial incentives, and voluntary agreements, may serve as important vehicles to assist in the reconstitution or recovery of a critical infrastructure necessitated by a physical or cyber event. Accordingly, it may prove productive for:

- The Administration and Congress to consider amending the DPA Declaration of Policy to include a finding that critical infrastructures are essential to national security;
- Lead Agencies associated with the critical infrastructures to study the energy provision for priorities in contracts as a potential model for other critical infrastructures in order to ensure continued operation and clear authority for use in reconstituting such infrastructures after an event;
- Congress to support funding for the DPA Fund and financial incentives and to consider making such funds available for research and development related to the critical infrastructures; and
- The Administration to direct Federal agencies with authorities pertaining to the critical infrastructures to review the DPA's authorities and work with industry to make available such authorities when needed to respond to a critical infrastructure incident.

Stafford Act/Federal Response Plan

The Stafford Act and Federal Response Plan set out the parameters of the federal response to major disasters as declared by the President. The Federal Emergency Management Agency's authority to prevent, mitigate, and respond to incidents affecting the operation of the critical infrastructures is unclear under the triggering mechanism currently contained in the statute.

A study should be undertaken of the Stafford Act and Federal Response Plan mechanisms to determine their applicability and suitability to cyber-induced disasters as well as their current implementation with regard to prevention and mitigation. Such a study would also take into account other disaster recovery authorities and their potential impacts on infrastructure assurance goals as well as the desirability of direct assistance to infrastructure owners and operators.

War Powers Resolution

The War Powers Resolution is one part of the legal and policy structure that enables the federal government to defend the nation. The Resolution was written before the emergence of information warfare and cyber threats. As these threats evolve, the mechanisms that enable defense of our nation will need to be reviewed and updated to ensure they enable an adequate and effective response, and instill adequate deterrence where appropriate.

Meanwhile, the Administration, including the National Security Council, Department of Defense, Department of State, Department of Justice, and other interested agencies, could ensure that the U.S. strategy for responding to an information warfare attack addresses the potential legal issues associated with current definitions of "attack" contained in important domestic and international legislation, including, but not limited to, the War Powers Resolution.

Nunn-Lugar-Domenisci Legislation

The Nunn-Lugar-Domenici legislation created a partnership involving many Federal agencies and state and local governments in an effort to increase the capability to respond to weapon of mass destruction incidents across the United States. This legislation focused on providing training, access to equipment, and information to local first responders. While these programs are in their initial phases, already state and local police, fire, and medical officials are requesting an expanded effort in this area. There may be more resources required for training and equipment and demand may also build that the effort be expanded in scope to address other infrastructure-related events.

Congress could consider whether the current Nunn-Lugar-Domenici program should be expanded to incorporate other critical infrastructure issues, including physical attacks on infrastructures by means other than Weapons of Mass Destruction, as well as the need for training, awareness, and information sharing efforts directed at state and local responders on the potential impact of disruptions of critical infrastructures, particularly information and communications, on emergency response efforts.

CHAPTER 9

ADEQUACY OF CRIMINAL LAW AND PROCEDURE (CYBER)

A "LEGAL FOUNDATIONS" STUDY

Report to the President's Commission on Critical Infrastructure Protection

1997

This report was submitted to the President's Commission on Critical Infrastructure Protection, and informed its deliberations and recommendations. This report represents the opinions and conclusions solely of its developers.

CONTENTS

PART ONE
INTRODUCTION

This paper studies the adequacy of current criminal law to provide adequate protection for the nation's critical infrastructures from cyber attack. It also addresses the adequacy of criminal law to deter or punish cyber attacks on infrastructures and if so, whether applicable laws should be amended, revised, or supplemented to fill the gap.

Research Issues

This paper addresses the following research questions:

- Do existing laws governing criminal procedure unduly hinder investigations of infrastructure threatening behavior?
- To the extent that they may unduly hinder such investigations, how should they be reconsidered or modified?

Research Findings

Upon reviewing various Federal, state and international statutes, the following observations were noted:

- In the United States, substantive computer crime laws (including sentencing laws) have been made by-and-large sufficient owing to recent amendments:
- Federal and state computer crime statutes are now broad enough to cover a wide range of behaviors, including unauthorized computer intrusion and propagation of malicious code.
- The Federal Sentencing Guidelines for these offenses have recently been amended to provide for increased and more comprehensive penalties for computer crime and are currently before Congress for approval.
- Procedural laws governing the investigation of computer-related offenses, however, have traditionally received far less attention, and may, in their current form, unduly hinder the conduct of investigations in electronic environments:
- U.S. law governing search and seizure requires that search warrants be obtained from courts "within a federal district . . . for a search of property . . . within the district." Fed. R. Crim. P. 41 (a). This formulation does not apply readily or predictably to evidence stored in networked environments, leading to confusion and delay in the conduct of time-sensitive investigations.
- U.S. law governing the issuance of "trap and trace" orders requires that the government apply for legal process in each of the jurisdictions

through which suspect electronic communications pass, thus hampering law enforcement's ability to quickly identify the source of an attack.

- Intrusions emanating from outside the United States compound these legal difficulties and create others–substantive and procedural:
  - In order to enlist assistance from foreign law enforcement services, mutual legal assistance treaties (MLATs) often require "dual criminality," that is, the behavior subject to investigation must also be a crime in the country in which the intruder is physically located. The United States thus has an interest in seeing that other countries also have adequate substantive computer crime laws.
  - In order to obtain timely and effective assistance from foreign law enforcement services, it is desirable for those services to be equipped with adequate procedural laws to foster the quick identification of the source of an attack.
  - In order to obtain evidence from foreign law enforcement services for use in U.S. criminal prosecutions, U.S. investigators currently must appeal to the courts to engage in a formal process (letters rogatory). Streamlined procedures would aid time-sensitive investigations.
- The legal channels necessary for maximally efficient international cooperation are just beginning to develop.

Assumptions

This paper uses certain assumptions in its analyses, and they are as follows:

- Computer crimes will continue to rise in frequency and severity, with (at least) a proportionate increase in the number of intrusions emanating from foreign jurisdictions.
- Although strong domestic enforcement will carry a significant deterrent effect on some classes of computer intruders, maximum deterrence can only be achieved through an effective international computer crime response.
- Enhancing the effectiveness of local, state, Federal and international law enforcement response to computer crime will achieve additional deterrence.
- Enhancing deterrence will reduce the cyber-threat to critical infrastructures.

PART TWO
HISTORICAL BACKGROUND:
COMPUTER CRIME LEGISLATION & INTERNATIONAL AGREEMENTS

As described more completely in the next section, legal reform in the areas of computer crime is underway at the local, state, national and international levels. The states continue to revise and reform their computer crime legislation, and many allocate funds to develop dedicated units of computer crime investigators and prosecutors. States continue to develop the most progressive ways of dealing with juvenile offenders.

On the Federal level, Congress has made frequent revisions to applicable federal law, which now covers virtually all forms of computer fraud and abuse. In late 1996, Congress enacted a federal law to address economic espionage and theft of trade secrets. The U.S. Sentencing Commission recently studied and proposed amendments to their guidelines for computer crimes. The new Guidelines are expected to go into effect in November 1997. Congress and the Administration are also reconsidering the current federal approach to crimes committed by juveniles. Despite its willingness to reconsider and revisit substantive computer crime laws, Congress has not been as receptive to changes in procedural criminal laws, particularly where revisions may be interpreted to impinge upon privacy and civil liberties. (Much of current criminal procedure is dictated not only by Congress, but by the Constitution and the Supreme Court.)

Work is also underway in a number of international forums, including the OECD, the Council of Europe, the European Union, and the G-7/P8, to insure the adequacy of criminal laws, procedures and resources among the respective member nations. In addition, the U.S. continues to pursue bilateral treaties and agreements to cement in place predictable and effective international response mechanisms.

The bulk of the items in this paper are premised on the work already undertaken or accomplished by these various bodies.

State Computer Crime Laws

Florida, Arizona and Virginia were at the forefront of state and federal efforts to craft legislation specifically to combat computer crime. As early as 1978, these states had computer crime statutes in place, and by 1979, California, Colorado, Illinois, Michigan, New Mexico, North Carolina, Rhode Island, and Utah had joined their ranks. By 1990, all states except Vermont had enacted some form of computer-related criminal law. Today, these statutes take a variety of forms and are, in some instances, more flexible and responsive to the unique character of computer crime than their federal counterparts. State statutory schemes often address unauthorized access or use, information abuse, and intrusive code (e.g. viruses).

States have also shown flexibility in responding to the challenges of investigating computer crimes. For instance, "Nebraska's computer crime statute provides incentives for potential victims of computer crimes to implement their own security measures. California's computer crime statute encourages victims to come forward by providing a civil cause of action for compensatory damages."[1] Civil actions are growing in popularity among states. A number of states have developed civil provisions either to supplement a criminal prosecution, act as an alternative to criminal prosecution, or fill a void left unfilled by the criminal law.[2] State civil actions not only cover a wider range of computer abuses, they also provide greater relief for the victim through the availability of injunctions, damages, court costs and attorney fees. Civil actions have the added advantages of a lower standard of proof and a higher degree of privacy.[3]

A number of states are also experimenting with novel sentencing provisions that afford the courts flexibility to craft penalties to minimize recidivism and achieve maximum deterrence, while at the same time providing additional monetary relief to the victim. Restitution and forfeiture provisions are becoming increasingly common.[4] Fines may be set in proportion to the loss of the victim or the financial gain of the defendant.[5]Judges may also impose restrictions on a defendant's use of computers following conviction.[6]

While there are benefits to state schemes, they do have legal and practical limitations. State investigators are often hampered in investigations of computer crime by jurisdictional boundaries. The nature of the Internet is not only interstate, but also international. States are beginning to develop special venue provisions that allow cases to go forward where any part of the criminal act took place, not only where the computer is located.[7] This is only

---

1 Project, *Eighth Survey of White Collar Crime,* 30 Am. Crim. L. Rev. 495, 516 (1993).

2 *See, e.g.,* Cal. Penal Code § 502(e) (permits recovery of compensatory damages, including expenses incurred to verify the extent of damage done by the access, and a reasonable attorney's fee to the prevailing party); Ga. Code Ann. § 16-9-93(g) (victim may recover damages, including lost profits and victim expenditure, plus cost of the suit); Tex. Civ. Prac. & Rem. Code Ann. §§ 143.001-143.002 (requires knowing or intentional violation of criminal provisions); Conn. Gen. Stat. Ann. § 52-570b (civil recovery independent of criminal actions for acts done recklessly, rather than intentionally or knowingly).

3 Arkansas, Virginia, and Wisconsin provide specifically for some measure of privacy for the victims of computer crime. *See* Ark. Code Ann. § 5-41-106(b); Va. Code Ann. § 18.2-152.12; Wis. Stat. § 943.70(5).

4 *See, e.g.,* N.M. Stat. Ann. § 30-45-7 (Restitution to be paid *in addition* to incarceration, forfeiture or fine); Cal. Penal Code § 502(g) (Forfeiture of computer equipment used during an offense); Ill. Ann. Stat. Ch. 38 para. 16D-6 (Forfeiture of "monies, profits, or proceeds"). *See also,* N.M. Stat. Ann. § 30-45-7; Ohio Rev. Code Ann. § 2925.44.

5 *See* Mon. Code Ann.§ 45-6-311(c)(2); 1984 Conn. Gen. Stat. Ann. § 53a-257; Del. Code Ann. tit. 11, § 937(f).

6 *See* Wis. Stat. § 943.70(4); Cal. Penal Code § 502(e)(3).

7 *See* Ark. Stat. Ann. § 5-41-105; Conn. Gen. Stat. Ann. § 53a-260; Ga. Code Ann. § 16-9-94; Ky. Rev. Stat. Ann. § 434.860; see also Md. Ann. Code art. 27, § 146(e); Miss. Code Ann. § 97-45-11; N.J. Stat. Ann. § 2C:20-34; S.C. Code Ann. § 16-16-30; S.D. Codified Laws Ann. § 43-43B-8; Tenn. Code Ann. § 39-14-603.

a small first step. Over the next several years, states can be expected to continue confronting these difficulties as well as to continue refining their substantive statutes.

Federal Computer Crime Legislation

In the early 1980's, there was no specific federal legislation in the area of computer crime. To prosecute a computer-related crime, one had to rely on federal statutes designed for other offenses, such as mail fraud[8] or wire fraud.[9] However, as evidence of computer-related crimes became more abundant, computer crime began to receive the attention of law enforcement personnel, the media, and the general public. Specifically, financial institutions (e.g. MasterCard International, VISA, etc.) reported large losses due to the exploitation of stolen and counterfeited credit cards and account numbers, in which exploitation was intertwined with the use of computers.

As a result, the House Committee on the Judiciary held hearings (Sept. 29 and Nov. 10, 1983, March 28, 1994) documenting society's dependence on credit cards and other financial instruments, many of which involved the use of computers and other electronic devices subject to criminal abuse. It became apparent that traditional theft/larceny statutes were not the proper vehicle to control computer abuse and computer assisted crimes. The Committee concluded that criminal laws needed 1) to be brought up to date, 2) to account for rapidly advancing technology, and 3) to shift attention from concepts such as "tangible property" to concepts of "information" and "access to information." The hearings led to passage of the Counterfeit Access Device and Computer Fraud and Abuse Act of 1984 (a part of the Comprehensive Crime Control Act).[10] This Act created 18 U.S.C. §§ 1029 and 1030 which made unauthorized access of classified information in a computer a felony and unauthorized access to financial records or credit histories in financial institutions or to trespass into a government computer a misdemeanor.[11]

In 1985, both the House and the Senate introduced legislation to expand the Computer Fraud and Abuse Act, 18 U.S.C. § 1030. Hearings were held before the Committee on the Judiciary to introduce variations. The Committee noted that, although financial institutions relied heavily on computer communications and were therefore especially vulnerable to attack, computer crime posed threats outside of the financial realm. After reviewing the American Bar Association Task Force on Computer Crime's Report

---

8 18 U.S.C. § 1341.

9 18 U.S.C. § 1343.

10 Pub. L. No. 98-473 (1984) (codified at 18 U.S.C. § 1030).

11 S. Rep. No. 357, 104th Cong., 2d Sess. 4 (1996).

on Computer Crime ("ABA Report"), which concluded that the most effective means of preventing computer crime was "more comprehensive and effective self-protection" by businesses, the Committee recommended the development of education programs for computer users and the general public to make people aware of the ethical and legal implications of using such technology. The Committee also considered issues of jurisdiction, and decided that, instead of enacting sweeping federal legislation, federal jurisdiction should be limited to cases in which there is a compelling federal interest. The result of the hearings was the Computer Fraud and Abuse Act of 1986 (CFAA).[12]

Congress amended the CFAA again in 1994. The 1994 amendments focused specifically on subsection (a)(5) of section 1030. The intent of the amendment was to expand the prohibition on damage to computers to cover not only "federal interest computers," but any "computer used in interstate commerce." This revision thus technically eliminated federal protection for government and financial institution computers which were not used in interstate commerce.

The 1996 amendments corrected this oversight in the revision of section 1030(a)(5) and also expanded the definition of damage required to trigger the prohibition. The statute, as amended, makes it a crime to:

- Knowingly access a computer without, or in excess of, authorization to obtain classified information.[13]
- Knowingly access a computer without, or in excess of, authorization to obtain (1) financial information from a credit reporting company or a financial institution, (2) any information in the possession of the government, and (3) any private information where the defendant's conduct involves interstate commerce.[14]
- Access, without authorization, a government computer–even if no damage is done or nothing is stolen.[15]
- Use a computer in a scheme to defraud victims of property.[16]
- Knowingly transmit a program (as well as information, code, or a command) and intentionally cause damage to a protected computer.[17]

---

12 Pub. L. No. 99-474 (1984).

13 18 U.S.C. 1030(a)(1) (1994).

14 18 U.S.C. 1030(a)(2) (1994). Obtaining information includes reading the information. Congress did not require that defendants copy or physically move the information. *Id.*

15 18 U.S.C. 1030(a)(3) 1994.

16 18 U.S.C. 1030(a)(4) 1994.

17 18 U.S.C. 1030(a)(5) 1994. Congress amended only this provision in 1995, clarifying

- Traffic in passwords or other similar information.[18]

Through the National Information Infrastructure Protection and Economic Espionage Acts of 1996, Congress expanded coverage of 18 U.S.C. § 1030 to:

- Protect civilian, state and local government computers from unauthorized access.
- Prohibit abuse of access privileges by government employees to obtain confidential or sensitive information.
- Include lost computer time within the fraud provision;
- Protect intrastate government computers and computers used in foreign communications and commerce from damage from viruses and other harmful computer programs. (A copy of the current law is attached as Appendix B.)

The 1996 amendments also added a new provision to section 1030. 18 U.S.C. § 1030(a)(7) prohibits the use of computers for blackmail or extortion.[19] The amendments also expanded the civil remedy by making it available for all seven provisions of 18 U.S.C. § 1030(a) and by expanding the definition of damage. "Damage" is either significant financial loss, potential impact on medical treatment, causing physical injury to any person, or threatening the public health or safety.[20]

While the CFAA is the major piece of federal legislation for prosecuting computer crimes, other federal legislation also touches on computer crime. The Electronic Communications Privacy Act of 1986 (ECPA)[21] updates the federal wiretap statute to protect the privacy and security of information transmitted through new computer and telecommunications technologies. The ECPA addresses the interception of wire, oral, and electronic communications, the unauthorized access to stored wire and electronic communications, and the use of pen registers and "trap and trace" devices. The ECPA protects the privacy of personal and proprietary information while at the same time satisfying the Federal government's legitimate law enforcement needs. In 1994, Congress supplemented the government's ability to conduct wiretaps through the Communications Assistance for Law Enforcement Act (CALEA) which sets out requirements and procedures for

---

that computer viruses and other similar attack mechanisms were included in the orbit of the statute.

18 18 U.S.C. 1030(a)(6) 1994.

19 The legislative history provides an example of the sort of act Congress intended to criminalize in subsection (a)(7)— "hackers penetrate a system, encrypt a database and then demand money for the decoding key." S. Rep. No. 357, 104th Cong., 2d Sess. 12 (1996).

20 *See*18 U.S.C. § 1030(e)(8).

21 Pub. L. No. 99-508 (1986) (codified at 18 U.S.C. § 2505 *et seq.*).

communications carrier cooperation with court ordered wiretaps.[22] In spite of this step forward, the procedural aspects of computer crime investigation and prosecution have not been as active an area for Congressional initiatives and lag well behind the substantive statutes.

Congress additionally passed legislation in 1996 primarily to address foreign government-sponsored economic espionage or theft of trade secrets.[23] The Economic Espionage Act of 1996 (attached as Appendix C) criminalizes various computer-related acts by any person, as well as persons "intending or knowing that the offense will benefit any foreign government, foreign instrumentality, or foreign agent[. . .]"[24] Prohibited acts to obtain trade secrets include:

- stealing or fraudulently obtaining;
- copying or downloading;
- knowingly purchasing or obtaining; and
- conspiring with others to obtain the trade secret.

The United States Sentencing Commission has recently revised the Sentencing Guidelines for computer crime and economic espionage. The amendments are set to go into effect in November, 1997. Unless the Guidelines are modified by Congress, they will provide for, among other things, an expanded definition of loss. When a defendant is convicted of trespass or theft under 18 U.S.C. § 1030(a)(3) or § 1831-32, upward departures in sentencing may be considered based not only on monetary losses, but also costs to the victim for damage assessments, restoring the system or data, and any lost revenue due to interruption of service. An upward departure may also be warranted in situations where the computer crime does not cause a loss per se, but results in a substantial invasion of privacy. If convicted under 18 U.S.C. § 1030(a)(5), an enhanced sentence will be appropriate if the computer crime caused a delay in medical treatment or disruption of important government, or private, services.

International Computer Crime Initiatives

With the advent of the Internet as a global resource, computer crimes are committed increasingly across national boundaries, potentially invoking different legal systems and traditions. While significant work has been done during the past 25 years by several multilateral organizations on substantive computer-related crimes, international organizations have only recently been devoting significant time and attention to procedural

---

22 Pub. L. No. 103-414 (1994) (codified at 18 U.S.C. § 2601 *et seq.*).

23 Economic Espionage Act of 1996, P.L. 105-15, 18 U.S.C. §§ 1831-1839 (1996).

24 *Id.* at 18 U.S.C. § 1831 (a)(1) - (a)(5).

issues raised by the need for cross-jurisdiction investigations. The most pressing issues being considered today include:

- mutual legal assistance/"dual criminality" requirements;
- extradition and prosecution;
- "rapid trace" capabilities;
- cross jurisdictional search and seizure; and
- evidence collection and computer forensics.

The United States has been an active participant in many of these efforts and has the potential to act as a leader in bringing many of these organizations' recommendations to fruition.

Organization for Economic Development (OECD)

OECD involvement in addressing computer crime and the use and abuse of computer networks began with several peripheral issues. In 1980, the OECD formally adopted its *Guidelines Governing the Protection of Privacy and Transborder Data Flows of Personal Data.* By 1992, the OECD had adopted well-known Guidelines for computer networks in the *Security of Information Systems.* Today, the OECD's focus is primarily on information systems, international telecommunications policy and the Global Information Infrastructure (GII). Currently, an OECD ad hoc Group of Experts is drafting Cryptography Policy Guidelines to assist governments and the private sector in developing guidelines for the use of cryptography technology in order to advance the development of the GII. The OECD hopes to balance these developments with traditional policies covering national security and public safety.

Several members of the OECD hope to focus the organization on the Internet. In particular, France and England are now arguing that the OECD should study, and later pass, international legal standards for the Internet. The French are circulating their own proposal. The OECD is also considering other areas of expansion, including computer crimes, however no specific work has been done to date.

Council of Europe (COE)

The Council of Europe has passed several well-known Recommendations in the area of computer crime. During the past ten years, the COE's European Committee on Crime Problems (CDPC)–a standing committee-commissioned two ad hoc groups to address issues of information technology and criminal law.

The first group focused on substantive criminal issues and produced *Computer-Related Crime,* an extensive report which lead to the well-known COE Recommendation No. R. (89) 9. This Recommendation, which the whole

Council of Ministers adopted in 1989, lists various acts that nations should criminalize as computer crimes.[25]

In 1992, the COE commissioned a second ad hoc Group to focus on procedural issues associated with computer-related crimes. The Committee of Experts on Problems of Criminal Procedure Law Connected with Information Technology produced a second Recommendation.[26] The whole Council of Ministers adopted Recommendation No. R. (95) 13 in September, 1995. The Appendix to Recommendation no. R (95) 13 recommends that countries create various procedural mechanisms to facilitate investigation and prosecution of crimes involving information technology. The Appendix specifically recommends that rules for search and seizure, technical surveillance, electronic evidence, cooperation with investigators, and international cooperation be reviewed and revised to ensure that they take into adequate account the unique nature of information technology.

Informally, the COE is indicating that it will soon commission a new Group of Experts on Crimes in Cyberspace. The agenda is broad and includes both substantive and procedural issues.

The European Union (EU)

The EU is undertaking several projects that complement the work done by the OECD and the COE. These projects cover topics including:

- Training
- Harmful content on the Internet
- Improving Mutual Legal Assistance Treaties (MLATs) and producing a new European MLAT
- Internet and EU Telecom Issues

G7/P8

Of all the multilateral and international organizations, the G7/P8 is perhaps the most focused in the area of procedural mechanisms to facilitate international cooperation.

The P8 evolved when Russia joined the Group of Seven, or G-7 nations. The P8 then adopted an international agenda that reached beyond the historically economic focus of the G-7, addressing both political and global issues. The Eight–Canada, England, France, Germany, Italy, Japan, Russia, and the

---

25 These acts included computer fraud, computer forgery, damage to computer data or programs, computer sabotage, unauthorized access, and unauthorized interception. There were also a number of optional crimes such as alteration of data or programs, computer espionage, and unauthorized computer use.

26 Recommendation No. R(95)13.

United States–recently embarked on two major initiatives–terrorism and transnational organized crime.

The P8's approach is both aggressive and practical. In the Communiqué from the recent Summit of the Eight in Denver (June 20-22, 1997), aggressive action was called for with regard to two areas of concern:

> *"First, the investigation, prosecution, and punishment of high-tech criminals, such as those tampering with computer and telecommunications technology, across national borders; Second, a system to provide all governments the technical and legal capabilities to respond to high-tech crimes, regardless of where the criminals may be located."*[27]

The High-tech Crime Subgroup of the Senior Experts' Group of the Eight on Transnational Organized Crime are currently working to implement "forty recommendations" relating to transnational organized crime, including computer crime, adopted by the Group in April 1996. The Subgroup is focusing on three areas specifically–(1) locating and identifying the perpetrator of a computer offense ("trap and trace"); (2) gaining access to the content of communications; (3) international cooperation in the collection of evidence (in particular "transborder searches").

The Group has since (in December 1997) issued a communiqué setting forth objectives and an Action Plan for improving response to computer and other high technology crime.

U.S. Involvement/Next Steps

In addition to these multilateral organizations, many foreign nations are working independently to improve the international environment for the investigation and prosecution of computer crime. Neighbors and allies, such as Canada and the United Kingdom, are actively working the problem. The United States government is an active participant and follows the efforts of these international bodies and individual nations closely. The Department of State, Department of Justice, and Federal Bureau of Investigation are just a few of agencies involved in the international battle against computer crime. These agencies and the interest of the Administration and Congress provide conduits for directing and forwarding the efforts of our international partners.

PART THREE
LEGAL AND LEGISLATIVE OPTIONS

The sheer number of options and permutations offered by bodies studying these issues has necessitated something of a novel approach. Possible approaches identified in this paper are organized by the jurisdiction to which

---

27 Communiqué, Summit of the Eight, § 40 (June 22, 1997).

they apply–state and local, Federal, and international–and then further by the aspect of criminal law on which they focus (whether substantive, procedural or pertaining to resources). To present the analysis better, each option has a "Pro" and "Con" element.

State and Local (Substantive)

Pursue Alternatives to Incarceration and Fines in Their Sentencing Schemes

Although a thorough study of state sentencing schemes was beyond the scope of this paper, it may be feasible for states to continue to create novel remedies that would increase deterrence and provide compensation for victims.

- **Pro:** Recognizes efforts by states that have adopted novel sentencing schemes to deal with unique nature of the crime and those who commit it. Restitution and other remedies, if uniformly provided, may increase the deterrent value of prosecutions by minimizing financial gain.
- **Con:** The value of innovative sentencing schemes has not adequately been demonstrated by states enacting them.

Consider Adoption of Computer Crime Laws Specifically Tailored to Juvenile Offenders

While the increased "detention" of juvenile offenders may be unpopular, some recognition of the need to address the problem they present may be warranted. Ethics training in schools or other vehicles may provide a more appropriate sentencing vehicle through which to "refocus" potential computer criminals.

- **Pro:** By drawing attention to volume of computer crimes perpetrated by juveniles and the weaknesses of some criminal schemes to prosecute or deter them, with further study, recommendations may ultimately be made that will be successful in reducing the number of intrusions by "recreational hackers." This would allow law enforcement to focus on truly destructive computer criminals.
- **Con:** Legal planning related to juvenile offenders is arguably best led by the states themselves, as Federal law enforcement is significantly less experienced in dealing with juvenile justice issues than are the states.

State and Local (Resources)

Devote Increased Resources to State and Local Enforcement of Computer Crime

Computer crime investigations place a tremendous drain on the resources of state and local authorities. Violent crimes are currently–and properly–given priority by strapped departments. An increase in resources would allow states and local jurisdictions to train officers or hire experienced computer security experts, to purchase equipment, and to expend additional man-hours on computer crime investigations. It may be prudent to examine increased training for or hiring of computer-literate prosecutors.

- **Pro:** A more effective state and local response may relieve pressure on federal law enforcement.
- **Con:** With the advent of the Internet, exclusively "local" computer crimes are likely to be increasingly rare. Jurisdictional restrictions often prevent state and local law enforcement from investigating cross-jurisdictional offenses on their own, often requiring inter-state cooperation or federal law enforcement assistance. Thus, resources might be better spent on a coordinated federal response.

Conclusions (State and Local)

The Administration, through the Department of Justice, Should Sponsor, a Comprehensive Study Aimed at Discovering Effective Ways of Deterring and Responding to Computer Crime and Abuse by Juveniles

The study would focus on the effectiveness of state-level handling of computer crime incidents, state computer crime laws, procedures for handling juvenile offenders, and sentencing regimes. It would determine whether modifications to applicable federal laws might be warranted. Study participants might include representatives of federal, state and local law enforcement; prosecutors offices; criminal defense attorneys; social services/juvenile justice experts; and representatives from the educational community. The study's aim would be to gather statistics on juvenile offenders (ages, recidivism rates, punishments, criminal activity as adults); examine how well juvenile issues are being covered by federal and state law enforcement efforts (*i.e.*, are there jurisdictional gaps allowing juveniles to "slip through the cracks"?); evaluate the effectiveness of sentencing provisions specifically tailored to juveniles (*e.g.*, community service requirements or extraordinary remedies such as expulsion from school); and promote education and awareness (e.g., ethics training, promote awareness of parents and educators). The study panel should be instructed to offer recommendations to state and federal governments for enhancing efforts to discourage hacker activity among juveniles and to promote responsible computer use.

Federal (Substantive)

Expand Sentencing Guidelines Treatment of "Harm" and "Loss" as Used for 18 U.S.C. § 1030 to Other Electronic Crimes

The U.S. Sentencing Commission, in new guidelines for computer crime, has taken the approach of increasing the deterrent effect of prosecutions by expanding the definitions of harm and loss to include interruptions in service; disruptions or delays in delivery of vital services endangering lives; invasions of privacy; and the cost to the victim of damage assessment, restoration of service and data, and loss of business or revenue due to interruption of service. This approach to sentencing could be adopted for other electronic crimes, including violations of the Wiretap statute and the Electronic Communications Privacy Act.

- **Pro:** The new amendments to sentencing guidelines tend to make punishment more proportional to overall damage done which could be an important consideration in deterring cyber attacks that disrupt critical infrastructures.
- **Con:** Amendments generally take time to be studied by the Sentencing Commission and may have to come as a result of Congressional prompting, which has to date caused a considerable backlog. Sentencing as a deterrent is reliant on reporting of an incident, successful investigation and prosecution. This approach may have limited deterrent value on its own.

Increase Attention to the Issue of Juvenile Computer Crime Offenders

Juvenile computer crime may merit study at the Federal level. Advocates can capitalize on existing federal initiatives which could be used to promote studies and recommendations for dealing with the issue (e.g., legislative initiatives, grant programs, education and awareness programs, research and study bodies). Timing of issue may be influenced by the current Federal juvenile justice statute which is set to expire in September, 1997.

- **Pro:** Federal attention may result in greater awareness of parents, media, and educators to the issue of computer abuse by juveniles. Federal attention may result in a more coordinated and well-rounded approach to dealing with juvenile hackers.
- **Con:** The Federal government has a limited role in criminal prosecutions of juveniles. The issue may be more appropriately handled at the state level. If the issue is not framed carefully, it may receive a negative public reaction from quarters already concerned about the treatment of the juveniles under the criminal law.

Federal (Procedural)

Improve Law Enforcement's Ability to Investigate Through Enhanced Cooperation of Victims and Witnesses

Recognizing difficulties inherent in bolstering law enforcement effectiveness through changes in criminal procedure, it may be preferable to encourage measures that would improve victim and witness cooperation and reporting of criminal incidents. Current victim/witness assistance legislation may provide a model to the extent it provides for a right to restitution and a right to be kept abreast of investigation and prosecution efforts. As an incentive for increased reporting, restitution will only work to the extent the defendant is able to pay. Mandatory requirements for reporting intrusions of a certain magnitude would be included in this general category.

- **Pro:** Additional reporting would provide information that law enforcement might otherwise have to acquire through burdensome procedures, and may obviate the need for procedural reform.
- **Con:** Any form of mandatory reporting is likely to be met with equal or greater opposition than would revisions to procedural criminal law. Incentives for reporting or reporting requirements can be costly and difficult to administer.

Identify And Endorse Viable Alternatives to Fortification of Criminal Enforcement Regime

Due to the relative intransigence of existing criminal procedures and the unlikely accomplishment of meaningful reform, Congress might consider viable alternatives to criminal enforcement to instill enhanced deterrence. (*See* PCCIP supplemental report *Approaches to Cyber Intrusion Response* for discussion and options).

Identify Procedural Changes

Congress may wish to consider revisions to procedural laws and rules governing how investigations can be performed in electronic environments. Specific recommendations could include new legislation on performing surveillance in an electronic environment (analogous to the current Wiretap Act, but written specifically for the Internet); a national "trap and trace" capability; or increased ability to monitor systems under 18 U.S.C. § 2511(2)(a)(i) (which currently provides for limited monitoring by service providers to operate and maintain their systems).

- **Pro:** Certain to stir immediate and vigorous discussion.
- **Con:** An immediate and vigorous debate will ensue, due to concerted opposition from civil libertarians, privacy advocates, and the web-surfing public. It is unclear whether significant improvements can be obtained without running afoul of constitutional considerations.

Federal (Resources)

Fortify Law Enforcement Resources Under Existing Substantive and Procedural Regime

Recognizing the relative adequacy of federal substantive criminal law and the limited likelihood of achieving meaningful modification to procedural criminal law, additional resources can be directed to overcoming difficulties inherent in the investigation and prosecution of crimes that arise in electronic environments.

- **Pro:** Avoids staking out potentially controversial position between law enforcement and civil liberties concerns, instead resorting to relatively uncontroversial, "more cops on the beat" approach.
- **Con:** Law enforcement response not cost-effective. Advocating enhancement of law enforcement resources to address high technology crime can be controversial, as it carries potential implication that fewer law enforcement resources might be available for other pressing concerns (violent crime, narcotics offenses, etc.).

Conclusions (Federal)

The Administration Should Promote Current Efforts to Develop Specific Procedural Changes to Assist Law Enforcement in the Investigation of Computer Crime.

The Department of Justice is currently exploring ways to ease undue administrative burdens on law enforcement officers investigating various forms of computer and high technology crimes that cross federal jurisdictional boundaries. There is a need for a national "trap and trace" capability and an analogous national search warrant capability that would allow for process obtained in one jurisdiction to be used in multiple jurisdictions involved in the offense under investigation. The Administration and Congress should also be sensitive to such needs, given the value of a rapid law enforcement response. Where currently law enforcement investigators are required to obtain court orders or warrants for each jurisdiction they will enter, a national capability would allow electronic searches or "trap and trace" efforts to be conducted across jurisdictional boundaries with the authorization of only one judge. This would not constitute an expansion of law enforcement's current ability to perform these vital functions, but would merely increase the speed and efficiency of their efforts in the computer crime investigation arena.

International (Procedural)

Clarify and Improve Current Procedures

The United States and other countries could pursue clarifications and improvements to their current procedures for investigating computer crime by, for example, a) clearly delineating and applying a distinction between searching computer systems and seizing data stored therein and intercepting data in the course of its transmission; b) granting investigators the authority, while executing a computer search, to extend the search to any computer system in their jurisdiction which is connected by a network and to seize the data therein, provided immediate action is necessary; c) enabling interception of telecommunications and collection of traffic data in investigations of serious offenses against the confidentiality, integrity, and availability of telecommunications or computer systems; and d) reviewing legal standards for collection of service data to ensure that they are appropriate to the level of personal intrusion.

- **Pro:** Clarifying procedural rules would benefit law enforcement both domestically and internationally. The U.S. may choose to be the leader and urge other nations to follow.
- **Con:** Many of the improvements and clarifications recommended by international bodies implicate complex statutory, constitutional, and policy questions in the United States.

Work to Create a Network of International Law Enforcement Agencies and Telecommunications Carriers to Draw Upon During Investigations

The creation of such a network of dedicated individuals would facilitate prompt "trap and trace" and identification of foreign sources of intrusions. Network members may also be expected to exchange information, engage in training exercises, and share technological innovations. (Note: P8 working groups have consistently highlighted the importance of cooperative networks for addressing law enforcement issues.)

- **Pro:** A network of trained personnel will provide an in-place capability to respond to computer intrusions quickly and effectively.
- **Con:** Developing a network alone will not facilitate investigations without clarification of procedures for obtaining international cooperation and resolution of dual criminality requirements in advance.

Promote International Agreements to Facilitate Cooperation in Computer Investigations

The U.S. currently has Mutual Legal Assistance Treaties (MLATs) in place with a number of countries to facilitate investigations of computer crimes. Various international organizations are analyzing alternative legal models to MLATs, which are typically variations on traditional and cumbersome procedures for international law enforcement cooperation. The U.S. could nonetheless encourage widespread adoption of agreements, or other vehicles, to improve international cooperation in computer crime, especially among countries who are not members of the primary international organizations working in this area (e.g., P8, OECD, COE).

- **Pro:** Well-crafted agreements will allow investigations to proceed at the speed necessary for successful identification of intruders and still observe international conventions for cooperation.
- **Con:** It may take a considerable amount of time to put in place all necessary agreements to allow a successful investigation. Hackers may be able to use countries not parties to such agreements to shield themselves from investigations.

International (Resources)

Improve Awareness of Sound Computer Forensic Principles and Applications

Gathering of computer evidence is a time-sensitive process which requires careful attention to procedures in order to assure authenticity of the evidence. Use of foreign investigators to collect evidence in U.S. intrusion investigations greatly forwards investigations and chances for prosecution provided the evidence is properly collected. It may be in the U.S. law enforcement interest to educate the international community of investigators, or even to work together to determine guidelines for proper computer forensic procedures. (Note: This option is based on the considerable work of P. Sommer in computer forensics.[28])

- **Pro:** This approach can be expected to increase the availability of evidence of foreign sources of intrusions to be introduced in U.S. courts and may contribute to successful prosecutions.
- **Con:** May be difficult to harmonize procedures internationally. The U.S. will have to work carefully with other participating countries to ensure that U.S. criminal procedure and evidentiary requirements are met.

Improve Technological Tools Available for Locating the Sources of Intrusion

The U.S. may want to stimulate an international effort at developing tools to assist in locating the sources of unauthorized intrusions. (Note: The P8

---

28 *See* P. Sommer, Forensic Computing, Computer Security Research Centre, London School of Economics & Political Science (1995), reprinted at http://csrc.lse.ac.uk/csrc/forncomp.htm.

has noted similar concerns about the tools of law enforcement in its study of computer crime.)

- **Pro:** New detection and location tools may have the capability to be tailored to enhance privacy by being less intrusive into the systems passed through during the course of an investigation. Coordination of research and development would bring the international community together and draw on skills and resources of a variety of countries. International guidelines for using such tools may develop in concert with development of the technology.
- **Con:** Research and development is costly and new tools may bring limited returns. Tools will still have to conform to U.S. laws governing tracking down intruders. Technology is moving so quickly that hackers will likely develop tricks to elude law enforcement as quickly as law enforcement develops new tools.

Create Specialized Computer Crime Units on an International Scale

Units would be specially trained and well-versed in the complexities of international investigations. Units would provide an existing network to carry out international investigations. (Note: This recommendation has been put forward by the Council of Europe.)

- **Pro:** Specialized units could be quickly mobilized for an international investigation. They could receive specialized training and would be expected to develop close working relationships across international borders.
- **Con:** Supporting such operations may detract from resources available domestically. There may be problems associated with close investigatory relationships with foreign investigators surveilling U.S. citizens. It would require resources to establish and maintain excellence.

Conclusions (International)

The Administration and Congress Should Continue to Support the Efforts by U.S. Delegations to Several International Bodies to Further Infrastructure Assurance Objectives and Highlight Potential Areas for Continued or Expanded Efforts in the International Area.

The U.S. has been a forerunner in efforts to clarify and improve current procedures for investigating computer crimes. The Administration and Congress should continue to recognize and support these efforts. This may require, for example, careful review of existing standards and procedures for searching and intercepting electronic data during an investigation, as well as support of internationally focused efforts to assure that other countries have the legal and technical capability to conduct and assist in computer crime investigations. Second, the United States should continue with efforts to establish a sound network of international law enforcement agencies and telecommunications carriers to draw on in investigations. The creation and maintenance of such a network can facilitate prompt "trap and trace" and identification of foreign sources of intrusions. Third, the United States should continue to promote efforts to enhance international cooperation in computer crime investigations whether through new international arrangements or traditional mutual legal assistance treaties. The U.S. should ensure that existing and future efforts are broad enough to cover the broadest range of potential sources of attack (i.e. where possible, agreements should cover more than members of P8, OECD, and COE.)

PART FOUR
A UNIFIED APPROACH TO CRIMINAL DETERRENCE

Perhaps the most powerful deterrent to cyber crime is a maximally effective criminal enforcement regime. A maximally effective response to computer crime requires the availability of adequate laws and resources at the state, federal and international level.

The effectiveness of a criminal enforcement regime depends in large part on the adequacy of underlying laws and enforcement resources. Adequate laws and enforcement resources. Adequate laws and resources should ideally support both the prosecutorial and investigative functions. Substantive criminal law should clearly define the prohibited behavior and set forth appropriate punishment. Procedural law should facilitate the conduct of investigations, within clearly defined legal boundaries, while respecting the rights of the accused and of the public. Investigative and prosecutive resources should be allocated in such a way as they remain balanced–they should grow in proportion to one another to avoid gluts and bottlenecks.

Owing to the unique, multi-jurisdictional nature of computer-related offenses, it is insufficient for any single jurisdiction to maximize the effectiveness of only its own law enforcement regime. Savvy computer criminals can choose targets and the routes to those targets, and can, in this way, take advantage of "weak links." Difficulties in investigating and prosecuting offenses occurring across state lines has fueled a need for a more robust national response. Difficulties in investigating and prosecuting offenses across national boundaries has likewise fueled the need for a more robust international response.

It is not enough to strengthen any one of the areas discussed in this paper without seeking to strengthen the others. The enhancement of procedural elements must be done in parallel with enhancement of substantive elements. If one is left with the status quo, improvements in the other will lose their effectiveness. Likewise, focusing on any one of the three major levels of government shown here without the other two will simply permit criminals to focus their activities at the level which provides the least amount of risk for them. Thus, a unified approach is the only one with a high probability of success.

To help understand the scope and complexity of the options, the following chart summarizes them. The reader should note that none of them are mutually exclusive.

Components of Criminal Deterrence

State & Local	LAWS	RESOURCES
INVESTIGATION	procedural	$$, personnel, training, equipment, etc...
PROSECUTION	substantive	$$, personnel, training, etc...

Federal	LAWS	RESOURCES
INVESTIGATION	procedural	$$, personnel, training, equipment, etc...
PROSECUTION	substantive	$$, personnel, training, etc...

International	LAWS	RESOURCES
INVESTIGATION	procedural	$$, personnel, training, equipment, etc...
PROSECUTION	substantive	$$, personnel, training, etc...

APPENDIX A

UNITED STATES CODE ANNOTATED
TITLE 18. CRIMES AND CRIMINAL PROCEDURE
PART I—CRIMES
CHAPTER 47—FRAUD AND FALSE STATEMENTS

Current through P.L. 105-15, approved 5-15-97

§ 1030. Fraud and related activity in connection with computers

(a) Whoever—

(1) having knowingly accessed a computer without authorization or exceeding authorized access, and by means of such conduct having obtained information that has been determined by the United States Government pursuant to an Executive order or statute to require protection against unauthorized disclosure for reasons of national defense or foreign relations, or any restricted data, as defined in paragraph y of section 11 of the Atomic Energy Act of 1954, with reason to believe that such information so obtained could be used to the injury of the United States, or to the advantage of any foreign nation willfully communicates, delivers, transmits, or causes to be communicated, delivered, or transmitted, or attempts t communicate, deliver, transmit, or cause to be communicated, delivered, or transmitted the same to any person not entitled to receive it, or willfully retains the same and fails to deliver it to the officer or employee of the United States entitled to receive it;

(2) intentionally accesses a computer without authorization or exceeds authorized access, and thereby obtains—

(A) information contained in a financial record of a financial institution, or of a card issuer as defined in section 1602(n) of title 15, or contained in a file of a consumer reporting agency on a consumer, as such terms are defined in the Fair Credit Reporting Act (15 U.S.C. 1681 et seq.);

(B) information from any department or agency of the United States; or

(C) information from any protected computer if the conduct involved an interstate or foreign communication;

(3) intentionally, without authorization to access any nonpublic computer of a department or agency of the United States, accesses such a computer of that department or agency that is exclusively for the use of the Government of the United States or, in the case of a computer not exclusively for such use, is used by or for the Government of the United

States and such conduct affects that use by or for the Government of the United States;

(4) knowingly and with intent to defraud, accesses a protected computer without authorization, or exceeds authorized access, and by means of such conduct furthers the intended fraud and obtains anything of value, unless the object of the fraud and the thing obtained consists only of the use of the computer and the value of such use is not more than $5,000 in any 1-year period;

(5)(A) knowingly causes the transmission of a program, information, code, or command, and as a result of such conduct, intentionally causes damage without authorization, to a protected computer;

(B) intentionally accesses a protected computer without authorization, and as a result of such conduct, recklessly causes damage; or

(C) intentionally accesses a protected computer without authorization, and as a result of such conduct, causes damage;

(6) knowingly and with intent to defraud traffics (as defined in section 1029) in any password or similar information through which a computer may be accessed without authorization, if—

(A) such trafficking affects interstate or foreign commerce; or

(B) such computer is used by or for the Government of the United States;

(7) with intent to extort from any person, firm, association, educational institution, financial institution, government entity, or other legal entity, any money or other thing of value, transmits in interstate or foreign commerce any communication containing any threat to cause damage to a protected computer;

shall be punished as provided in subsection (c) of this section.

(b) Whoever attempts to commit an offense under subsection (a) of this section shall be punished as provided in subsection (c) of this section.

(c) The punishment for an offense under subsection (a) or (b) of this section is—

(1)(A) a fine under this title or imprisonment for not more than ten years, or both, in the case of an offense under subsection (a)(1) of this section which does not occur after a conviction for another offense under this section, or an attempt to commit an offense punishable under this subparagraph; and

(B) a fine under this title or imprisonment for not more than twenty years, or both, in the case of an offense under subsection (a)(1) of this section which occurs after a conviction for another offense under this section, or an attempt to commit an offense punishable under this subparagraph;

(2)(A) a fine under this title or imprisonment for not more than one year, or both, in the case of an offense under subsection (a)(2), (a)(3), (a)(5)(C), or (a)(6) of this section which does not occur after a conviction for another offense under this section, or an attempt to commit an offense punishable under this subparagraph; and

(B) a fine under this title or imprisonment for not more than 5 years, or both, in the case of an offense under subsection (a)(2), if—

(i) the offense was committed for purposes of commercial advantage or private financial gain;

(ii) the offense was committed in furtherance of any criminal or tortious act in violation of the Constitution or laws of the United States or of any State; or

(iii) the value of the information obtained exceeds $5,000;

(C) a fine under this title or imprisonment for not more than ten years, or both, in the case of an offense under subsection (a)(2), (a)(3), or (a)(6) of this section which occurs after a conviction for another offense under this section, or an attempt to commit an offense punishable under this subparagraph; and

(3)(A) a fine under this title or imprisonment for not more than five years, or both, in the case of an offense under subsection (a)(4), (a)(5)(A), (a)(5)(B), or (a)(7) of this section which does not occur after a conviction for another offense under this section, or an attempt to commit an offense punishable under this subparagraph; and

(B) a fine under this title or imprisonment for not more than ten years, or both, in the case of an offense under subsection (a)(4), (a)(5)(A), (a)(5)(B), (a)(5)(C), or (a)(7) of this section which occurs after a conviction for another offense under this section, or an attempt to commit an offense punishable under this subparagraph; and

[**(4)** Repealed. Pub. L. 104-294, Title II, § 201(2)(D), Oct. 11, 1996, 110 Stat.3493]

(d) The United States Secret Service shall, in addition to any other agency having such authority, have the authority to investigate offenses under subsections (a)(2)(A), (a)(2)(B), (a)(3), (a)(4), (a)(5), and (a)(6) of this section. Such authority of the United States Secret Service shall be exercised in accordance with an agreement which shall be entered into by the Secretary of the Treasury and the Attorney General.

(e) As used in this section—

(1) the term "computer" means an electronic, magnetic, optical, electrochemical, or other high speed data processing device performing logical, arithmetic, or storage functions, and includes any data storage facility or communications facility directly related to or operating in conjunction with such device, but such term does not include an automated typewriter or typesetter, a portable hand held calculator, or other similar device;

(2) the term "protected computer" means a computer—

(A) exclusively for the use of a financial institution or the United States Government, or, in the case of a computer not exclusively for such use, used by or for a financial institution or the United States Government and the conduct constituting the offense affects that use by or for the financial institution or the Government; or

(B) which is used in interstate or foreign commerce or communication;

(3) the term "State" includes the District of Columbia, the Commonwealth of Puerto Rico, and any other commonwealth, possession or territory of the United States;

(4) the term "financial institution" means—

(A) an institution with deposits insured by the Federal Deposit Insurance Corporation;

(B) the Federal Reserve or a member of the Federal Reserve including any Federal Reserve Bank;

(C) a credit union with accounts insured by the National Credit Union Administration;

(D) a member of the Federal home loan bank system and any home loan bank;

(E) any institution of the Farm Credit System under the Farm Credit Act of 1971;

(F) a broker-dealer registered with the Securities and Exchange Commission pursuant to section 15 of the Securities Exchange Act of 1934;

(G) the Securities Investor Protection Corporation;

(H) a branch or agency of a foreign bank (as such terms are defined in paragraphs (1) and (3) of section 1(b) of the International Banking Act of 1978); and

(I) an organization operating under section 25 or section 25(a) of the Federal Reserve Act.

(5) the term "financial record" means information derived from any record held by a financial institution pertaining to a customer's relationship with the financial institution;

(6) the term "exceeds authorized access" means to access a computer with authorization and to use such access to obtain or alter information in the computer that the accessor is not entitled so to obtain or alter;

(7) the term "department of the United States" means the legislative or judicial branch of the Government or one of the executive departments enumerated in section 101 of title 5; and

(8) the term "damage" means any impairment to the integrity or availability of data, a program, a system, or information, that—

(A) causes loss aggregating at least $5,000 in value during any 1-year period to one or more individuals;

(B) modifies or impairs, or potentially modifies or impairs, the medical examination, diagnosis, treatment, or care of one or more individuals;

(C) causes physical injury to any person; or

(D) threatens public health or safety; and

(9) the term "government entity" includes the Government of the United States, any State or political subdivision of the United States, any foreign country, and any state, province, municipality, or other political subdivision of a foreign country.

(f) This section does not prohibit any lawfully authorized investigative, protective, or intelligence activity of a law enforcement agency of the United States, a State, or a political subdivision of a State, or of an intelligence agency of the United States.

(g) Any person who suffers damage or loss by reason of a violation of this section may maintain a civil action against the violator to obtain compensatory damages and injunctive relief or other equitable relief. Damages for violations involving damage as defined in subsection (e)(8)(A) are limited to economic damages. No action may be brought under this subsection unless such action is begun within 2 years of the date of the act complained of or the date of the discovery of the damage.

(h) The Attorney General and the Secretary of the Treasury shall report to the Congress annually, during the first 3 years following the date of the enactment of this subsection, concerning investigations and prosecutions under subsection (a)(5).

APPENDIX B

UNITED STATES PUBLIC LAWS
104th Congress - Second Session
Convening January 3, 1996

PL 104-294 (HR 3723)
October 11, 1996
ECONOMIC ESPIONAGE ACT OF 1996

An Act to amend title 18, United States Code, to protect proprietary economic information, and for other purposes.

Be it enacted by the Senate and House of Representatives of the United States of America in Congress assembled,

SECTION 1. SHORT TITLE.

This Act may be cited as the "Economic Espionage Act of 1996".

TITLE I—PROTECTION OF TRADE SECRETS

SEC. 101. PROTECTION OF TRADE SECRETS.

(a) IN GENERAL.—Title 18, United States Code, is amended by inserting after chapter 89 the following:

"CHAPTER 90—PROTECTION OF TRADE SECRETS

"Sec.

"1831. Economic espionage.

"1832. Theft of trade secrets.

"1833. Exceptions to prohibitions.

"1834. Criminal forfeiture.

"1835. Orders to preserve confidentiality.

"1836. Civil proceedings to enjoin violations.

"1837. Conduct outside the United States.

"1838. Construction with other laws.

"1839. Definitions.

"§ 1831. Economic espionage

"(a) IN GENERAL.—Whoever, intending or knowing that the offense will benefit any foreign government, foreign instrumentality, or foreign agent, knowingly—

"(1) steals, or without authorization appropriates, takes, carries away, or conceals, or by fraud, artifice, or deception obtains a trade secret;

"(2) without authorization copies, duplicates, sketches, draws, photographs, downloads, uploads, alters, destroys, photocopies, replicates, transmits, delivers, sends, mails, communicates, or conveys a trade secret;

"(3) receives, buys, or possesses a trade secret, knowing the same to have been stolen or appropriated, obtained, or converted without authorization;

"(4) attempts to commit any offense described in any of paragraphs (1) through (3); or

"(5) conspires with one or more other persons to commit any offense described in any of paragraphs (1) through (3), and one or more of such persons do any act to effect the object of the conspiracy,

shall, except as provided in subsection (b), be fined not more than $500,000 or imprisoned not more than 15 years, or both.

"(b) ORGANIZATIONS.—Any organization that commits any offense described in subsection (a) shall be fined not more than $10,000,000.

"§ 1832. Theft of trade secrets

"(a) Whoever, with intent to convert a trade secret, that is related to or included in a product that is produced for or placed in interstate or foreign commerce, to the economic benefit of anyone other than the owner thereof, and intending or knowing that the offense will, injure any owner of that trade secret, knowingly—

"(1) steals, or without authorization appropriates, takes, carries away, or conceals, or by fraud, artifice, or deception obtains such information;

"(2) without authorization copies, duplicates, sketches, draws, photographs, downloads, uploads, alters, destroys, photocopies, replicates, transmits, delivers, sends, mails, communicates, or conveys such information;

"(3) receives, buys, or possesses such information, knowing the same to have been stolen or appropriated, obtained, or converted without authorization;

"(4) attempts to commit any offense described in paragraphs (1) through (3); or

"(5) conspires with one or more other persons to commit any offense described in paragraphs (1) through (3), and one or more of such persons do any act to effect the object of the conspiracy, shall, except as provided in subsection (b), be fined under this title or imprisoned not more than 10 years, or both.

"**(b)** Any organization that commits any offense described in subsection (a) shall be fined not more than $5,000,000.

"§ 1833. Exceptions to prohibitions

"This chapter does not prohibit—

"(1) any otherwise lawful activity conducted by a governmental entity of the United States, a State, or a political subdivision of a State; or

"(2) the reporting of a suspected violation of law to any governmental entity of the United States, a State, or a political subdivision of a State, if such entity has lawful authority with respect to that violation.

"§ 1834. Criminal forfeiture

"**(a)** The court, in imposing sentence on a person for a violation of this chapter, shall order, in addition to any other sentence imposed, that the person forfeit to the United State—

"(1) any property constituting, or derived from, any proceeds the person obtained, directly or indirectly, as the result of such violation; and

"(2) any of the person's property used, or intended to be used, in any manner or part, to commit or facilitate the commission of such violation, if the court in its discretion so determines, taking into consideration the nature, scope, and proportionality of the use of the property in the offense.

"**(b)** Property subject to forfeiture under this section, any seizure and disposition thereof, and any administrative or judicial proceeding in relation thereto, shall be governed by section 413 of the Comprehensive Drug Abuse Prevention and Control Act of 1970 (21 U.S.C. 853), except for subsections (d) and (j) of such section, which shall not apply to forfeitures under this section.

"§ 1835. Orders to preserve confidentiality

"In any prosecution or other proceeding under this chapter, the court shall enter such orders and take such other action as may be necessary and appropriate to preserve the confidentiality of trade secrets, consistent with the requirements of the Federal Rules of Criminal and Civil Procedure, the Federal Rules of Evidence, and all other applicable laws. An interlocutory appeal by the United States shall lie from a decision or order of a district court authorizing or directing the disclosure of any trade secret.

"§ 1836. Civil proceedings to enjoin violations

"(a) The Attorney General may, in a civil action, obtain appropriate injunctive relief against any violation of this section.

"(b) The district courts of the United States shall have exclusive original jurisdiction of civil actions under this subsection.

"§ 1837. Applicability to conduct outside the United States

This chapter also applies to conduct occurring outside the United States if—

"(1) the offender is a natural person who is a citizen or permanent resident alien of the United States, or an organization organized under the laws of the United States or a State or political subdivision thereof; or

"(2) an act in furtherance of the offense was committed in the United States.

"§ 1838. Construction with other laws

"This chapter shall not be construed to preempt or displace any other remedies, whether civil or criminal, provided by United States Federal, State, commonwealth, possession, or territory law for the misappropriation of a trade secret, or to affect the otherwise lawful disclosure of information by any Government employee under section 552 of title 5 (commonly known as the Freedom of Information Act).

"§ 1839. Definitions

"As used in this chapter—

"(1) the term 'foreign instrumentality' means any agency, bureau, ministry, component, institution, association, or any legal, commercial, or business organization, corporation, firm, or entity that is substantially owned, controlled, sponsored, commanded, managed, or dominated by a foreign government;

"(2) the term 'foreign agent' means any officer, employee, proxy, servant, delegate, or representative of a foreign government;

"(3) the term 'trade secret' means all forms and types of financial, business, scientific, technical, economic, or engineering information, including patterns, plans, compilations, program devices, formulas, designs, prototypes, methods, techniques, processes, procedures, programs, or codes, whether tangible or intangible, and whether or how stored, compiled, or memorialized physically, electronically, graphically, photographically, or in writing if—

"(A) the owner thereof has taken reasonable measures to keep such information secret; and

"(B) the information derives independent economic value, actual or potential, from not being generally known to, and not being readily ascertainable through proper means by, the public; and

"(4) the term 'owner', with respect to a trade secret, means the person or entity in whom or in which rightful legal or equitable title to, or license in, the trade secret is reposed.".

(b) CLERICAL AMENDMENT.—The table of chapters at the beginning part I of title 18, United States Code, is amended by inserting after the item relating to chapter 89 the following:

"90. Protection of trade secrets 1831".

(c) REPORTS.— Not later than 2 years and 4 years after the date of the enactment of this Act, the Attorney General shall report to Congress on the amounts received and distributed from fines for offenses under this chapter deposited in the Crime Victims Fund established by section 1402 of the Victims of Crime Act of 1984 (42 U.S.C. 10601).

SEC. 102. WIRE AND ELECTRONIC COMMUNICATIONS INTERCEPTION AND INTERCEPTION OF ORAL COMMUNICATIONS.

Section 2516(1)(c) of title 18, United States Code, is amended by inserting "chapter 90 (relating to protection of trade secrets), "after" chapter 37 (relating to espionage),".

TITLE II—NATIONAL INFORMATION INFRASTRUCTURE PROTECTION ACT OF 1996.

SEC. 201. COMPUTER CRIME.

Section 1030 of title 18, United States Code, is amended—

(1) in subsection (a)—

(A) in paragraph (1)—

(i) by striking "knowingly accesses" and inserting "having knowingly accessed";

(ii) by striking "exceeds" and inserting "exceeding";

(iii) by striking "obtains information" and inserting "having obtained information";

(iv) by striking "the intent or";

(v) by striking "is to be used" and inserting "could be used"; and

(vi) by inserting before the semicolon at the end the following: "willfully communicates, delivers, transmits, or causes to be communicated, delivered, or transmitted, or attempts to communicate, deliver, transmit or cause to be communicated, delivered, or transmitted the same to any person not entitled to receive it, or willfully retains the same and fails to deliver it to the officer or employee of the United States entitled to receive it";

(B) in paragraph (2)—

(i) by striking "obtains information" and inserting "obtains—

"(A) information"; and

(ii) by adding at the end the following new subparagraphs:

"(B) information from any department or agency of the United States; or

"(C) information from any protected computer if the conduct involved an interstate or foreign communication;";

(C) in paragraph (3)—

(i) by inserting "nonpublic" before "computer of a department or agency";

(ii) by striking "adversely"; and

(iii) by striking "the use of the Government's operation of such computer" and inserting " that use by or for the Government of the United States";

(D) in paragraph (4)—

(i) by striking "Federal interest" and inserting "protected"; and

(ii) by inserting before the semicolon the following: "and the value of such use is not more than $5,000 in any 1-year period";

(E) by striking paragraph (5) and inserting the following:

"(5)(A) knowingly causes the transmission of a program, information, code, or command, and as a result of such conduct, intentionally causes damage without authorization, to a protected computer;

"(B) intentionally accesses a protected computer without authorization, and as a result of such conduct, recklessly causes damage; or

"(C) intentionally accesses a protected computer without authorization, and as a result of such conduct, causes damage;"; and

(F) by inserting after paragraph (6) the following new paragraph:

"(7) with intent to extort from any person, firm, association, educational institution, financial institution, government entity, or other legal entity, any money or other thing of value, transmits in interstate or foreign commerce any communication containing any threat to cause damage to a protected computer;";

(2) in subsection (c)—

(A) in paragraph (1), by striking "such subsection" each place that term appears and inserting "this section";

(B) in paragraph (2)—

(i) in subparagraph (A)—

(I) by inserting", (a)(5)(C)", after "(a)(3)"; and

(II) by striking "such subsection" and inserting "this section";

(ii) by redesignating subparagraph (B) as subparagraph (C);

(iii) by inserting immediately after subparagraph (A) the following:

"(B) a fine under this title or imprisonment for not more than 5 years, or both, in the case of an offense under subsection (a)(2), if—

"(i) the offense was committed for purposes of commercial advantage or private financial gain;

"(ii) the offense was committed in furtherance of any criminal or tortious act in violation of the Constitution or laws of the United States or of any State; or

"(iii) the value of the information obtained exceeds $5,000;"; and

(iv) in subparagraph (C) (as redesignated)—

(I) by striking "such subsection" and inserting "this section"; and

(II) by adding "and" at the end;

(C) in paragraph (3)—

(i) in subparagraph (A)—

(I) by striking "(a)(4) or (a)(5)(A)" and inserting "(a)(4), (a)(5)(A), (a)(5)(B), or (a)(7)"; and

(II) by striking "such subsection" and inserting "this section"; and

(ii) in subparagraph (B)—

(I) by striking "(a)(4) or (a)(5)" and inserting "(a)(4), (a)(5)(A), (a)(5)(B), (a)(5)(C), or (a)(7)"; and

(II) by striking "such subsection" and inserting "this section"; and

(D) by striking paragraph (4);

(3) in subsection (d), by inserting "subsections (a)(2)(A), (a)(2)(B), (a)(3), (a)(4), (a)(5), and (a)(6) of" before "this section.";

(4) in subsection (e)—

(A) in paragraph (2)—

(i) by striking "Federal interest" and inserting "protected";

(ii) in subparagraph (A), by striking "the use of the financial institution's operation or the Government's operation of such computer" and inserting "that use by or for the financial institution or the Government"; and

(iii) by striking subparagraph (B) and inserting the following:

"(B) which is used in interstate or foreign commerce or communication;";

(B) in paragraph (6), by striking "and" at the end;

(C) in paragraph (7), by striking the period at the end and inserting "; and"; and

(D) by adding at the end the following new paragraphs:

"(8) the term 'damage' means any impairment to the integrity or availability of data, a program, a system, or information, that—

"(A) causes loss aggregating at least $5,000 in value during any 1-year period to one or more individuals;

"(B) modifies or impairs, or potentially modifies or impairs, the medical examination, diagnosis, treatment, or care of one or more individuals;

"(C) causes physical injury to any person; or

"(D) threatens public health or safety; and

"(9) the term 'government entity' includes the Government of the United States, any State or political subdivision of the United States, any foreign country, and any state, province, municipality, or other political subdivision of a foreign country."; and

(5) in subsection (g)—

(A) by striking ", other than a violation of subsection (a)(5)(B),"; and

(B) by striking "of any subsection other than subsection (a)(5)(A)(ii)(II)(bb) or (a)(5)(B)(ii)(II)(bb)" and inserting "involving damage as defined in subsection (e)(8)(A)".

TITLE V—USE OF CERTAIN TECHNOLOGY TO FACILITATE CRIMINAL CONDUCT

SEC. 501. USE OF CERTAIN TECHNOLOGY TO FACILITATE CRIMINAL CONDUCT.

(a) INFORMATION.—The Administrative Office of the United States courts shall establish policies and procedures for the inclusion in all presentence reports of information that specifically identifies and describes any use of encryption or scrambling technology that would be relevant to an enhancement under section 3C1.1 (dealing with Obstructing or Impeding the Administration of Justice) of the Sentencing Guidelines or to offense conduct under the Sentencing Guidelines.

(b) COMPILING AND REPORT.—The United States Sentencing Commission shall—

(1) compile and analyze any information contained in documentation described in subsection (a) relating to the use of encryption or scrambling technology to facilitate or conceal criminal conduct; and

(2) based on the information compiled and analyzed under paragraph (1), annually report to the Congress on the nature and extent of the use of encryption or scrambling technology to facilitate or conceal criminal conduct.

CHAPTER 10

ADEQUACY OF CRIMINAL LAW AND PROCEDURE (PHYSICAL)

A "LEGAL FOUNDATIONS" STUDY

Report to the
President's Commission
on Critical Infrastructure Protection
1997

This report was prepared for the President's Commission on Critical Infrastructure Protection, and informed its deliberations and recommendations. This report represents the opinions and conclusions solely of its developers.

CONTENTS

PART ONE
INTRODUCTION

This paper examines the adequacy of current criminal law to provide protection for the nation's critical infrastructures from physical attack. It also addresses the adequacy of criminal law to deter or punish attacks on infrastructures, how it might be amended, revised, or supplemented to close the gap.

Research Issues

This paper addresses the following research questions:

- Do existing laws governing criminal procedure unduly hinder investigations of infrastructure threatening behavior?
- To the extent that they may unduly hinder such investigations, how should they be reconsidered or modified?

Research Findings

Upon reviewing Federal, state and international legal authorities, the following observations were noted:

- Critical infrastructures may prove attractive targets for a variety of attackers, whether disgruntled employees, terrorists (foreign or domestic), or agents of nation-states. Federal criminal law may not adequately differentiate in terms of punishment based on the severity of threat or the identity of perpetrator.
- There are no Federal criminal statutes specifically prohibiting attacks on "critical infrastructures" or denials of "critical infrastructure services" per se. Constitutional considerations (*e.g.*, vagueness, overbreadth, ex post facto concerns) might even preclude successful enforcement of such a broad statutory proscription.
- Nonetheless, Department of Justice experts have suggested that the vast majority of physical acts that could damage critical infrastructures are anticipated by and covered under general Federal criminal law, and that there are no significant legal deficiencies that would hinder the pursuit of serious cases.
- Statutory penalties for these general Federal crimes that would be used to prosecute physical attacks on critical infrastructures generally range from 10 to 20 years' imprisonment.
- Infrastructure assurance is not currently being taken into account by judges when handing down sentences for physical acts against infrastructures. The *U.S. Sentencing Guidelines* encourage courts to make upward departures in sentences for crimes that involve violence against

persons, but not necessarily for disruption of critical infrastructure services or extensive physical damage.

- Current legislation providing for Rewards-For-Information in terrorism cases could be more actively and efficiently implemented by Federal agencies.
- Agency-specific regulations containing prohibitions on acts against critical infrastructures (which do provide for minor criminal penalties) are seldom enforced by the agencies themselves or by the Department of Justice for a number of reasons. More comprehensive criminal statutes provide better avenues for improving protection of critical infrastructures.
- International treaties are in place that address physical crimes. The United States is a party to many of these treaties. The treaties, however, do not appear to take into account or adequately address infrastructure assurance issues.

Assumptions

This paper uses certain assumptions in its analyses, and they are as follows:

- Critical infrastructures serve as potential targets of physical attacks; such attacks will likely persist, if not increase, in the future.
- Enhancements to the criminal law, including stiffer penalties for physical acts against critical infrastructures will act as a deterrent to such acts, and thus contribute to infrastructure assurance.

Background

Federal criminal statutes have been put in place over time largely in reaction to specific events. Recent acts of domestic terrorism, abuse of firearms, and attacks on certain types of facilities (e.g., trains) have resulted in a series of laws that enable prosecutors to charge defendants for similar offenses should they recur. These newer laws, such as the Antiterrorism and Effective Death Penalty Act of 1996, in addition to the more established criminal statutes provide a rich menu of options for prosecutors faced with devastating acts of physical violence. However, because of the piecemeal nature of these statutes, they do not appear to provide blanket protection for critical infrastructures from physical attack. Neither the substantive criminal prohibitions nor the sentencing guidelines have been put in place specifically with infrastructure assurance objectives in mind.

Substantive Prohibitions on Physical Crimes Against Infrastructures

Current Federal statutes which provide protection to infrastructures are difficult to categorize. As a preliminary matter, Federal investigators must be establish that the "victim"(i.e. the critical infrastructure) is involved in interstate commerce. This finding then allows the Federal government to exercise jurisdiction over a crime that otherwise would fall to the states to investigate and prosecute.[1] Once Federal jurisdiction is established, prosecutors have a range of options to choose from in charging a physical crime against an infrastructure. 18 U.S.C. § 844 is a statute of general applicability that may be used for crimes involving explosives targeted at instrumentalities of interstate commerce. Weapons of mass destruction, which include nuclear, chemical and biological agents, also have specific statutes addressed to their use in the U.S. Code.[2] Specific critical infrastructures, such as pipelines, airports, maritime vessels, and electric utilities have statutes that are aimed at protecting them from physical damage. However, to the extent a more general provision is available it will generally be preferred to a specific statute because of the higher penalties available. (*See* Appendix A for further discussion of Federal statutes prohibiting physical crimes).

Sentencing for Physical Crimes Against Infrastructures

While there may be some holes in this statutory framework when it comes to a prosecutor charging a defendant for a crime against an infrastructure, the overall scheme is fairly comprehensive. One of the larger obstacles Federal law enforcement faces with respect to enhancing deterrence is in the area of sentencing for violations. Penalties and sentencing are controlled by two sources: Congressional statutes and the *Sentencing Guidelines* published by the United States Sentencing Commission.

The United States Sentencing Commission is an independent agency in the judicial branch of government. Its principal purpose is to establish sentencing policies and practices for the Federal courts, including detailed guidelines prescribing the appropriate form and severity of punishment for offenders convicted of Federal crimes. While the development, monitoring, and amendment of the sentencing guidelines is the centerpiece of the agency's work, the Commission provides training, conducts research on sentencing-related issues, and serves as an information resource for Congress, criminal justice practitioners, and the public.

In many cases, Congress will specifically set forth a punishment in the statute which contains the substantive prohibition. In almost every criminal offense associated with critical infrastructure protection, Congress

---

1 State laws protecting critical infrastructures were not studied for the purposes of this issue paper. To the extent this paper expresses a preference for Federal investigation and prosecution, it reflects not necessarily a deficiency in state law, but a lack of uniformity which may ultimately adversely affect the deterrent value of prosecutions. (e.g., state policies on the death penalty vary widely).

2 *See* 18 U.S.C. § 2332a & 2332c (1997).

identifies a range of punishment options—such as from two years to 20 years with or without a particular fine. Because courts are given a range of punishments, the U.S. Sentencing Commission publishes Guidelines to assist Courts in delivering the most appropriate sentence. In general, the Guidelines suggest stricter punishments where criminal acts result in death, are committed with explosives, or are intended to cause harm. Amendments made to the Guidelines this year set forth more stringent punishments for terrorist-related acts. These are addressed further in the examples below.

Example - Terrorism and the U.S. Sentencing Guidelines

Use of U.S. Sentencing Guidelines to Supplement Legal Authorities: U.S. Sentencing Guideline Enhancement

- Prior to passage of the Antiterrorism and Effective Death Penalty Act of 1996, Federal court judges were not provided any guidance with regard to acts that could be seen as "acts of terrorism."
- Sentencing Guidelines associated with the Antiterrorism and Effective Death Penalty Act of 1996 accounted for criminal acts such as the Oklahoma bombing.
- The U.S. Sentencing Commission amended the Guidelines in May 1997 [3] to increase the sentencing level for any offense intended to promote the "Federal crime of terrorism."

While there are some provisions in the current Sentencing Guidelines that would allow for enhanced prison terms for certain acts against critical infrastructures, the Guidelines as written do not take into account extensive physical damage to infrastructures or disruptions in services due to physical attack on infrastructures. Amendment to the Guidelines to allow upward departures for these types of harms may be appropriate in light of the recent changes to the Sentencing Guidelines for computer crimes (as discussed in the companion paper to this effort on the adequacy of criminal law and procedure (cyber)).

One specific weakness of the current sentencing regime was uncovered with respect to biological substances. Under existing laws, crimes involving biological substances may result in unusually low prison sentences.

3 Amendments to the Sentencing Guidelines (U.S.S.G.), United States Sentencing Commission, May 13, 1997 at 58 (Amendment 19).

Congressional laws do adequately address crimes involving toxic biological substances. In particular, legal authorities set forth penalties for a wide range of activities–acquiring, developing, producing, stockpiling, etc.[4] Congress sets the prison term for any of these, and other related activities at "life or any term of years."[5] However, while Federal court judges may sentence an offender for life, there is no mandatory sentence in the statute. The Sentencing Guidelines set forth the following criteria:

- Base offense is equivalent to "weapons possession"—less than two years; [6]
- Where the Court finds that there was no intent to commit harm, the Sentencing Guidelines suggest only a minor increase in sentencing (*e.g.*, six months); [7]
- Where the Court finds intention was to commit murder, the prison term can be raised by 27 to 33 months; [8]
- Where the Court finds that the biological substance was used as a "weapon of mass destruction," then the Guidelines increase the sentence considerably.[9]

Federal courts then have the discretion to apply the statute strictly or to find for an extremely low prison sentence. Where such crimes involve critical infrastructures, an unusually low prison sentence is–at least theoretically–possible based on relevant precedent.[10] Sentencing would depend, in large part, on the court's sensitivity to critical infrastructure harm and related issues. In other areas, the Sentencing Guidelines provide strict implementation options–such as where the offender uses the substances in a terrorist-related act or as a weapon of mass destruction. Given the extreme toxicity of biological substances, and the level of threat they pose to certain infrastructures, Congress may wish to recommend (1) reducing the discretion now available to judges through Congressionally mandated sen-

---

4 18 U.S.C. 175.

5 *Id.*

6 The base offense is equivalent to 12 points under U.S.S.G. 2K2.1.

7 *See* U.S.S.G. 2N2.1.

8 *See* U.S.S.G. 2A. 1 series for particulars.

9 *See* U.S.S.G. 2M6.1. The Court may also find here that the act was terrorist-related and increase the term to life; where death result, the court can find for the death penalty. *See* USSG3A1.4.

10 Recently, in a Minnesota case, an offender was convicted for manufacturing ricin. Ricin is one of the most toxic substances discovered to date, and small amounts could taint a large supply of treated water. The Court was faced with the options listed above and chose a middle course —which increased the base offense by approximately 27 to 33 months (*See* U.S.S.G. 2K2.1.)

tencing, or (2) raising the courts awareness of these issues through sentencing enhancement with the existing Sentencing Guidelines framework.

Alternatives to Traditional Law Enforcement Efforts

While it is assumed that successful prosecution and severe sentences will deter many of the possible criminal acts that could be targeted at infrastructures, recent cases (e.g., suicide bombers targeting N.Y. subways) suggest that there is a segment of society that will remain undeterred by even the most successful law enforcement efforts. For this reason, it may be important to consider other means of detecting and defeating these acts before they become realities. Neither of the alternatives suggested below should be taken as independent of law enforcement efforts, but rather as supplements to current or enhanced law enforcement efforts.

The first area where there is potential for improvement is in the administration of the current award program for information about terrorist activities. Through this Congressionally-mandated program, tips that lead to the capture and arrest of terrorists are to receive monetary awards. Such tips have played important roles in many recent law enforcement investigations such as the New York subway case. However, the current administration of the system may not be adequate to provide the necessary incentive to bring all who possess such crucial information forward. Improvements in education and awareness, scope of coverage, and the clarity of guidelines for receiving awards may contribute to the success of such a program.

In addition, preventive measures could be taken by many of the critical infrastructures to increase the physical security of the truly critical elements of their infrastructures. While it may be infeasible to make specific recommendations to each of the infrastructures as to the proper security measures to have in place at identified sites, it could still be possible to recognize the value of physical security as a deterrent and a deterrent that prevents damage rather than react to it.

PART TWO
POSSIBLE APPROACHES TO ENHANCING DETERRENCE

Enact "Critical Infrastructure" Crimes Legislation

Congress could re-examine existing legislation covering critical infrastructures and consider revising and adding additional provisions to cover all of the critical infrastructures. Legislation would then be packaged as a "Critical Infrastructure Protection" statute. Specific guidance could also be provided to the Sentencing Commission on how to draft the guidelines for the new and revised provisions in the package.

- **Pro:** New legislation would allow Congress and the Sentencing Commission to focus on infrastructure assurance issues not previously addressed. A detailed study would allow an infrastructure-by-infrastructure approach which is probably necessary to avoid the Constitutional overbreadth issues that would plague a generic "damage to critical infrastructures" statute. Packaging of provisions together would raise the level of awareness of critical infrastructure issues. This approach may provide a useful model for states to use in evaluating their criminal statutes.

- **Con:** Congress is traditionally reluctant to consider creation of any broad criminal statutes where narrow statutes effectively address most of the problem. Congress is likely to be very reluctant to try new and untested approach where the old approach produces satisfactory results. Where there are holes, other less drastic means are available to "fill the gaps," such as sentencing enhancement. Congress may be reluctant to pursue new legislation without an identified and urgent need (i.e. without a well-publicized event).

Enhanced Sentencing Guidelines for Infrastructure-Related Harm and Damage

Several Federal criminal laws protect critical infrastructures from physical attacks. These authorities consist primarily of Federal statutes from Title 18 of the United States Code. There are over 30 major criminal laws that that may be invoked in response to physical attacks on critical infrastructures–statutes that confer adequate investigative jurisdiction and authority in the vast majority of instances. Thus, the legislative scheme is comprehensive and provides an adequate "legal fortification" from physical attacks. However, there are several potential shortfalls relating to the abilility of this complicated patchwork of laws to deter crimes against critical infrastructures.

Critical infrastructure deterrence may be undercut for several reasons. The first is that judges are likely not yet aware of, nor have they been trained to identify, critical infrastructure issues. Also, the Sentencing Guidelines support lighter sentencing for attacks against property, however serious. More severe sentences are reserved for attacks against people and attacks using traditional, violent means–such as explosives and firearms. In addition, the Sentencing Guidelines may not adequately take into account the severity of consequential damages arising from attacks on critical infrastructures, for example, damage resulting from the "downstream" effects of a denial of service attack. As a consequence, the Sentencing Guidelines may not adequately embody critical infrastructure assurance goals and interests, thus creating a possibility of disproportionately light sentences for some forms of attack on critical infrastructure.

Either the Administration or Congress could instruct the U.S. Sentencing Commission to enhance existing Sentencing Guidelines to account for harm to critical infrastructures. Using the Guideline on terrorist acts as a model, a specific guideline for used in sentencing physical crimes against critical infrastructures could be developed.[11] In particular, a recommendation could be offered to increase the range for prison sentencing for harm to critical infrastructure-related property. (*See* U.S.S.G. 2B1.3.) Disruptions in services that harm individuals or have substantial costs should be taken into account. (Sentencing Guidelines for computer crime may provide a model.) It may also be appropriate to recommend that the U.S. Sentencing Commission address infrastructure assurance objectives in its training program for Federal Court judges.

- **Pro:** A recommendation to enhance sentencing is an efficient legal method to protect critical infrastructures from physical threats; currently, the Guidelines do not account for, or in any way contemplate, damage to critical infrastructures.
- **Con:** Enhancing the Sentencing Guidelines will not lead to dramatically higher penalties for crimes against critical infrastructures. Only substantial changes to the underlying statutes, by Congress, will force dramatic changes from current sentencing practices.

There are three forms of "attack" against critical infrastructures which may not adequately be addressed by existing guidelines and which may require some degree of special attention. They include attacks—

1. With biological weapons or chemical weapons where no deaths occur.
2. Where there is only property damage.
3. Where no explosives or firearms are used.

There are three ways to amend the Sentencing Guidelines. This first is that Congress may instruct the Sentencing Commission to amend the Guidelines immediately to coincide with an enacted statute. Second, the U.S. Sentencing Commission may amend the Guidelines in accordance with a specific delegation of authority from Congress. Under both of these methods, Congress explicitly *empowers* the Sentencing Commission to amend the Guidelines. The third way is that any interested party may propose a Sentencing Guideline amendment.

The first two methods may result in Guideline amendments in a relatively short period of time. The only way to evade the "normal amendment cycle" is via express authority delegated by Congress. The third method—initiation by an interested party—requires at least one cycle (one full year); U.S. Sentencing Commission history suggests that two to three years is more likely required to amend a Guideline.

---

11 Such a provision could read: "*If the offense is a felony that involved, or was intended to affect a critical infrastructure, as defined, then increase offense level by________.*"

- **Pro:** These options complement Congressional statutes that penalize physical threats to critical infrastructures, but do so within the parameters of existing law and sentencing mechanisms. They accordingly should not require additional government structures or significant expenditures of resources. By instructing the President and Congress to delegate authority to the Sentencing Commission, infrastructure assurance goals will be incorporated outside the "normal U.S. Sentencing Commission cycles," which may last as long as several years. Finally, sentencing enhancement achieves important consciousness-raising goals, providing information on critical infrastructure protection to a vital portion of the law enforcement and criminal justice network of professionals, judges, legislators, probation officers, local law enforcement, and academicians.
- **Con:** Working through the U.S. Sentencing Commission, as opposed to changing or developing new legislation, is not necessarily the most visible or comprehensive method for filling gaps in existing law. Critical infrastructure issues are complex and do not translate easily into U.S. Sentencing Commission Guideline Amendments. Real deficiencies in jurisdictional authority, for example, can be corrected only through specific criminal legislation.

Include Within Infrastructure-Related Legislation a Finding that Critical Infrastructures are "Instrumentalities of Interstate Commerce"

In order to have investigative or prosecutorial jurisdiction over physical acts against critical infrastructures, the affected infrastructure must have some relationship with "interstate commerce."[12] Although not common, there are examples of purely intrastate infrastructure facilities. Various rail lines, natural gas pipe networks, and state-run and operated subway systems may be deemed "intrastate." The recent attempted pipe-bomb attack in the New York subway reveals a legal vulnerability; according to the Department of Justice, defense attorneys could defeat application of important Federal laws because the subway system is not clearly an "instrumentality in interstate commerce." Congress could deem certain infrastructures to be "instrumentalities of interstate commerce." Such a finding will ensure that crimes associated with critical infrastructures will be under Federal jurisdiction.

- **Pro:** Such a finding is necessary to ensure that all crimes associated with critical infrastructures are handled in Federal court, and are addressed with Federal criminal legislation. A Congressional finding is not overly dramatic or burdensome to Congress. It can be easily included in any piece of infrastructure-related legislation. Application of Federal crimi-

---

12 Please refer to the Listing of Legal Authorities for examples (*See* Appendix A).

nal legislation to all critical infrastructure related crimes best protects these infrastructures by providing uniformity, especially in sentencing.

- **Con:** A Congressional finding that critical infrastructures are all instrumentalities of interstate commerce may be overly broad and may lead to unintended results. Such an attempt to secure Federal jurisdiction in all criminal matter affecting critical infrastructures may be premature without further study of state handling of such crimes and may be seen by state lawmakers as a usurpation of power.

The comprehensiveness of existing criminal law to cover a full range of physical attacks on critical infrastructure was examined as part of developing this paper. Deficiencies that might pose undue limitations on the exercise of Federal jurisdiction in such instances were also looked at. Two potential deficiencies were identified (discussed below) with respect to purely *intrastate* attacks against critical infrastructures—even when such an attack may result in severe damage. In these instances, in order legally to assume jurisdiction over an investigation or prosecution, the Federal government must demonstrate on a case-by-case basis that the incident affects interstate commerce. This may be a particularly difficult determination to make at the earliest stages of an investigation, before the scope of an attack or its effects are well known.

Also, there are several hypothetical scenarios involving purely intrastate facilities that may preclude Federal investigative and prosecutorial involvement.[13]

A *Congressional Finding* that all critical infrastructures (suitably defined) were "instrumentalities of interstate commerce" may be an efficient way of ensuring that Federal jurisdiction exists to respond to appropriate instances. Subsequent research, including a review of related cases and discussions with legislative counsel about these issues, lead to the following conclusions pertaining to issuance of a Congressional Finding:

- A Congressional Finding must be made part of a legislative enactment; a Congressional Resolution or statement read into the Congressional Record would not be sufficient to support a subsequent claim to Federal jurisdiction.[14]

---

13 Possible examples of attacks against certain purely intrastate railroad and subway lines and natural gas pipe networks fit the profile.

14 *See e.g.*, U.S. v. Lopez, 2 F.3d 1342 (5th Cir. 1993), *affirmed* 115 S.Ct. 1624, 514 U.S. 549, 131 L.Ed.2d 626 (Gun-Free School Zones Act prohibiting any individual from knowingly possessing firearm in school zone is invalid as beyond power of Congress under commerce clause. Neither Act nor legislative history reflected any congressional determination that possession denounced by Act was in any way related to interstate commerce or its regulation, or that Congress was exercising its powers under commerce clause).

- The standard used by Courts to test Congressional Findings is low: Courts demand only a "reasonable" nexus between the facts offered (in the Congressional finding) and interstate commerce.[15] For threats to critical infrastructures, this standard would likely be attainable.
- The Congressional Finding must be sufficiently specific to identify the critical infrastructures.[16]

Mandatory Prison Sentences

Congress could establish new, mandatory prison sentencing for all crimes involving biological substances; or in the alternative, recommend that the U.S. Sentencing Commission set forth stricter sentencing guidelines for crimes involving biological substances and train Federal court judges in effective implementation.

Biological substances are very serious agents which in small amounts can do serious damage. Congressional statutes are in place to charge crimes involving such substances; however, judges are currently given wide discretion in sentencing. Some past cases suggest that judges failed to appreciate the seriousness of the crimes in handing down their sentences.

---

15 When Congress itself determines that particular conduct adversely affects interstate commerce, the courts defer to Congress' judgment so long as there is a rational basis for finding a chosen regulatory scheme necessary to the protection of commerce. Rasmussen v. American Dairy Ass'n, 472 F.2d 517 (9th Cir. 1972), cert. denied, 93 S.Ct. 3014, 412 U.S. 950, 37 L.Ed.2d 1003. When reviewing Congress' exercise of power under commerce clause, courts apply a rational basis test under which court must defer to a congressional finding that a regulated activity affects interstate commerce if there is any rational basis for such a finding. U.S. v. Wilson, 880 F.Supp. 621 (E.D. Wis. 1995), reversed 73 F.3d 675, rehearing and suggestion for rehearing en banc denied, cert. denied 117 S.Ct. 46, 136 L.Ed.2d 12, cert. denied 117 S.Ct. 47, 136 L.Ed.2d 12; Freedom of Access to Clinic Entrances Act (FACE), providing criminal and civil penalties against persons using or threatening to use force to interfere with reproductive health services, was validly enacted under Commerce Clause; congressional findings of nationwide violence against abortion clinics showed a substantive impact on interstate commerce. U.S. v. McMillan, 946 F. Supp. 1254 (S.D. Miss. 1995).

16 *See* U.S. v. Dawson, 467 F.2d 668 (8th Cir. 1972), cert. denied, 93 S.Ct. 1427, 410 U.S. 956, 35 L.Ed.2d 689. Congressional failure to make formal findings of fact in connection with drafting of this section which among other things makes it unlawful to receive, transport, sell, etc., explosive materials knowing or having reasonable cause to believe that they were stolen did not render invalid underlying legislation despite claim of defendant that this section is infirm because its enactment was not accompanied by congressional findings concerning manner in which misuse of explosives affect interstate commerce, as there was a rational basis upon which Congress could properly have determined that the misuse of explosive materials is one activity which as a class affects commerce; U. S. v. Perez, 426 F.2d 1073 (2nd Cir. 1970), cert. granted 91 S.Ct. 175, 400 U.S. 915, 27 L.Ed.2d 154, affirmed 91 S.Ct. 1357, 402 U.S. 146, 28 L.Ed.2d 686 (In view of rational congressional finding that even purely intrastate extortionate credit transactions affect interstate commerce, this section is a constitutional exercise of Federal power over interstate commerce).

Congress or the Sentencing Commission could study and propose stricter penalties, with less discretion for judges, for cases involving biological agents. In addition, training in such new types of crimes should be given to judges to increase their sensitivity to the nature of the substances and their potential criminal uses.

- **Pro:** This approach would compel awareness of judges to relatively new crimes such as those involving biological agents. It would ensure strict sentencing in cases involving biological substances and may increase deterrence.
- **Con:** Congress has addressed the issue in the statute and may be reluctant to reopen the issue to consider sentencing. The Sentencing Commission may resist efforts to limit judicial discretion any further–as may the judicial community.

Monetary Awards

The authors also reviewed legislation that offers reward monies for information leading to the capture of terrorists.[17] Under these legal authorities, Congress authorizes the Attorney General and the Secretary of State to administer rewards and payment-for-information programs. These laws appear to effectively supplement other Federal crimes legislation to protect critical infrastructures.

Few Federal agencies that control critical infrastructures have programs to protect infrastructure facilities–whether in the form of funds-for-information, rewards, or training and awareness. One of the few programs designed and administered by a Federal agency is Bonneville Power Administration's Crime Witness Program.[18] Formed by BPA in 1994, the Crime Witness Program is a sophisticated blend of public relations,[19] grass roots training,[20] and cooperation with other Federal agencies to reduce losses and protect facilities.[21] BPA funds reward monies as a percentage of vandalism-related savings.

---

17 18 U.S.C. §§ 3071-3077.

18 Bonneville Power Administration Journal, June 1997 at 2 (reprinted at Bonneville Power Administration Website, *www.bpa.gov/Corporate/ACS/jl/97jl/jl0697x.htm/.*

19 BPA commissioned a professional video to advertise the program as a Public Service Announcement. Television stations throughout areas that BPA maintains transmission lines broadcast the message at no charge; BPA produced similar radio spots as well.

20 BPA additionally prints posters, pamphlets, and signs to advertise the Program at schools, community centers, and on-site throughout the Northwest United States. Dissemination of materials is especially heavy in sparsely populated areas. BPA has approximately 15,000 miles of transmission lines.

21 The Bureau of Reclamation and the Army Corps of Engineers recently agreed to coordinate efforts with BPA as part of the Crime Witness Program. After calls come in

Government agencies could review and promote participation in Congressional programs that provide monetary awards for tips leading to the capture and arrest of terrorists (*see* 18 U.S.C. §§ 3071-3077). Such programs may be one of the best ways to prevent crimes from successfully occurring. In order for such programs to be truly effective, especially with regard to crimes against critical infrastructures, they must be well-publicized and provide clear guidelines (as little red-tape as possible) for obtaining the reward. There is evidence from recent investigations of the promise such programs have for preventing criminal acts (*e.g.*, the New York subway case). The utility of such programs could be highlighted and expanded to include the critical infrastructures. Their administration could be streamlined and visibility heightened to improve participation.

- **Pro:** This approach is consistent with Congress' intent in establishing the reward money program. It focuses on preventing the crime (and subsequent damage). It supplements law enforcement efforts and contributes to successful prosecution. It publicizes low tolerance and proactive stance of the Federal government with regard to crimes directed against critical infrastructures.
- **Con:** Such programs are expensive and difficult to administer. Expansion to critical infrastructure issues may be resisted by atgencies that administer the programs (State and DOJ).

Critical Infrastructure Owners and Operators Pursue Increased Physical Security for Particularly Vital and Vulnerable Elements of their Infrastructures

This approach would allow the agencies, trade associations and owners and operators to work together to determine the areas that need additional protection, best level of protection, and appropriate means to supply it. Focusing on prevention may provide an element of deterrence for small-time criminals without access to powerful weapons or sophisticated equipment.

- **Pro:** This approach focuses on preventing damage rather than reacting to criminal act. It involves infrastructure owners and operators in protecting their infrastructures, and supplements enhanced law enforcement responses (*e.g.*, stiffer sentences).
- **Con:** Without a powerful mandate, infrastructure owners and operators may be reluctant to sink additional funds into security. Increased security will not deter the most sophisticated and determined criminals.

---

advising of a potential crime, operators direct calls to the agency that controls the relevant infrastructure facility. The Bureau of Reclamation provides funding directly to BPA.

PART THREE
CONCLUSIONS

If the President and Congress so chose, they could encourage the Sentencing Commission to amend the Sentencing Guidelines to adequately and proportionately address physical threats to critical infrastructures. Specifically, they could ensure that–

- The U.S. Sentencing Commission ensure that the applicable guideline ranges for a defendant convicted of a crime against a critical infrastructure are sufficient to deter such crime, and provide an appropriate vehicle through which the sentencing court may consider the full range of consequential damages, including but not limited to economic damages, that may have been incurred by the victim and others as the result of the crime;
- The U.S. Sentencing Commission study and report to Congress on sentencing issues–with emphasis on sentencing deficiencies for damage to critical infrastructure property and attacks with biological and chemical weapons that do not result in any deaths and are not currently covered as "terrorist" acts for guidelines purposes;
- The U.S. Sentencing Commission train Federal Court judges to identify and incorporate critical infrastructure assurance goals into sentencing; and
- Sentencing Commissions and Councils within each state, and state legislatures, study Amendments to the U.S. Sentencing Guidelines intended to further critical infrastructure assurance, and to consider similar changes to their state sentencing guidelines or criminal statutes.

Congress may find it prudent to consider the propriety of including in appropriate infrastructure-related legislation a Congressional Finding that "critical infrastructures" are "instrumentalities of interstate commerce."

Such a finding may be necessary to ensure that all serious attacks on critical infrastructures can be investigated by Federal law enforcement officials and prosecuted in Federal court. Rendering a Congressional Finding under such circumstances would likely not prove burdensome or inappropriate. It can be included in any piece of infrastructure-related legislation, and need not be tied to "new" criminal legislation.

Some issues militate against this approach however. There may be considerable delay before passage of specific "critical infrastructure" legislation. Without a Congressional statute, there can be no Congressional Finding. The Finding itself must be stated with adequate specificity as to put a defendant on notice as to what might constitute a "critical infrastructure," and the adequacy of this definition may provide grounds for appeal of a conviction. Thus, Congress should carefully consider whether a suitable definition can be drawn, and whether the benefits might justify the effort.

The President and Congress could review and promote Federal government agencies' participation in programs that provide monetary rewards for information leading to the capture and arrest of vandals and terrorists. Congress could further encourage agencies to explore the creation of such novel programs aimed at protecting critical infrastructure facilities from damage.

This position is wholly consistent with Congress' intent in establishing rewards and payment for information programs. Further, at the grass roots level, there seems to be movement in creating these programs to meet individual agency and infrastructure needs. It should be noted however, that these programs are likely expensive and difficult to administer. Expansion of programs, other than the existing programs for the Department of Justice and the State Department, may be too onerous for other agencies.

APPENDIX A
LISTING OF PRIMARY LEGAL AUTHORITIES

(Note: Headings provided by editors)

Federal Statute	Comments	Dept. of Justice	U.S. Sentencing Guidelines
Explosives/ incendiary devices 18 U.S.C. 844	All critical infrastructures. Not applied to crimes associated with infrastructure elements not in "interstate commerce."	Criminal code legal authority to prosecute use, or threat of use of explosives. Involves, government and private property. Used extensively by Department of Justice to prosecute use of explosives in criminal acts.	Must be in "interstate commerce"; if not, the provision will not apply. Not all infrastructures are necessarily instrumentalities in interstate commerce (e.g., certain rail lines, gas pipe networks, etc.) Penalties vary based on offense: 844h -Damage to government property with explosives mandatory 10 years; if private property then mandatory 5 years. If is a second offense, penalties are far stricter. Where death results, then courts may sentence for any term of years up to the death penalty.

Federal Statute	Comments	Dept. of Justice	U.S. Sentencing Guidelines
18 U.S.C. 2332a	All critical infrastructures. Any U.S. Government property, or against any person (must have results in interstate commerce).	Criminal code legal authority to prosecute for use of any biological, chemical, or conventional explosives. Department of Justice applies broadly— use of any type of explosives. Part of 1996 Antiterrorism and Effective Death Penalty Act (1996 Terrorism Act).	Would not apply if criminal act to infrastructure were solely "intrastate." Penalty is term of years or life. Sentencing Guidelines increase time of sentence if: -against property - U.S.S.G. 2B1.3; -harm is caused to a person U.S.S.G. 2A1.1; -where death results, then may be punished by death, or life in prison.
18 U.S.C. 2332c	Attacks against all critical infrastructures.	Criminal code legal authority to prosecute for use of any biological, chemical, or conventional explosives. Department of Justice applies broadly—use of any type of explosives. (Note 18 U.S.C. 2333 and provisions for civil penalties). Part of 1996 Antiterrorism and Effective Death Penalty Act (1996 Terrorism Act).	Would not apply if criminal act were solely "intrastate." No specific Sentencing Guideline covers this provision. Congress calls for penalties similar to 2332a—term of years to life; where act causes death, then death penalty may be invoked.

Federal Statute	Comments	Dept. of Justice	U.S. Sentencing Guidelines
18 U.S.C. 831	All critical infrastructures.	Criminal code - legal authority to prosecute for use of any nuclear materials.	Use of nuclear materials: -Congress lists range of penalties from not more than 20 years (no harm is caused) to life (where indifference to life in manifest by criminal act). -Where death is caused, term of years to life, and the death penalty is an Option. *See* U.S.S.G. Amendment 3A1.4 allows for the death penalty. If the intent is to "aid a foreign government" or to cause harm to the U.S., then the Sentencing Guidelines suggest a higher term of years U.S.S.G. 2M6.1.

Federal Statute	Comments	Dept. of Justice	U.S. Sentencing Guidelines
18 U.S.C. 175-178 Biological Weapons Anti-Terrorism Act of 1989	All critical infrastructures.	Criminal code legal authority to prosecute for use of any biological weapon.	No "interstate commerce" requirement. However, there are no specific U.S. Sentencing Guidelines for this provision and Congress indicates a "for life or any other term." The absence of a specific Sentencing Guideline might prevent courts from applying this provision strictly—especially in the absence of any harm. DOJ suggests that three possible U.S. Sentencing Guideline applications
		Recently, in a Minnesota case, a Federal judge sentenced someone who was producing ricin to "minimum sentence." The Judge ultimately based the decision on U.S. Sentencing Guidelines for unlawful weapons possession (U.S.S.G. 2K.2.1) rather than the more serious "weapons of mass destruction" guideline. The Court could have set forth an even lighter sentence based on possession of a biological weapon.	Enhancing the Sentencing Guidelines might be appropriate. Production and stockpiling ofricin, for example, would likely lead to a 0 to 6 month sentence based on available Sentencing Guideline 2N2.1 or 2K2.1: -2N2.1 (use of biological product). -2A2.1 (attempted murder),, -2K2.1 (weapons possession); or -2M6.1 -(weapon of mass destruction).

Federal Statute	Comments	Dept. of Justice	U.S. Sentencing Guidelines
18 U.S.C. 2332b	All critical infrastructures.	Aimed at prosecuting terrorism "beyond U.S. boundaries." Provision transcends U.S. boundaries. Mini-RICO—Additional teeth for conspiracies— more than one person.	Penalties are severe for acts that lead to death. May be charged with term of years to life, or the Death Penalty under the 1996 Terrorism Act and U.S. Sentencing Guideline 3A1.4. Long-arm jurisdiction: Congress attempt to catch criminal "wherever and whenever."
18 U.S.C. 1363	All critical infrastructures.	Buildings of property within maritime or territorial jurisdiction.	Generic: destruction of buildings and property, whether government or private. Sentencing Guidelines force increase in term based on: -property damage -weapons used (gun v. explosives) -planning involved See differences between U.S.S.G. 2B1.3 and U.S.S.G. 2K1.4. According to DOJ, this provision used to catch "prankster" more than terrorist. Statute limits prison term to 20 years.

Federal Statute	Comments	Dept. of Justice	U.S. Sentencing Guidelines
18 U.S.C. 1361	All critical infrastructures.	Government property or contracts	This provision applies to government only. DOJ considers this provision to be the most generic provision for destruction of Federal property. Penalties of zero to ten years, based on property damage and harm to human life. Statute limits maximum prison term to ten years.
18 U.S.C. 1992	Transportation facilities—railroad and railroad facilities.	"Wrecking trains" statute. If act is to intrastate facility, DOJ applies 18 U.S.C. 844I—five to 15 years for first offense is prison range.	Does not cover mass transit. Must be in interstate commerce; where facility is solely intrastate, statute might not apply. Maximum prison term is 20 years. Sentencing Guidelines set forth increase terms where act involves: -homicide (U.S.S.G. 2A1.1) terrorism -(U.S.S.G. 3A1.4). Where train is carrying high-level radioactive waste or spent nuclear fuel, term increases from 30 years to life. Under this scenario, same U.S. Sentencing Guidelines apply to calculate actual term.

Federal Statute	Comments	Dept. of Justice	U.S. Sentencing Guidelines
18 U.S.C. 32	Airport facilities.	Destruction of aircraft facilities.	For destroying or an act of violence against an aircraft facility. Includes aircraft and facilities. Congress sets penalty at maximum of 20 years. Where the act involves kidnapping, homicide, or terrorism, U.S. Sentencing Commission instructs courts to increase sentences accordingly. Does not apply to "international airports."
18 U.S.C. 37	International airport facilities.	Any act of violence at international airport facilities.	Applies only to airport facilities at international airports. Does not cover aircraft (see 18 U.S.C. 32, above). Applies to acts outside the U.S. if offender is subsequently found in the U.S. Congress sets limit at 20 years.
18 U.S.C. 33	Motor vehicles or vehicle facilities.	Destruction of motor vehicles or motor vehicle facilities.	Congress sets limit at 20 years. DOJ also uses 844I (compare with application of 844I to intrastate criminal act to a railroad, discussed above).

Federal Statute	Comments	Dept. of Justice	U.S. Sentencing Guidelines
18 U.S.C. 2271-2281	Vessels; mixed platforms.	Vessels.	Wide range of Congressional statutes covering crimes regarding vessels. Comprehensive. Specific as to party committing act. U.S. Sentencing Guidelines cover adequately.
49 U.S.C. 60123 and 49 USCA 60114(c)	Pipelines; pipeline facilities.	Acts of destruction to pipelines.	Congress sets forth different penalties depending on act: -damage or destroy facility - 5 to 15 years -destroy sign or natural gas marker - not more than one year; -failure of owner/operator to mark adequately - not more than 5 years (See 49 USCA 60114(c). Also sets forth penalties from not following marking rules. NOTE: Application of the U.S. Sentencing Guidelines to these statutes suggests that Act lacks any real "teeth." U.S. Sentencing Guidelines for destruction to property suggest extremely low penalties–often zero to six months. See U.S.S.G. 2B1.3.

Federal Statute	Comments	Dept. of Justice	U.S. Sentencing Guidelines
18 U.S.C. 1366 49 U.S.C. 60101	Energy facilities; power utilities; interstate gas pipeline facility.	Destruction of energy facility.	Similar to the pipeline criminal statute discussed above, Congress' attempt to penalize destruction to energy facilities might not adequately punish offenders.
18 U.S.C. 924c			According to the statute, Congress sets penalties of up to 10 years (attempt or actual damage of over $100,00) and up to five years (attempt or actual damage of over $5,000). Because the U.S. Sentencing Guideline applied to property damage–2B1.3 suggests extremely low prison terms, the destruction of such critical infrastructures could occur with little or no actual prison sentences. One method DOJ uses to obtain more appropriate results is to apply 18 U.S.C. 924c to similar crimes. This provision may be used to obtain Federal jurisdiction and a mandatory, consecutive sentence. The criminal act must involve, however, use of a "firearm" or a destructive device.
42 U.S.C. 2284	Nuclear fuel facility.	Any act of destruction, violence	

Federal Statute	Comments	Dept. of Justice	U.S. Sentencing Guidelines
18 U.S.C. 1362	Telecommunication lines.	Any act of destruction.	This provision, which Congress intended to protect telecommunication lines and stations, may lack necessary teeth. Similar to the energy facility and pipeline examples above, an offender who does not use a firearm or other destructive device, could end up causing substantial damage and not be subject to a commensurate prison term. The basis for this is U.S. Sentencing Guideline 2B.1.3. Congress sets forth a maximum term of ten years. The Guidelines will suggest a higher term where weapons are involved (U.S.S.G. 2K1.4) or if part of a terrorist scheme (U.S.S.G. 3A.1.4) However, where no weapons are used, and the offender has no previous crimes, the Guidelines will militate toward a far shorter term—perhaps zero to six months—where no "weapons" are used.

APPENDIX B
The United States Sentencing Commission

1. What is the United States Sentencing Commission's mission?

2. Why did Congress create the Sentencing Commission?

3. How do the Sentencing Guidelines Work?

4. How is the Sentencing Commission Structured?

5. Who are the Commissioners?

6. What is the process for amending the Sentencing Guidelines and how much time would the Sentencing Commission require to amend the existing Guidelines?

1. What is the United States Sentencing Commission?

The United States Sentencing Commission is an independent agency in the judicial branch of government. The Commission's principal purpose is to establish sentencing policies and practices for the Federal courts, including detailed guidelines prescribing the appropriate form and severity of punishment for offenders convicted of Federal crimes. While the development, monitoring, and amendment of the sentencing guidelines is the centerpiece of the agency's work, the Commission provides training, conducts research on sentencing-related issues, and serves as an information resource for Congress, criminal justice practitioners, and the public.

The U.S. Sentencing Commission was created by the *Sentecing Reform Act* provision of the *Comprehensive Crime Control Act of 1984*, and its authority and duties are set out in chapter 58 of title 28, United States Code; procedures for implementing guideline sentencing are prescribed in a new chapter 227 of Title 18. [22]

Congress charged the Commission with three significant ongoing responsibilities:

> *1. Evaluating the effects of the sentencing guidelines on the criminal justice system,* [23]

---

22 *See* 18 U.S.C. § 3553(a)(2); 28 U.S.C. § 991(b).

23 The Sentencing Guidelines are designed to incorporate the purposes of sentencing:
-just punishment, deterrence, incapacitation, and rehabilitation;
-certainty and fairness in meeting the purposes of sentencing by avoiding unwarranted disparity among offenders with similar characteristics convicted of similar criminal conduct;

2. Recommending to Congress appropriate modifications of substantive criminal law and sentencing procedures, and

3. Establishing a research and development program on sentencing issues.

In addition to creating the Sentencing Commission, the *Sentencing Reform Act* abolished parole for offenders sentenced under the guidelines so that the sentence received would be basically the sentence served.[24]

2. Why did Congress Create the Sentencing Commission?

Disparity in sentencing has long been a concern for Congress, the criminal justice community, and the public. After decades of research and debate, Congress decided that the previously unfettered sentencing discretion accorded Federal trial judges needed to be structured. Congress created a permanent commission charged with formulating national sentencing guidelines to define the parameters for Federal trial judges to follow in their sentencing decisions.

Organized in October 1985, the U.S. Sentencing Commission submitted its initial set of Guidelines and Policy Statements to Congress on April 13, 1987. Prior to this submission, the Commission held 13 public hearings, published two drafts of guidelines for public comment, and received more than 1,000 position papers from individuals and organizations. After the required period of congressional review, the guidelines became effective on November 1, 1987, and apply to all offenses committed on or after that date.[25]

3. How do the Sentencing Guidelines Work?

The United States Sentencing Commission has authority to submit guideline amendments each year to Congress between the beginning of a regular congressional session and May 1. Amendments automatically take effect 180 days after submission unless a law is enacted to the contrary.

-sufficient judicial flexibility to take into account relevant aggravating and mitigating factors; and

-advancement in the knowledge of human behavior as it relates to the criminal justice process.

24 Under the law, inmates may earn up to only 54 days of credit a year for good behavior.

25 Shortly after implementation of the guidelines, Defendants throughout the country began challenging the constitutionality of the *Sentencing Reform Act*, claiming improper legislative delegation and violation of the separation of powers doctrine. On January 18, 1989, in the landmark case of *Mistretta v. United States*, the Supreme Court rejected these challenges and upheld the constitutionality of the Act and the Commission as an independent judicial branch agency.

The Guidelines take into account both the defendant's count of conviction and actual nature of the criminal conduct by assigning a base offense level (a number) that serves as a starting point in assessing the seriousness of an offense. This base offense level can increase or decrease based on the circumstances of the particular case. The factors that modify the base offense level ("specific offense characteristics") are enumerated in the guidelines. A base offense level, modified by specific offense characteristics and general adjustments, forms one axis of the table used to determine sentencing ranges. The sentencing table's offense axis extends from level 1 (least serious) to level 43 (most serious).

The other axis reflects the defendant's criminal history as expressed in one of six categories (Category I-Category VI). The point at which the offense level and criminal history category intersect on the sentencing table determines an offender's guideline range. In order to provide flexibility, the top of each guideline range exceeds the bottom by no more than six months or 25 percent (whichever is greater). Ordinarily, the judge must choose a sentence from within the guideline range unless the court identifies a factor that the Sentencing Commission failed to consider that should result in a different sentence. However, the judge must in all cases provide the reasons for the sentence. Sentences outside the guideline range are subject to review by the courts of appeals for "reasonableness," and all sentences can be reviewed for incorrect guideline application or sentences imposed in violation of law.

4. How is the Sentencing Commission Structured?

Unlike many special-purpose "study" commissions within the executive branch, Congress established the U.S. Sentencing Commission as an ongoing, independent agency within the judicial branch. The seven voting members on the Commission are appointed by the President and confirmed by the Senate. The Attorney General is an ex officio member of the Commission, as is the Chairman of the U.S. Parole Commission.

The Commission staff of approximately 100 employees is divided into five offices with the director of each office reporting to the Staff Director who in turn reports to the Chairman. The five offices are: General Counsel, Monitoring, Training and Technical Assistance, Policy Analysis, and Administration.

In addition to developing the initial guidelines, the Commission:

- monitors the guidelines' operation through a comprehensive data collection system;
- conducts sentencing-related research;
- collects and disseminates information on Federal sentencing practices;

- provides reports and technical assistance to Congress and the judiciary on sentencing issues;
- revises the guidelines as necessary;
- trains thousands of judges, probation officers, prosecutors, defense attorneys, and others involved in the sentencing process;
- operates hotlines to assist judges, probation officers, and prosecuting and defense attorneys with guideline application;
- assesses the guidelines' effectiveness in achieving the sentencing purposes of just punishment, deterrence, incapacitation, and rehabilitation; and
- responds to requests for information from the courts, inmates, the general public, the defense bar, students, professors, the business community, and Congressional offices.

5. Who are the Current Commissioners?

Congress set forth slots for nine commissioners: seven voting and two non-voting, *ex-officio.* Commissioners are appointed by the President. Currently, there are only five sitting Commissioners: two others are waiting for confirmation by the Senate. The following is a listing of the current U.S. Sentencing Commission commissioners:

Richard P. Conaboy of Scranton, Pennsylvania, was sworn in as Chair of the U.S. Sentencing Commission on October 11, 1994. He currently serves as a United States District Judge for the Middle District of Pennsylvania. In 1989, he became Chief Judge; he took senior status in 1992. Formerly Judge of the Court of Common Pleas of Lackawanna County (PA), Hearing Examiner for the Pennsylvania State Liquor Control Board, Pennsylvania Deputy Attorney General, and a practicing attorney in Pennsylvania. From 1977 to 1980, he served as Chairman of the Pennsylvania Commission on Sentencing. He also served as Chairman of the Pennsylvania Joint Council on Criminal Justice, Chairman of the Pennsylvania Conference of State Trial Judges, and Vice-Chairman of the Pennsylvania Governor's Justice Commission.

Michael S. Gelacak of Centreville, Virginia, formerly practicing attorney in Washington, D.C. and Buffalo, NY. He served as Chief Counsel of the United States Senate Subcommittee on Penitentiaries, Staff Director of the Senate Subcommittee on Criminal Laws, and Minority Staff Director of the Senate Judiciary Committee. He also served as Legislative Director for Senator Joseph R. Biden, Jr., from 1983 to 1985.

Michael Goldsmith of Salt Lake City, Utah, Professor of Law at Brigham Young University. Formerly Assistant Professor of Law at Vanderbilt Law School, Counsel to the New York State Organized Crime Task Force, Assistant U.S. Attorney for the Eastern District of Pennsylvania, Senior Staff Counsel to the House Select Committee on Assassinations, and Deputy State's Attorney for Chittenden County, Vermont. Professor Goldsmith received the Order of the Coif from Cornell University Law School. He has written numerous articles on a wide range of criminal law topics, including sentencing and Federal guidelines.

Deanell R. Tacha of Lawrence, Kansas, United States Circuit Judge for the Tenth Circuit. Formerly Chair of the Judicial Conference's Committee on the Judicial Branch, Vice Chancellor for Academic Affairs at the University of Kansas, Associate Dean at the University of Kansas School of Law, Professor of Law at the University of Kansas, and a practicing attorney in Kansas and Washington, D.C. Listed in Outstanding Young Women in America. Received Jaycee's "Outstanding Young Woman in Kansas" Award in 1980 and Girl Scout's "Women of Distinction" Award in 1994 among numerous other awards.

Ex-Officio Members:

Michael J. Gaines of Little Rock, Arkansas, was designated as Chairman of the U.S. Parole Commission on February 4, 1997. Formerly, he served as Commissioner of the Parole Commission's National Appeals Board, Chairman of the Arkansas State Board of Parole and Community Rehabilitation, Executive Director of the State Supreme Court Committee on Professional Conduct, and Criminal Justice Liaison and Pardon and Extradition Counsel to then Governor Clinton.

Mary Frances Harkenrider of Chicago, Illinois, Counsel to the Assistant Attorney General for the Criminal Division, United States Department of Justice. Formerly a Federal prosecutor in the Northern District of Illinois and a Senior Supervising Attorney for the City of Chicago. She has taught Trial Practice and Pretrial Procedure classes as an Adjunct Professor at Northwestern University School of Law.

6. What is the Process for Amending the Sentencing Guidelines and how much time would the Sentencing Commission Require to Amend the Existing Guidelines?

Congress did not legislate minimum or maximum time requirements in the Sentencing Reform Act.[26] Rather, the Sentencing Commission is left with discretion to select relevant issues and topics for sentencing reform and enhancement. In principle, changes proposed by Congress or the Sentencing Commission may take effect on an expedited basis–in some cases, as quickly as several months' time. All other amendments require at least six months to one year, assuming there is no opposition.

---

26 28 U.S.C. § 1991(b) (1995).

The Sentencing Commission operates under its Rules of Practice and Procedure.[27] Pursuant to the Rules, Amendments are introduced into the process in three ways:

(1) by Congress, as a mandated Emergency Guideline Amendment (Rule 4.1);

(2) by the Sentencing Commission, pursuant to authority delegated from Congress and approved by four of the Commissioners (Rules 2 and 4.1);

(3) by any party, via submission, as a proposed Guideline Amendment (Rules 4.1 -4.5).

Congress may direct the Sentencing Commission to amend the Guidelines where a new statute requires an addition to the Guidelines. Congress may also force an amendment where Guidelines are inconsistent with existing Statutes. In both cases, Congress notifies the Commissioners, provides the necessary language, and the amendment occurs without public debate or Sentencing Commission deliberations. This is the most efficient method and requires the least amount of time.

The Sentencing Commission may draft the Guideline Amendment when instructed to by Congress. In contrast to Congress' unilateral action to amend the Guidelines, the Sentencing Commission has authority to discuss and to draft Amendments, but only on an "emergency amendment basis". *See* 28 U.S.C. § 994(a); Rules of Practice 2.2 (1997).[28] Under both scenarios, Congress must instruct or delegate the power to the Sentencing Commission to amend the Guidelines outside of the normal process.

Other interested parties may propose amendments to the Sentencing Guidelines. The proposals take the form of submissions and set forth all arguments and rationale for the proposed amendments. The cycle–which lasts one full year—typically follows the following time frame:

- December to February—Received Submissions for Amendments.
- March - April—Debates and considers public reaction; votes on Amendments.
- May 1—Sends Amendments to Congress.
- May - November—Congress Reviews.

---

27 Rules of Practice and Procedure, U.S. Sentencing Commission (July 1997) (the "Rules of Procedure"). The Sentencing Commission promulgates the Rules of Procedure pursuant to Congressional mandate. 28 U.S.C. § 995(a)(1). The Sentencing Commission amended its Rules of Procedure in June 1997. For background information on the recent changes, refer to 61 Fed. Reg. 39493-39496 (July 19, 1996) and 61 Fed. Reg. 52825-52826 (October 8, 1996).

28 Note that the two ex-officio Commissioners do not have voting authority. Rules of Practice 2.1 (1997).

- Enacted in November, unless Congress explicitly rejects.

Under this third scenario, any organization may propose an Amendment to the Guidelines. The cycle technically lasts one year. In reality, however, the Commission often takes up to two to three cycles before hashing out a particular change to the Guidelines.[29]

The Sentencing Commission may also invite the airing of an issue, or series of issues. According to the Rule of Practice, the Sentencing Commission publishes a "Notice of Priorities" in the Federal Register. The purpose of this Notice is to–

> *"make available to the public a notice of the tentative priorities for future Commission inquiry and possible action, including areas for possible amendments to guidelines, policy statements, and commentary."*[30]

It may be appropriate that the Sentencing Commission immediately list critical infrastructure assurance as a priority issue.

---

29 Based on conversations with John Steer, General Counsel to the U.S. Sentencing Commission. *See* Angeli, *A "Second Look" at Crack Cocaine Sentencing Policies: One More Try for Federal Equal Protection,* 34 AM. CRIM. L. REV. 1211 (1996) (Demonstrates the political and legal complexities of amending the Sentencing Guidelines).

30 Rules of Practice, 5.2 (May 1997).

CHAPTER 11

PRIVACY LAWS AND THE EMPLOYER-EMPLOYEE RELATIONSHIP

A "LEGAL FOUNDATIONS" STUDY

Report to the
President's Commission
on Critical Infrastructure Protection
1997

This report was submitted to the President's Commission on Critical Infrastructure Protection, and informed its deliberations and recommendations. This report represents the opinions and conclusions solely of its developers.

CONTENTS

PART ONE
INTRODUCTION

This paper examines what the Federal government should do, if anything, to insure that appropriate legal means exist by which governments and the private sector can collect, retain and disseminate data or information integral to achieving infrastructure assurance objectives while respecting fundamental privacy concerns of individuals and the interests of state legislatures in protecting citizens' privacy. The paper explores the possibility of making available to owners and operators, for use in filling certain sensitive positions within the critical infrastructures, some of the techniques and methods currently used by the Federal government to screen employees for employment in certain sensitive positions.

Research Findings

- "Insider" misconduct is the most immediate and, to some, the most predominant security concern of infrastructure owners and operators. Insiders pose a potential threat along physical and cyber dimensions.
- Insider harm can prove more devastating than outsider harm by virtue of the insider's superior knowledge and access.
- Employee screening and employee monitoring are two techniques by which employers can impose additional security in the workplace. Of the two, employee screening appears to be the least intrusive.
- In 1994, Congress attempted to strike the delicate balance between the employer's right to monitor and employees' privacy interests in a series of comprehensive legislative proposals. The failure of those efforts demonstrates the difficulty and sensitivity of the issue.
- Enhanced employee screening, resembling measures taken by the Federal government in issuing security clearances, may prevent or deter some forms of insider misconduct.
- Efforts to prevent insider harm through more stringent employee screening implicate employee privacy issues, necessitating that a balance be struck between privacy and security:
  - Employers may wish to take into account in employment decisions factors such as criminal history, employment history, credit history, or even the results of a polygraph examination.
  - Individuals have a strong interest in controlling collection and dissemination of such highly personal (and potentially unreliable) information.
- Information technology exacerbates this potential clash of interests by increasing the ease with which such information might be collected, stored and disseminated.

- State lawmaking bodies have been responsive to citizens' privacy concerns, particularly where electronically stored information is concerned, and have enacted measures designed to protect sensitive information from collection or disclosure.
- Though responsive, the states have not adopted uniform approaches to these information privacy issues. This has resulted in a rich and varied "patchwork" of laws governing, among other things, the employer-employee relationship.
- This patchwork of privacy protections applies predominantly to private sector employees. Federal, state and local government employees are often expressly exempt.
- Federal laws have been established in some instances within this environment to unify divergent approaches, or to protect or restore vital national security concerns.

Assumptions

- Some state and Federal employee "privacy" laws, though enacted with best intentions, may unduly restrict the collection or dissemination of information that might otherwise be used by an owner or operator of the critical infrastructures to enhance infrastructure assurance.
- The elimination of some undue restrictions on collecting and sharing job-related information (at least for the owners and operators of critical infrastructures seeking to fill key positions) would permit and encourage owners and operators to avail themselves of employee screening procedures akin to those used by the government in issuing security clearances.
- Allowing owners and operators to more carefully pre-screen employees or prospective employees for certain sensitive positions within the critical infrastructures will enhance infrastructure assurance.

PART TWO
BACKGROUND

One the most pressing and complex problems facing our nation's critical infrastructures is the threat posed by their own employees. In a recent survey, 87 percent of respondents cited "disgruntled employees" as the likely source of computer attacks on their company.[1] Despite the alarming percentage of computer-related incidents, and the persistence of traditional criminal activity by insiders ($120 billion in employee theft each year),[2] few

---

1 CSI/FBI 1997 Computer Security Survey.

2 Rochelle B. Ecker, *To Catch a Thief: The Private Employer's Guide to Getting and Keeping*

recommendations have been made in the area to date. The reluctance to address this problem may be traced to concerns among legislators, at the Federal and state levels, over infringing on the legitimate privacy interests of upstanding citizens. "The employer's interest in maintaining an honest workplace is not always consistent with the employee's interest in privacy."[3]

In fact, the trend in state legislation has been toward increasing the protections of employees and prospective employees from employer incursions into their personal privacy. These state statutes may forbid anything from inquiring about personal relationships to investigating criminal histories. While many of these statutes are well-crafted and serve the interests of employees and employers alike, others may be over-inclusive and ultimately inhibit infrastructure owners and operators from taking necessary security precautions.[4]

Not only is addressing the "insider threat" difficult because of the privacy interests at stake, but there are also difficult federalism concerns that must be considered. The Federal government has jurisdiction over the critical infrastructures through the interstate commerce power, and has in the past regulated some of the infrastructures quite heavily. However, the area of privacy has traditionally been left to the states to develop the appropriate balance of interests. This is consistent with the constitutional authority of the states to exercise general police powers, including legislating for the public health, safety, morals and welfare of their citizens. The result has been varied and inconsistent approaches among the states.

In each of the specific areas identified by the research into privacy laws and their impact on critical infrastructure owners and operators, issues of Federal-state relations dictate the options that are available. Federal law is only possible in those areas that touch on interstate commerce as a legal matter. As a matter of policy, Federal legislation, trumping existing state laws, may only be appropriate in those areas where the lack of uniformity among the states or failure of the states to recognize security-related issues rise to the level of national security concerns. The available approaches include:

1. **The Status Quo:** The status quo in this context amounts to a recognition and tacit acceptance of the existence and continued development of fifty very different genres of state privacy and employer-employee-related legislation.

---

an Honest Employee, 63 U.M.K.C.L. REV. 251, 252 (1994).

3 *Id.*

4 In contrast, recent state legislation promoting the on-line availability of public records may provide more access than is needed to protect the interests of critical infrastructure employers and employers generally, to the detriment of the employee. *See, e.g.*, H.R. 2112, 55th Leg., Reg. Sess., (Washington 1997) (providing for criminal history information to be available via the Internet).

2. **Highlighting Critical Infrastructure Issues to the States and Suggest Reconsideration of Privacy Legislation:** This is a deferential approach that will not on its own achieve uniformity, but may succeed in achieving revisions to current legislation or new legislation that is more sensitive to the needs of critical infrastructure owners and operators.

3. **Creation of a Federal Privacy Baseline:** Congress may draft and enact legislation that sets the minimum for privacy protection in a given area. States would then be free to allow for greater privacy protection as necessary. This approach, while it does articulate a Federal position, is not preemptive of state efforts (unless the state currently falls below the threshold of the Federal baseline).

4. **Federal Preemption:** In areas that touch interstate commerce, Congress may exercise its authority to pass legislation that, rather than setting a minimum which states must adhere to, completely overtakes the area of law from state control. While this is a highly intrusive approach, it creates a level uniformity across Federal and state levels that is otherwise unachievable.

5. **Study of the need for and advisability of any or all of the above approaches:** While yet another study delays action, it allows the appropriate interests to be considered and incorporated into any change, whether merely recommended to the states or mandated by Federal preemption.

These approaches may be applicable to the following areas identified as potential impediments to the achievement of infrastructure assurance objectives. These impediments tend to restrict the ability of employers, or potential employers, from screening or monitoring employees or applicants, thus constraining the ability of private owners and operators to protect critical infrastructures from insider threats by conducting the sort of background investigations routinely performed on Federal employees in sensitive positions.

Considerations for Employees in Sensitive Positions

Critical infrastructures, whether run by private sector owners and operators or the government, must utilize the services of employees. Prudence or due diligence may compel operators to explore basic questions about a current or prospective employee:[5]

- Does the employee have the necessary skills?

---

5 Fear of tort liability may also prompt employers to conduct background investigations. *See* discussion *infra* p. 15 of "negligent hiring."

- Is there any background information that would militate against relying on the employee's skills?
- Is the employee honest and, does he or she possess integrity?
- Has the employee disclosed relevant job-related information, such as his or her criminal, financial, or employment history—especially background data that goes to the issue of trust, reliability, and predictability?

These inquiries are routinely made by the Federal government in screening employees for sensitive positions. The authority for these background investigations was specifically granted to the President in the Civil Service Act of 1883.[6] Federal privacy-related statutes may affect the scope of Federal background investigations. However, state privacy laws do not affect the authority of the Federal government to conduct background investigations pursuant to this Federal legislation. Without similar Congressional authorization, critical infrastructure owners and operators must observe the various Federal and state laws that govern the individual elements of employee screening. Although such inquiries may be central to an owner or operator's ability to protect critical infrastructure, Federal and state laws passed to protect personal privacy may prohibit such inquiries, or make inquiries difficult and legally suspect. In this important sense, some privacy legislation may actually serve to undercut critical infrastructure assurance.

In addition, Federal and state laws, whether statutes or common law causes of action, limit the ability and incentives for employers to share information relating to employees. For example, in 1994, an airplane crashed killing fifteen people on board. The National Transportation Safety Board concluded the cause of the crash was pilot error. Investigations revealed that before joining his current airline, the pilot had resigned from another airline to avoid being fired for failing a critical flight test. The airlines did not share that information. In fact, most airlines have specific policies in place preventing sharing of that type of information to avoid liability from suit.[7] The fear of liability, whether or not it is justified, operates in many instances to prevent adequate screening of potential employees or sharing of information about prospective employees.

Relevant Law and Legislation

The following types of state laws may present infrastructure assurance issues:

---

6 Currently codified at 5 U.S.C. § 3301 (1997). This authority is implemented by Exec. Order 10450, Civil Service Investigations.

7 Robert Adler & Ellen Pierce, *Encouraging Employers to Abandon Their "No Comment" Policies Regarding Job References: A Reform Proposal*, 53 WASH. & LEE L. REV. 1381, 1383 (1996).

Criminal History Information

The criminal history of a prospective employee, or information on any subsequent arrests or convictions once hired, is an important tool used by employers in determining the suitability of an individual to a sensitive position. As a legal matter, obtaining criminal history information is primarily governed by state law. States regulate the collection and maintenance of criminal justice record systems and the rules under which such information may be made available to individuals and third parties. State laws regarding civil rights, employment, or (in the case of New York state) corrections, set out the rights of applicants and employers with respect to requesting and reviewing criminal history information. In addition, both Federal and state equal employment laws provide guides for the proper use of criminal history information in hiring decisions. It is important for purposes of employment issues to maintain a crisp distinction between information disclosing a *conviction* and information reporting a mere *arrest*. While arrest information may be available through state means, Federal law strictly prohibits potential employers from requesting such information or using the information in making hiring decisions.

Access to Criminal History Records

Criminal history information is maintained both by the Federal and state governments. According to the rules set out at 28 C.F.R. § 20.1 *et seq.*, the Federal Bureau of Investigation administers the National Criminal Information Center (NCIC). States may receive funds and participate in the NCIC if willing to conform to the regulations governing the database. Individuals may request and review their NCIC records to verify their accuracy. Federal regulations place substantial restrictions on the availability of arrest (without conviction) information, but allow states full discretion to permit access to conviction information by third parties. In fact, the regulations specifically recognize the public nature of much of the criminal history information related to convictions and does not regulate that information to the extent it is available from an independent source (i.e., other than the centralized state criminal information system or the NCIC). Federal officials, through the FBI, will only release criminal history information for employment purposes to requesting third parties who are banks, state and local governments, registered securities exchanges or nuclear power providers.[8]

If a state does not participate in the NCIC, they are free to regulate their centralized criminal history information system as they choose. In addition, to the extent that they do participate, states may still regulate access to conviction information by third parties. Some states, such as Colorado,

---

8 *See* 28 C.F.R. §§ 20.33 & 50.12.

have taken a liberal approach to the availability of such information allowing complete access to criminal history information except for the names of victims of sexual offenses. Delaware releases conviction information of job applicants and employees to employers directly without any requirement of employee consent.[9] Other states limit the information available to third parties to information relating to convictions within a certain time frame.[10] As a general rule, individuals have access to their criminal history records at state and Federal levels.[11] Records may, in some instances, also be released to a third party based on the individual's consent.[12] Only a few states do not allow copies of criminal history records to be released to individuals or their agents.[13]

Employer Inquiries into Criminal History

A more contentious issue than simply the availability of criminal history records, is employer inquiries into criminal histories as part of a hiring process. At the Federal and state levels, this is regulated as a matter of fair employment and civil rights practices. Both the Federal government and 41 states prohibit requests for *arrest* information.[14] Some states do allow inquiries into current arrests and pending criminal matters. States have set additional limitations on inquiries relating to criminal backgrounds. Massachusetts, for example, prohibits questions relating to misdemeanor convictions.[15] Still other states set time limitations to allow access only to "current" offenders. Other states only allow inquiries into criminal histories for certain statutorily enumerated job positions, usually involving the supervision of minors (e.g., teachers, day care operators, etc.). The California Penal Code prohibits employers from requiring applicants to provide a

---

9 DEL. CODE ANN. tit. 11, § 8513(c)(1) (1996).

10 *See, e.g.,* ALASKA STAT. § 12.62.160 (1996).

11 *See, e.g.,* ARK. CODE ANN. §§ 12-12-211 through 12-12-213 (1995 Repl. Vol.); CAL. PENAL CODE § 13323 (1996); GA. CODE ANN. § 35- 3-34 - 35-3-37 (Supp. 1996).

12 *See, e.g.,* KAN. STAT. ANN. § 22-4710 (1996) (employer access to criminal history based on signed release by applicant).

13 *See, e.g.,* LA. REV. STAT. ANN. § 15:588 (West 1996); MINN. STAT. ANN. § 13.87 (West 1996) (allows any member of the public to view criminal history information on a computer monitor); MISS. CODE ANN. § 45-27-11 (1996) (copy of record only available to individual or their attorney if contesting information therein).

14 *See* Commission Decision No. 74-02, CCH EEOC Decision (1973); Rochelle B. Ecker, *To Catch a Thief: The Private Employer's Guide to Getting and Keeping an Honest Employee,* 63 U.M.K.C. L. REV. 251, 255 (1994). Inquiring about arrest histories has been determined to be employment discrimination in violation of Title VII of the Civil Rights Act of 1964 on a theory of disparate impact. Disparate impact is generally applicable when a criterion, while neutral on its face, has a statistically greater impact on one segment of the population than others.

15 *See, e.g.,* MASS. GEN. LAWS ANN. ch. 151B, § 4(9) (1991).

copy of their criminal records.[16] As a general rule, inquiries regarding *convictions* are permissible to the extent job-related.

Use of Past Convictions in the Hiring Decision

Even more controversial than access to or inquiries into criminal histories are the uses of such information by an employer in making a hiring decision. In general, both as a matter of Federal and state law, employers may take conviction information into account, to the extent it is available (see above), in the hiring process provided the employer considers: (1) the seriousness of the offense or offenses; (2) the age of the person at the time of the offense; (3) the time which has elapsed since the offense or offenses; and (4) the bearing the conviction of such an offense has on the ability of the individual to perform the duties or responsibilities of the position for which they applied.[17]

While such considerations should be taken into account by infrastructure owners and operators, in the vast majority of cases, individuals, particularly those in sensitive positions, may be screened for criminal convictions. And the closer the relationship to public safety (which in the critical infrastructures is often a very close nexus), the greater the assurance of the employer that they will not be subjected to liability for violations of employment-related laws. However, care must still be given to properly observe the state law requirements with regard to use of criminal conviction information. Some states do require written explanations of the grounds for denying an applicant a position based on criminal history.

Credit History Information

Availability of credit history information is governed primarily by the Federal Fair Credit Reporting Act (FCRA).[18] The FCRA sets out rules for requesting and receiving information, procedures for use of the information in making hiring decisions, and provisions for applicants to seek recourse for inaccurate and damaging information. Individuals may request and receive information concerning their own credit history from credit bureaus. In addition, employers may directly, and with no requirement of applicant consent, request and receive credit history information from credit bureaus as long as they are using the information to make employment decisions.[19] If an applicant is not hired based on information contained in their credit report, they must be notified of the reason in writing with the name and address of the credit reporting agency. Appli-

---

16 CAL. PENAL CODE § 13326 (1996).

17 *See, e.g.*, N.Y. CORRECTION LAW § 753 (McKinney 1997).

18 15 U.S.C. §§ 1681-1681t (1996).

19 15 U.S.C. § 1681b.

cants can then request the information and review it for discrepancies. Procedures are in place for disputing information contained in credit reports. This framework established by the FCRA is the model for much of the state legislation in the area.[20] In fact, to the extent any state law is inconsistent with the Federal provisions in the FCRA, the state law is preempted.

Polygraph Examinations

Federal and state laws limit the use of polygraph testing by employees, including the owners and operators of critical infrastructures. Federal law does not apply to state and local governments; absent state legislation, there is no general prohibition against use for state and local government employees.

State Law

The need for Federal polygraph protection surfaced in the 1980s after employers forced polygraph testing on unsuspecting employees in record numbers, often circumventing state laws to protect employees. In the early part of the 1980s, employee theft resulted in losses ranging from $9.2 billion to $50 billion per year.[21] Employers responded with over two million polygraph tests annually to screen job applicants, investigate specific incidents of theft, and to uncover employee misconduct.[22] Forty-one state legislatures responded with legislation after employees protested that the examinations were unduly intrusive and did not detect deception.[23] However, because the state laws lacked uniformity, employers were able to force employees to submit to tests in neighboring states, where laws were relatively lax.

Federal Law

Congress responded by passing strong Federal legislation. The Employee Polygraph Protection Act (EPPA)[24] makes it unlawful for an employer

---

20 *See, e.g.*, ARIZ. REV. STAT. ANN. § 44-1691 through 1696 (1996); CAL. CIVIL CODE §§ 1785-1786.56 (1996); KAN. STAT. ANN. §§ 50-701 through 722 (1996); MASS. GEN. LAWS ANN. ch. 93 §§ 50-68 (1997); N.M. STAT. ANN. §§ 56-3-1 through 56-3-8 (1997).

21 *See* C. Cullen, The Specific Incident Exemption of the Employee Polygraph Protection Act: Deceptively Straightforward, 65 NOTRE DAME L. REV. 262 (1990).

22 *Id., citing to* Privacy Protection Study Commission, Personal Privacy in an Information Society (1997) at n.. 7 (reporting that in 1977, by contrast, only 300,00 lie detector tests were given).

23 *See id.* at 263-264. For an excellent summary of the Act, please refer to L. Pincus and C. Trotter, *The Disparity Between Public and Private Sector Employee Privacy Protections: A Call for Legitimate Privacy Rights For Private Sector Workers*, 33 AM. BUS. LAW JOURNAL, 50, 68-70 (1995) (hereinafter "Employee Protections").

"directly or indirectly, to require, request, suggest, or cause any employee or prospective employee to take or submit to any lie detector test."[25] Violations may be punished by a civil penalty of up to $10,000.[26] The Act specifically identifies: who may be asked to take a polygraph; for what purpose or limited set of positions it may be requested; the types of questions employers may ask in polygraph testing; the allowable uses of the results; and whether the results may be disseminated.

According to the Act, information obtained from the polygraph examination cannot be disclosed except to the examinee, the employer requesting the test, courts or agencies pursuant to a court order.[27] The Act includes several major exemptions to the general rule that no one may be subject to a polygraph test. With respect to the critical infrastructures, these exemptions are significant, and include: all Federal, state, local governments and employees;[28] certain DOD, DOE, intelligence community and FBI contractors and employees.[29] Exemptions also appear for employers to use testing as a tool in ongoing employment-related investigations, which require the employer to show reasonable suspicion and follow carefully delineated investigative procedures.[30]

Exemptions written into the Act carve out several areas where employers may continue to use the polygraph.[31] One pertinent exception applies to employers who provide security-related services. For these employers, polygraph testing for applicants and existing employees is permitted. This exemption covers employers (1) whose primary business purpose consists of providing security, such as armored car personnel, personnel who install or maintain alarm systems, or other uniform or plainclothes security personnel; *and* (2) whose function includes protection of facilities, materials, or operations having a significant impact on the health or safety of any State or political subdivision, including electric power plants, public water supply systems, public transportation, and the protection of currency, negotiable securities, precious commodities or proprietary information.[32] The

---

24 29 U.S.C. §§ 2001 - 2009; Pub. L. No. 100-347, 102 Stat. 646.

25 29 U.S.C. § 2002.

26 29 U.S.C. § 2005.

27 29 U.S.C. §§2006 - 2008; 29 C.F.R. §§ 801, 801.10 - 801.14 (1997).

28 29 U.S.C. § 2006 (1988), 29 C.F.R. § 801.10 (1997). States may choose, however, to pass more restrictive polygraph legislation. The Act also would not preempt a more restrictive collective bargaining agreement provision.

29 29 U.S.C. § 2006e; 29 C.F.R. §§ 801.11 - 801.14 (1997) (Department of Labor regulations construing EPPA).

30 *Id.*

31 29 U.S.C. § 2006(e)(1)(A)(i) - (iv); *see also* 29 C.F.R. Part 801, 801.10 - 801.14 (1997).

32 *Id.* at 29 U.S.C. § 2006(e)(1).

exemption appears to extend to a wide range of *physical* security services, but its application to those who provide *information* security services remains unclear. Likewise, providers of telecommunications services are not included in the enumeration of protected facilities.

The Department of Labor's implementing regulations elaborate on this exemption. According to the regulations, both of these preconditions must be met for the employer to claim an exemption.[33] That is, the employer must be primarily in the business of protecting the facilities and the facilities may be publicly or privately owned, and may be construed broadly to include a wide range of infrastructure facilities.[34]

Interaction of State and Federal Law

It is important to note that the EPPA does not completely preempt state law on the use of polygraph tests. The EPPA does govern exclusively the exemptions allowing use of polygraph tests by government employers (Federal, state and local), for national defense and security reasons by the Federal government, and of FBI contractors.[35] The EPPA also sets a minimum level of protection for private-sector employees in all states. That is, states must observe the prohibition on use of polygraph tests set out in the EPPA, however, they may refuse to accept the exceptions to the prohibition recognized in Federal law, or craft more narrow ones. To the extent a state law is more restrictive than the Federal law, it will control in an employment matter in that state. For this reason, exemptions must be considered on a state by state basis. The result being that any modification to the Federal exemptions that broaden their scope may be of limited effect alone. Unless they are included among the preemptive provisions of the Act, or states opt to recognize the need for broader exemptions, they will not have wide application outside of Federal issues. Select state exemptions are highlighted in the table, below.

---

33 29 C.F.R. § 801.14(c)-(d)(1) (1997).

34 *Id.* (". . . These examples are intended to be illustrative, and not exhaustive.") *Id.*

35 29 U.S.C. §§ 2006 & 2009.

State	Polygraph law
Alaska	General rule is to prohibit; exemptions for law enforcement. Alaska Stat. § 23.10.037(a).
California	General rule is to prohibit; exemptions for government officials. Cal. Labor Code § 432.2 (West 1989) (Lie-Detector Testing).
District of Columbia	General rule is to prohibit; exemptions for criminal investigation by Metro police or fire departments, or Department of Corrections.
Kansas	Repealed 1989 law (Kan. Stat. Ann. § 75-744 (1989).
Maine	Pre-employment: General rule is to prohibit. Current employees: General rule is to prohibit. Exemptions for law enforcement. Employee can voluntarily submit.
Minnesota	General rule is to prohibit. If one is given, results can only be given to employee and persons authorized by employee to receive results.
Montana	General rule is to prohibit; exemptions in areas of "security", "public safety" and where there is a "fiduciary responsibility."
New York	General rule is to prohibit.
Pennsylvania	General rule is to prohibit. Exemptions for law enforcement or who dispense narcotics and dangerous drugs.

Employment History

Personnel records often contain details of an individual's employment history, such as reprimands or disciplinary actions, that could have a bearing on their fitness for a sensitive position within a critical infrastructure. Depending on whether an employer is Federal government, state government or private, a different set of rules governing access to employment records applies (*e.g.*, Federal and state Freedom of Information Acts). However, a few generalizations can be made. Employers are generally required to ensure the accuracy of the records they keep and to allow employees an opportunity to review the records to verify the information they contain. Provisions for inspection of personnel records are included in the Federal Privacy Act (which applies only to information held by the Federal government) and in various state statutes.[36] These records may be released to third parties only with the consent of the employee they concern or pursuant to advanced written notice.[37] Employers may not share employee personnel records with each other without the involvement of the employee.

---

36 *See* 5 U.S.C. § 552a(d); *see, e.g.*, CONN. GEN. STAT. § 31-128f (1997); D.C. CODE ANN. § 1-632.5 (1997); MICH. COMP. LAWS § 423.506 (1997); OR. REV. STAT. § 652.750 (1997); WIS. STAT. § 103.13 (1997).

37 *Id.*

Defamation as a Limit on Disclosure

In addition to the specific statutory limits on dissemination of information in employment records, there are also tort-based laws which may prevent employers from sharing relevant information. These tort laws are state level statutory or common law creations. They protect individuals from invasions of privacy and damage to reputation to varying degrees. To the extent this issue is governed primarily by case law rather than statutes, the individual predilections of judges and juries contribute to immense diversity among the states in possible acts that could create liability for an employer. Although other causes of action are available, defamation has most recently been invoked with regard to unfavorable references. Defamation is the publication of anything injurious to the good name or reputation of another, or which tends to bring him into disrepute.[38] A common, modern day dilemma exists for all employers who are called to discuss a former employee. What can the employer say if the employee's performance was poor? What if the employee was let go for a suspected theft? If a critical infrastructure facility is hiring for a sensitive position, can it seek and obtain accurate references on a job applicant?

Defamation is handled predominantly in state legislatures or the courts. Most states have handled defamation in widely different ways. Many states recognize defenses, such as "truth" and "employer immunity." Certain generalizations can be made, such as:

- *Truth* is always a defense to the common law claim of defamation;
- *Qualified privilege* is used if it is a true statement made in good faith serving a business interest or purpose. A harmed employee can defeat a qualified privilege by showing malice—actual knowledge of falsity or reckless disregard for the truth.
- Some states allow for employer immunity where information is offered in good faith (see chart below).[39]

---

38 *Western Union Telephone Co. v. Lesesne*, 198 F.2d 154 (4th Cir. 1952) (prima facie case defined).

39 *See, e.g.*, Florida, Delaware, and Kansas have also recently passed employer immunity statutes. *See* FLA. STAT. ANN. § 768.095 (West 1997) (employer immunity from liability); KANS. STAT. ANN. § 44-119(a) (Supp. 1996) (immunity from liability); DEL. CODE ANN. Tit. 19, § 708 (Supp. 1996) (immunity for employer). *See also* D. Scott Landry & Randy Hoffman, *Walking the Fine Line on Employee Job Reference Information*, 43 LA.B.J. 457 (1996). (Law review article on trend in South for employer immunity).

State	Approach to Defamation
Idaho	An employer who in good faith provides information about the job performance, professional conduct, or evaluation of a former or current employee to a prospective employer of that employee, at the request of the prospective employer of that employee, or at the request of the current or former employee, may not be held civilly liable for the disclosure or the consequences of providing the information. Idaho Code § 44-201 (Employer Duties).
Maine	26 M.R.S.A. § 598 MAINE REVISED STATUTES ANNOTATED TITLE 26. LABOR AND INDUSTRY CHAPTER 7. EMPLOYMENT PRACTICES SUBCHAPTER I. CONDITIONS FOR EMPLOYMENT § 598. Employment reference immunity An employer who discloses information about a former employee's job performance or work record to a prospective employer is presumed to be acting in good faith and, unless lack of good faith is shown by clear and convincing evidence, is immune from civil liability for such disclosure or its consequences. Clear and convincing evidence of lack of good faith means evidence that clearly shows the knowing disclosure, with malicious intent, of false or deliberately misleading information. This section is supplemental to and not in derogation of any claims available to the former employee that exist under state law and any protections that are already afforded employers under state law.
Indiana	§ 22-5-3-1 Disclosure of information after employee's discharge Sec. 1. (a) A person who, after having discharged any employee from his service, prevents the discharged employee from obtaining employment with any other person commits a Class C infraction and is liable in penal damages to the discharged employee to be recovered by civil action; but this subsection does not prohibit a person from informing, in writing, any other person to whom the discharged employee has applied for employment a truthful statement of the reasons for the discharge. (b) An employer that discloses information about a current or former employee is immune from civil liability for the disclosure and the consequences proximately caused by the disclosure, unless it is proven by a preponderance of the evidence that the information disclosed was known to be false at the time the disclosure was made. (c) Upon written request by the prospective employee, the prospective employer will provide copies of any written communications from current or former employers that may affect the employee's possibility of employment with the prospective employer. The request must be received by the prospective employer not later than thirty (30) days after the application for employment is made to the prospective employer.

State	Approach to Defamation
Louisiana	LSA-R.S. 23:291 § 291. Disclosure of employment related information; presumptions, causes of action; definitions A. Any employer that, upon request by a prospective employer or a current or former employee, provides accurate information about a current or former employee's job performance or reasons for separation shall be immune from civil liability and other consequences of such disclosure provided such employer is not acting in bad faith. An employer shall be considered to be acting in bad faith only if it can be shown by a preponderance of the evidence that the information disclosed was knowingly false and deliberately misleading. B. Any prospective employer who reasonably relies on information pertaining to an employee's job performance or reasons for separation, disclosed by a former employer, shall be immune from civil liability including liability for negligent hiring, negligent retention, and other causes of action related to the hiring of said employee, based upon such reasonable reliance, unless further investigation, including but not limited to a criminal background check, is required by law. C. As used in this Section, the following words and phrases shall have the meanings contained herein unless the context clearly requires otherwise: (1) "Employer" means any person, firm, or corporation, including the state and its political subdivisions, and their agents, that has one or more employees, or individuals performing services under any contract of hire or service, expressed or implied, oral or written. (2) "Employee" means any person, paid or unpaid, in the service of an employer. (3) "Prospective employer" means any "employer", as defined herein, to which a prospective employee has made application, either oral or written, or forwarded a resume or other correspondence expressing an interest in employment. (4) "Prospective employee" means any person who has made an application, either oral or written, or has sent a resume or other correspondence indicating an interest in employment. (5) "Job performance" includes, but is not limited to, attendance, attitude, awards, demotions, duties, effort, evaluations, knowledge, skills, promotions, and disciplinary actions.

Despite the availability of defenses and the likelihood they may prevail in a civil suit, many employers are nonetheless reluctant to do anything more than verify dates of employment, salaries and other factual information. Some states have recently enacted legislation expanding immunity from civil suits for employers who provide references in good faith. Prior to 1995 only five states had such statutes. By October of 1995, nine more states had enacted such legislation. Louisiana passed similar legislation in 1996.[40] The momentum of these legislative initiatives may increase due to other emerging civil actions relating to employee hiring.

---

40 43 LA. B.J. at n.3.

There is an additional incentive to share information about employees. As of 1991, 30 states, including California, recognized the tort of negligent hiring. In determining whether the tort of negligent hiring has been committed, "courts look to whether the employer reasonably investigated the employee's background before hiring the employee."[41] Access to information about prospective employees is important to all employers, if for no other reason than to avoid this type of liability. The current state of tort law with regard to employee references truly demonstrates the complexity of the interests at stake in the employer-employee relationship and the difficulty of balancing those equities.

PART THREE
APPROACHES FOR ADDRESSING PRIVACY ISSUES

General Recommendation By The Federal Government That States Individually Examine Potential Unintended Consequences Of Related Legislation And Consider Benefits Of Adopting Unified Approaches

Without enacting any particular approach, Congress or the Administration could acknowledge the importance of *certain* employers being permitted to gather *certain* types of job-related information for *certain* sensitive positions, and the difficulties and unpredictability brought about through multiple state approaches. The Federal government could suggest that states clarify their own laws regarding the types of records and information employers may acquire, for what positions, and under what circumstances. In this way, the Federal government can influence state legislatures to revisit the impact of their own laws in light of infrastructure assurance objectives. Critical infrastructure owners and operators can be encouraged to lobby state legislatures to incorporate specific exemptions in these privacy provisions—thereby allowing more careful screening of applicants for sensitive positions. Additional measures may include Federally-directed awareness campaigns at state legislatures and private industry executives to raise awareness of infrastructure assurance issues relating to employee security.

- **Pro:** Allows Federal government to "air views" while demonstrating maximum deference to states. Minimally intrusive approach impinges neither upon individual privacy concerns nor state sovereignty or federalism concerns, but encourages long-term conformity to infrastruc-

---

41 Janet Swerdlow, *Negligent Referral: A Potential Theory for Employer Liability*, 64 S. CAL. REV. 1645, 1646 (1991) (in the article, the author advocates the creation of a tort law duty of employers to provide accurate and substantive references on former employees).

ture assurance objectives. Allows each state legislature to determine the balance between privacy and security concerns *for that state.*

- **Con:** Unlikely to result in meaningful change absent more specific guidance backed by statute or regulation. Retains geographical differences between laws governing electronic data transmissions—resulting in lack of uniformity between the states with respect to "rules of the road" for information conveyed across state lines. Difficulty of achieving compliance with laws across more than 50 different jurisdictions may discourage owners and operators from seeking information they might otherwise be entitled to receive. Leaves legislative "spadework" to infrastructure owners and operators—but they are unlikely to rally at state and local levels for what may be perceived as an unpopular cause.

Federal Government Points Out Benefits of States Voluntarily Adopting Minimal Consensual "Baseline" for Acquisition of Background Information, At Leat For Owners and Operators of Critical Infrastructures

Congress and the Administration can acknowledge the importance of allowing critical infrastructure owners and operators to acquire sensitive background information for certain employment positions through consent of the employee or applicant. Assuring nationwide availability of a consensual "baseline" would represent a reasonable compromise between employees' privacy concerns and the security needs of owners and operators of critical infrastructures. This approach would allow implementation by the states in a narrow or broad manner—they could make legitimate, job-related information available to a broad or narrow class of employers, for a broad or narrow set of positions.

- **Pro:** This approach is minimally intrusive as most of the states already achieve this "baseline" under current law. Recognizes benefits of uniformity and importance of screening for employees in sensitive positions in critical infrastructures, while allowing state lawmakers to achieve balance based on regional values. Such an approach would allow the Federal government to point to the benefits of a predictable nationwide climate governing employer-employee relationships in this area, particularly where electronic record checks could one day be the norm.
- **Con:** Will fail to achieve uniformity.

The Administration can Name a Panel to Consider Best Approach to Achieving Legislative Balance Between Security Needs and Privacy Interests

The Administration can form a study panel constituted to adequately represent Federal, state, and local government interests, as well as private sector representatives of labor and management, to consider a balance between security needs and privacy interests. The study panel can be tasked to consider, among other things, some of the concerns and equities raised in this paper (e.g., balance between privacy and security concerns, employee and employer equities, and state and Federal interests). The balance they strike may be reflected in form of a general statement of objectives, or perhaps a model privacy statute or uniform law for states to consider. They may decide to issue a call for partially or wholly preemptive Federal legislation.

- **Pro:** Relatively unintrusive process, can be implemented with limited resources. Promotes open and balanced dialogue between numerous interested parties, and implementation by legal experts. Enhances awareness and broadens participation in the process. Allows careful consideration and crafting of recommendations.
- **Con:** Defers serious resolution of the issue indefinitely.

Congress and The Administration can Recommend Narrow Amendments to *Federal* Privacy Laws to Serve as Model for State Legislative Reform

The Federal government can identify potential security-privacy shortfalls in existing Federal legislation and recommend modest changes while at the same time demonstrating to state legislatures and owner-operators some of the benefits to be derived from reconsideration of legislation in light of infrastructure assurance objectives.

- Pro: Offers specific guidance to Congress to promote immediate action and discussion without need to await findings of study group. Offers useful example to states while at the same time respecting comity concerns.
- **Con:** Unlikely, on its own, to enhance uniformity. Could even promote further diversity by encouraging individualized state approaches.

The Administration can Recommend Limited Congressional Preemption of State Privacy Laws to Accommodate Data Collection Needs

The Administration can recommend that Congress consider the propriety of establishing a mandatory information collection and dissemination "baseline" with which relevant state privacy legislation must comport, thereby assuring owners and operators the ability to perform the equivalent of "background checks" on employees or prospective employees occupying certain sensitive positions. This might be accomplished, for example, through Congressional findings equating certain positions of employment within the critical infrastructures with Federal government positions that require security clearances.

- **Pro:** This approach would set a standard for privacy across all fifty states at the level required for infrastructure protection. It would tend to unify existing practices thus increasing predictability for business.
- **Con:** Congressional support is unpredictable if not unlikely. It is not clear that infrastructure owners and operators would be willing to shoulder additional expense associated with background checks absent subsidies or incentives unless they were in their own vested business interests. This approach may be polarizing in new and unfamiliar ways.

PART FOUR
CONCLUSIONS

Employers have a valid interest in acquiring information about current and prospective employees. This is certainly true of owners and operators of critical infrastructures—particularly with respect to employment in certain sensitive positions. Recently, however, the prevailing concerns of state and Federal lawmakers has been to solidify other valuable intereststhose of employee privacy. And while the Federal government has been the traditional arbiter of concerns relating to the national security, concerns relating to individual privacy are the traditional province of the States. The Administration and Congress should proceed cautiously to insure that a proper balance is struck between security and privacy concerns, a balance respecting the authority of the Federal and state governments. One way to achieve such a balance may be to allow greater employer access to job-related background information (criminal history, credit, past employment) for specific types of positions, while also ensuring that appropriate procedures are in place to allow employees to be certain of the accuracy and fair use of such information.

Diversity of State and Federal Law

A diversity of interests have given rise to varied state and Federal approaches to the acquisition and use of personal information for employment purposes (e.g., criminal history, credit, employment). These approaches often reflect regional values—but have resulted in a disparate framework for the nation. This creates complications for employers whose own operations may span several jurisdictions, or who may wish to inquire about employees who have themselves lived in a number of states. It is not vital to advocate resolution of these uncertainties for all employers and all job positions. Rather, the Federal government should only seek to make predictable procedures available for employers seeking to fill positions of high sensitivity within the critical infrastructures.

Insider Threats to the Critical Infrastructures

The Federal government recognizes that insofar as the critical infrastructures may provide attractive targets to physical and cyber attack, the owners and operators and their employees may be, in some instances, the first and only line of defense. This fact suggests that some of the positions held within the critical infrastructures may be relevant if not vital to national security. The Federal government would not likely fill a position of such gravity without conducting a background investigation, and would likely keep that information current. Some owners and operators of critical infrastructures, however, are prevented by the operation of state and Federal laws from taking analogous protective measures. Impediments to background inquiries have been put in place in the service of other interests—privacy, fair employment, post-conviction rehabilitation—without necessarily having taken into account the relatively novel concerns of infrastructure assurance. Similarly, the exemptions to those impediments are legislative responses to a different though often related set of concerns. Infrastructure assurance objectives may be achieved in some instances through minor modifications to the exemptions.

Employee Screening Procedures

The current practice of the Federal government is to conduct routine reinvestigations into employee backgrounds at regular intervals after the initial screening for many of its sensitive employment positions. General baseline principles for collecting and using criminal history, credit, polygraph and other types of information identified here can apply equally to initial hiring as well as to the subsequent "updating" of relevant, job-related information on an employee.

Such conclusions do not represent a mandate or endorsement of any particular employee screening or reinvestigation practice. It is far better to establish a consensual baseline between *certain* employers and *certain* applicants for sensitive employment. This baseline would serve to merely make available to a potential or current employer relevant job-related information, to be used in a manner consistent with prevailing law, upon consent of the applicant.

Fair Information Practices

It is very important to observe fair information practices in any aspect of these issues bearing on privacy and information sharing.[42] Fair information practices are at the foundation of Federal legislation such as the Fair Credit Reporting Act and the Privacy Act of 1974. Fair information practices are principles for collection, use and dissemination of information which should be considered by both Federal and state governments when drafting laws dealing with personal information.

Advocating a "Consensual Baseline" Approach

It may be appropriate to raise to state legislatures and to Congress the possibility that existing privacy laws may unduly hinder attempts by certain employers to take legitimate security precautions—precautions that may otherwise be justified in light of national security and economic concerns. Ideally, lawmakers would want to reexamine existing legislation and take into account these security concerns in the future. Naturally, careful consideration of all of the interests at stake should be given before striking an appropriate balance of those interests in law. Consideration should also be given to the benefits of uniformity among the states in this area.

Under such circumstances, an appropriate "baseline" might be constructed around employer-employee consent, whereby an applicant for certain sensitive positions within the critical infrastructures might grant, upon request, third party access to documents (under conditions that reflect fair information practices). This is an approach currently in place in several states, but a few states do not allow for individuals—much less third parties—to obtain copies of certain types of information. As a first step, states should allow individuals to obtain copies of their records and allow

---

42 Fair information practices are a set of global principles that define fair procedures for the collection and use of personal information. Personal information is any information that can be associated with an identifiable individual. These principles are the basis for all federal privacy laws and are reflected in the Clinton Administration policies for the Global Information Infrastructure. Fair information practices include the following principles:
-The information collected should be clearly relevant to the purpose for which it is being collected;
-People must be able to learn what personal information is in their records and how the information is being used. There should be no secret systems;
-People must be able to inspect their records and correct any errors;
-Personal information should not be collected for one purpose and used for other purposes without the individual's knowledge and consent; and
-Organizations that create, maintain, use or disseminate records containing personal information must assure the reliability of the information for its intended use and must ensure that the information is protected against misuse. Fair Information Practices were developed in 1973 by a Advisory Committee to the Department of Health, Education & Welfare (now HHS), chaired by Dr. Willis Ware.

employers to request that employees provide relevant information when applying for certain sensitive positions within the critical infrastructures.

Privacy - Security Study Group

A baseline of consent across U.S. jurisdictions will provide a degree of uniformity and consistency currently absent from employee screening laws. But this paper does not account for the role of labor unions and collective bargaining. the trend toward increased availability of personal information via the Internet, nor many of the other practical, financial or legal issues involved in the employee-employer relationship. A comprehensive study of these issues would best be performed by a study body of relevant professionals—lawyers; labor and management representatives; privacy advocates; Federal, state and local government representatives. They could explore the issues identified in greater depth in order to make findings and recommendations for the further modification of Federal and state legal regimes governing this area.

Ideally, the Attorney General could be directed to convene such a study group charged to address, among other related issues, the following:

- Identify current Federal and state legislation that may impede owners and operators of critical infrastructures from obtaining job-related background information on job applicants and current employees for sensitive positions (under conditions similar to those by which, for example, the Federal government obtains information before issuing security clearances);
- Identify other potential impediments to a private background investigation process (e.g., practical difficulties in obtaining criminal history information, financial issues, influence of collective bargaining agreements, etc.);
- Identify individual privacy and civil liberties issues implicated by a private background investigation process and existing legal protections for those interests;
- Analyze whether there are gaps in authority with respect to availability, collection, or dissemination of information relevant to an employee's suitability for a sensitive position, or protection of personal privacy interests that should be filled by legislation;
- Analyze the current tort litigation climate and its influence on employer-employee relations taking into account recent trends in legislation;
- Make recommendations to enable a private background investigation process that strike the appropriate balance between all of the stakeholders, taking into account:
    - federalism concerns;
    - privacy interests;

- civil rights/fair employment practices;
- labor law;
- tort liability;
- fair information practices;
- financial issues and other practical impediments.

- Specifically delineate the infrastructures and positions to which such recommendations may apply;
- Consider the propriety of producing model state legislation, or (if perceived as necessary under the circumstances) partially preemptive Federal legislation reflecting vital national security interests.

Illustrative Amendments to Federal Legislation

Given an adequate opportunity to review all the laws that are implicated by this issue, fairly specific recommendations for legislative reform may follow. But more modest changes can bring some degree of relief, and can illustrate the benefits of reconsidering prevailing laws in light of infrastructure assurance objectives. One statute in particular represents an example of the type of change a study body may suggest. This statute is the Employee Polygraph Protection Act (EPPA). The EPPA broadly prohibits private employers from subjecting employees to polygraph examinations. The statute also provides, however, some narrow exemptions. Included in the exemptions are employers who are in the business of providing security services for the protection of certain types of enumerated facilities, many of which are part of critical infrastructures.

In fact the current Department of Labor regulations implementing the EPPA appear to include all of the critical infrastructures within the scope of the exemption.[43] Through the operation of this exemption, for example, employers who provide employees to install alarm systems at electric power facilities may administer polygraphs to those employees under the limited conditions described in the Act. Interestingly, however, under current wording, an employer responsible for providing analogous "cyber" security services—such as the installation of firewalls or other protective technologies—appears not to be able to take similar precautions.

Research to date has shown that some distinctions—such as those between physical and information security—often break down where infrastructures are highly interdependent and where threats can be physical, cyber or both. Amendments could be made to the EPPA to include within

43 See 29 C.F.R. Part 801.

the scope of its exemptions those who are in the business of providing information security services. (*See* Appendix A). Such amendments do not make it incumbent upon covered employers to polygraph employees, but merely allows them to do so to the extent permitted under applicable state law.[44]

The insider threat is as important an issue as it is difficult to address. It is important to emphasize the importance of linking these efforts to other efforts, such as international policies, standards, and training and education, which also may touch on potential problems arising from an unfortunate but potentially significant insider threat.

---

44 The proposed amendments shown in Appendix A would include under the EPPA's exemptions most of the critical infrastructures and would also be an important step in amending extant legislation to reflect the growing importance of information security alongside physical security. However, the potential limitations to the effectiveness of such a revision further indicate the need for a careful study of these issues. The EPPA provides only limited preemption of state legislation. States remain free to create more restrictive exemptions and thus broadening the federal exemptions may leave state coverage unchanged. Nonetheless, the EPPA affords an important opportunity for Congress to demonstrate to state lawmakers its concern over information security, particularly as implemented within the critical infrastructures.

Appendix A
Proposed Amendment to EPPA

[Summary of 29 U.S.C. 2001-2009; Employee Polygraph Protection Act, incorporating proposed revisions to section 2006(e).]

Section 2001. Definitions

Section 2001 defines "lie detector" to include a "polygraph, deceptograph, voice stress analyzer, psychological stress evaluator, or any similar device (whether mechanical or electrical) that is used . . . [to] render a diagnostic opinion regarding the honesty or dishonesty of an individual."

Section 2002. Prohibitions on lie detector use

Section 2002 makes it unlawful for any employer in interstate commerce to (1) "require, request, suggest or cause" an employee or prospective employee to take or submit to a lie detector test; (2) to "use, accept, refer to or inquire concerning" the results of "any lie detector test of any employee or prospective employee"; (3) to discharge, discipline or otherwise discriminate against an employee or prospective employee for refusing to take the test or on the basis of the results of such a test; or (4) to discharge, discipline or otherwise discriminate against an employee or prospective employee for exercising rights under this section.

Section 2003. Notice of protection

Section 2003 requires the Secretary of Labor to produce and distribute, and for employers to post, summaries of the pertinent provisions of this chapter in conspicuous places.

Section 2005. Enforcement provisions

Section 2005 allows the Secretary of Labor to assess penalties of up to $10,000. The Secretary may bring suit to restrain violations. This provision also permits employees a private right of action to recover damages and seek reinstatement. It expressly disallows employee waiver of any of the provisions of the chapter.

Section 2006. Exemptions

Section 2006 makes the chapter inapplicable to (a) federal, state or local government employees, and to (b) certain DoD, DOE, intelligence community, and FBI contractors. It also *permits* an employer to ask an employee to submit to an examination if it is offered (d)(1) in connection with an ongoing investigation; (d)(2) if the employee had access to the property that is the subject of the investigation; (d)(3) if the employer reasonably suspects the employee was involved; and (d)(4) the employer complies with procedural requirements, such as providing prior written and signed notice of the offense under investigation, an estimation of the resulting loss or damage, and a statement describing the basis of the employer's reasonable suspicion.

Section 2006(e) contains an exemption for providers of security services

* * * * *

e) Exemption for security services

(1) In general

Subject to paragraph (2) and sections 2007 and 2009 of this title, this chapter shall not prohibit the use of polygraph tests on prospective employees by any private employer whose primary business purpose consists of providing armored car personnel, personnel engaged in the design, installation, and maintenance of security alarm systems, [*information security personnel*] or other uniformed or plainclothes security personnel and whose function includes protection of—

(A) facilities, materials, or operations having a significant impact on the health or safety of any State or political subdivision thereof, or the national security of the United States, as determined under rules and regulations issued by the Secretary within 90 days after June 27, 1988, including—

(i) facilities engaged in the production, transmission, or distribution of electric or nuclear power,

(ii) public water supply facilities,

(iii) shipments or storage of radioactive or other toxic waste materials, and

(iv) public transportation.

(B) currency, negotiable securities, precious commodities or instruments, or proprietary information.

(2) Access

The exemption provided under this subsection shall not apply if the test is administered to a prospective employee who would not be employed to protect facilities, materials, operations, or assets referred to in paragraph (1).

* * * * *

Section 2006(f) exempts authorized manufacturers, distributors or dispensers of controlled substances.

Section 2007. Restrictions on use of exemptions

Section 2007 sets forth permissible uses of test results when performed by employers under conditions permitted by section 2006. Subsection (a) provides that adverse test results cannot provide the sole basis for discharge, discipline, or other discriminatory employment action. Subsection (b) sets forth the rights of the examinee, including the ability to terminate the test at any time; the right not to be asked degrading or needlessly intrusive questions; and the right not to be asked about religious, racial, or political beliefs or practices, sexual behavior, or union affiliations. This section also sets forth the procedures to be followed. The employer must, for example, provide the employee with notice of the time and place of the test, the right to obtain legal counsel, and what type of monitoring is to occur. The employee must read and sign a written notice of these provisions, and must be provided with an opportunity to review all questions in advance. The section sets forth additional requirements regulating the content and duration of the tests themselves. Subsection (c) requires the examiner to have a current State license in states requiring same, and to maintain a minimum of a $50,000 bond or professional liability insurance. All conclusions from the tests must be rendered in writing.

Section 2008. Disclosure of information

Section 2008 allows information obtained during the test to be revealed only by the examinee, or the examiner to the examinee, the employer, or to any court with proper process. The employer is generally prohibited from disclosing results, unless the disclosed information is an admission of criminal conduct.

Section 2009. Effect on other laws and agreements

Section 2009 makes clear that this law is not intended to preempt any state law or collective bargaining agreement that is more restrictive with respect to lie detector testing.

CHAPTER 12

LEGAL IMPEDIMENTS TO INFORMATION SHARING

A "LEGAL FOUNDATIONS" STUDY

Report to the President's Commission on Critical Infrastructure Protection
1997

This report was submitted to the President's Commission on Critical Infrastructure Protection, and informed its deliberations and recommendations. This report represents the opinions and conclusions solely of its developers.

CONTENTS

PART ONE
INTRODUCTION

Data collection and information sharing are likely to play a key role in the protecting of our nation's critical infrastructures, whether it is to involve data and information originally compiled by government or by the private sector. The development of data collection and sharing capabilities between government (Federal, state and local) and the private sector can contribute, for example, to an effective threat and warning capability.

In attempting to forecast potential legal impediments to the broad sharing of information for the purpose of assuring the critical infrastructures, existing information-sharing activities and the legal issues implicated by them were studied.

The *Legal Foundations* study *Information-Sharing Models* [Chapter 2] is the companion volume to the *Legal Foundations* study *Legal Impediments to Information Sharing*. The two documents should be studied together to fully understand the issue of information sharing. *Information-Sharing Models* details the current approaches to information sharing and sets the stage for further understanding the concepts described in *Legal Impediments to Information Sharing*.

Research Issues

- What potential legal impediments (Federal or state statutes, regulations, or common law causes of action) could inhibit the flow of infrastructure assurance-related information between and among government bodies and the private sector?
- In light of these impediments, are there specific courses of action that could improve the information-sharing climate?
- How can an information-sharing mechanism best be structured to avoid these legal impediments?

Research Findings

Upon reviewing various Federal and state statutes, the following research findings were noted:

- Information collection, dissemination, and protection each implicate discrete legal issues, and hence pose potential legal impediments to information sharing. Relevant legal issues include, among others, antitrust, liability, national security concerns, and privacy.

- These impediments are generally manageable when they arise in conjunction with private sector-to-private sector or government-to-government transactions.
- These impediments are less manageable with respect to "mixed" transactions, i.e., government-to-private or private-to-government information collection, dissemination, and protection.
- An understanding of the legal issues implicated by these transactions is relevant, if not central, to the design and embodiment of an effective information-sharing capability.
- An appreciation for these legal issues can be gained through the study of existing information-sharing mechanisms, and models that can be abstracted from them.

Assumptions

This paper assumes that a successful information-sharing mechanism or mechanisms will contribute to infrastructure assurance.

Impediments to Information Sharing

Current data collection and information-sharing activities are underway in a number of sectors. These efforts range from informal discussions of industry executives about an emerging problem to statutorily required collections and mandated dissemination. Our study of these existing activities and mechanisms further suggested five theoretical models for information sharing.

(For further information on the individual models, *See Information Sharing Models*, a special study of the *Legal Foundations* series.) These models include:

1. Information Clearinghouse
2. Advisory Committee/Commission
3. Mandatory Provision of Information with Limited Exchange
4. Voluntary Cooperation or Exchange of Information
5. Threat and Hazard Detection and Notification

The models differ in terms of the degree of formality, the legal requirements relating to reporting and dissemination, and degrees of public and private participation. The attributes of a particular model may subject it to further legal impediments to information flow making some models more advantageous than others to achieve infrastructure assurance objectives.

The study of these models has revealed a series of legal issues that arise under each of these models to varying degrees. Antitrust, liability, national security, proprietary data and privacy concerns are potentially implicated no matter which option is employed. These issues may impede the sharing of information in the following ways:

- **Antitrust:** Federal antitrust and anti-competitive behavior statutes, and their state level counterparts are primarily intended to prevent price-fixing and market-splitting. Although many activities conducted by private companies to share information relating to critical infrastructure protection should not implicate antitrust concerns, many companies have raised antitrust as a potential "show-stopper" in information-sharing efforts.
- **Liability for Failure to Disclose or Inform:** Liability for failure to disclose information that could have prevented a critical infrastructure-related incident is a concern for government and private parties. Unless adequately addressed, concern over this form of liability may discourage robust collection of information, or may militate in favor of cautious dissemination among a narrow and well-defined field of participants.
- **Federal Government:** The Federal government can make itself subject to suit under the Federal Tort Claims Act. The statute can permit recovery if the government negligently or intentionally withholds information that could have prevented a loss. There is an exception to this liability that would apply when a Federal employee was acting with due care in the execution of a statute or regulation. Some information-sharing statutes and executive orders are explicit that they do not create a right of action for failure to receive information.
- **Private Parties:** To hold a private party liable for failure to provide information that could have prevented a loss, the plaintiff would have to show that the party had a duty— whether statutory, regulatory, or common law— to provide the information. This generally requires a relationship between the parties, a relationship that can be adequately established, (either express or implied), through information-sharing activities or agreements.
- **National Security Concerns:** National security issues relate to the protections that certain classes of information receive by virtue of their relation to national security issues (e.g., intelligence sources) or the concerns that arise in relation to foreign access to sensitive, but not necessarily classified, critical infrastructure information. The government has regulations and specific statutes that punish improper disclosure of classified information. These statutes and regulations are of particular concern as they apply to "foreign persons." Generally, release of information to a foreign person is contingent on an evaluation of

their country of origin and a satisfactory inspection of their background.

- **Government Protection of Private Sector Information:** The Federal and state governments operate under rules designed to allow openness and transparency in the democratic process. The most important of the legislation surrounding government disclosure of information is the Freedom of Information Act (FOIA). FOIA, and its state level counterparts, require the government to disclose all information that does not fall into a series of narrowly defined exemptions (e.g., on-going law enforcement investigations). Information disclosed by the private sector does receive some protection from disclosure under FOIA exemption b(4) if that information is either a trade secret or proprietary and voluntarily disclosed. Disclosure of information can also be prevented by independent statutory mechanisms. These may be tied to the mission of the organization collecting the information, to judicial procedures, or specific trade secret protections. However, without specific guarantees of confidentiality, the private sector may be unwilling to share information with government voluntarily.
- **Privacy:** Privacy issues are discussed in a separate paper in the Legal Foundations series: *Privacy Laws and the Employer-Employee Relationship.*

PART TWO
ADDRESSING LEGAL IMPEDIMENTS

Limiting Liability of the Federal Government

A specific statement of the Federal government's duties with respect to disclosure and notification could be included in executive orders or statutes relating to critical infrastructure information sharing in order to limit the liability of the Federal government under the Federal Tort Claims Act. By specifically delineating the government's responsibilities in an information-sharing mechanism, the government can limit its liability. If the legal vehicle by which the mechanism is established states in specific terms the parties with whom the government will share information, parties who are not enumerated will not be able to sue for failure to inform. However, the government will have some liability exposure with regard to the parties it has said that it will inform if it fails to share information due to negligence. The government may also avoid liability by specifically stating in the legal vehicle forming the information-sharing mechanism that it does not create a right of action against the government.

- **Pro:** This approach would allow the government to avoid the need to include *every* critical infrastructure owner and operator specifically in an information-sharing effort because of liability concerns, thus limiting the burden on the information-sharing process. The government could instead specifically decide the level of exposure to which it should be amenable and act accordingly.

- **Con:** The specific delineation of the government's responsibilities may raise opposition of those not included at the table. Eliminating liability concerns in this way may remove the incentive to try to include all interested parties in the information-sharing effort.

Antitrust Opinion From Department of Justice (DOJ) or Federal Trade Commission (FTC)

To respond to perceived antitrust concerns, the Department of Justice Antitrust Division or the Federal Trade Commission (FTC) could consider the propriety of rendering an opinion or providing guidelines clarifying for private industry the standards for communicating critical infrastructure information both through intermediaries and directly. Guidance from one of the agencies directly involved in enforcing antitrust violations as to appropriate processes for sharing information among private sector entities, particularly as it relates to infrastructure assurance issues, may alleviate apprehension of industry about engaging in such activities.

National Security Concerns

As part of its approach to infrastructure assurance, the Federal government will have to address the issue of sharing sensitive information with "foreign corporations" which may be either owned or controlled by non-U.S. citizens. The next two items describe the issues involved.

Developing Government-wide Guidelines for Sharing Sensitive Information with Foreign Corporations

The Federal government needs to study and consider developing comprehensive and centralized guidelines for sharing sensitive information with foreign corporations. Currently, many agencies have individual regulations defining and setting out rules for dealing with "foreign corporations." The definition is usually based on a percentage of foreign ownership. These percentages vary from agency to agency. Under this option, an interagency task force, perhaps under the guidance of the NSC, would study whether a uniform approach to defining who is and who is not a "foreign corporation" would be in the interest of national security and government efficiency. A government-wide study would allow foreign policy and economic implications to be taken into account. If found to be desirable, the body could also study where the proper percentage of foreign ownership for the entire government to share information should be set and whether there are any entities which should be exempt or for which the percentage should differ.

- **Pro:** This approach would bring the issue of foreign ownership to the Administration's and Congress' attention. All agency policies would be reviewed and potentially collapsed into one guideline. It would reduce complexity and uncertainty for foreign investors, and provide the Federal government an opportunity to revisit current regulations in light of critical infrastructure issues.
- **Con:** One set of regulations may not be able to adequately address the needs of all government agencies. Regulations may be too restrictive for some agencies and too lax for others. The study of foreign ownership is likely to be a difficult and controversial process, a process made exceptionally difficult to implement in an environment of rapid restructuring and globalization.

Developing Interim Guidelines for Sharing Information with Foreign Corporations

Notwithstanding resolution of the issue above, the information-sharing mechanism for critical infrastructure assurance should consider studying, and if necessary, setting guidelines for sharing critical infrastructure information with foreign corporations. If the immediately-preceding option is adopted, and uniform Federal guidelines are eventually created, the information-sharing mechanism may need to adopt those guidelines. Until such time, the mechanism should consider the proper way to include corporations with foreign owners in their process. This may result in guidelines that determine the percentage of foreign ownership that is allowed for participating entities to receive information. The guidelines could be specifically tailored by infrastructure or to provide for a sliding scale of ownership percentage according to the sensitivity of the information.

- **Pro:** This approach would allow the rules for sharing information with foreign corporations to be specifically tailored to critical infrastructures.
- **Con:** Current approaches of agencies connected to critical infrastructures will likely remain unchanged. This approach depends on the successful implementation of recommendations for an information-sharing mechanism or follow-on entity to come to fruition.

Government Protection of Private Sector Information

A potentially contentious set of issues having to do with infrastructure assurance relates to the Federal government's protection of private sector information. The next three items discuss the ramifications of different approaches and strategies.

Legal Vehicle to Protect Confidential Information

A legal vehicle could be established within the critical infrastructure information-sharing mechanism that would include specific provisions allowing the protection of confidential information. Under exemption b(3) of FOIA, information protected from disclosure under other statutes are also exempt from public disclosure under FOIA. If the information-sharing mechanism is established by Congressional legislation, a provision protecting information from disclosure would qualify for nondisclosure under exemption b(3). The provision would need to be carefully drawn to allow the information-sharing mechanism to disclose certain types of information or sanitized versions of information and also to meet the requirements for a b(3) statute.

- **Pro:** If a b(3) statute is in place, the private sector participants can be fairly certain that their information will not be disclosed under FOIA of otherwise. If Congress must act to create the information-sharing mechanism, it will be easy to insert the needed language to accomplish this sort of protection for confidential information.
- **Con:** Exemption b(3) requires a statute, thus an executive order will not be able to accomplish the same goal. If the information-sharing mechanism is to be established by executive order, this option will have to be accomplished separately and may not be addressed until well after that mechanism is up and running. The current administration stance on FOIA is very pro-disclosure, even when exemptions are available.

Impact of Classification

The critical infrastructure information-sharing mechanism might consider the need for classification of certain information and the impact classification would have on the information dissemination process. Classified information has specific handling and dissemination rules and regulations that must be followed. Classified information can be sanitized and declassified to share with the private sector, or shared with a select group of private sector representatives who are able to qualify for a clearance. However, such a process is time-consuming and would limit the circle of participants in the information-sharing activity. In order to address this potential impediment, the information-sharing mechanism should study whether there will be information collected that will need to be classified, whether adequate authority currently exists to classify that type of information, and the process by which such information can and should be shared with other participants.

- **Pro:** This approach raises classification as an issue and allows for further study of whether it will act as an impediment to information sharing. Such a study is likely to produce recommendations specifically tailored to the information-sharing process.
- **Con:** The study approach does not specifically and immediately address classification as an impediment to information sharing.

Proper Procedures for Exchanging and Treating Information

The critical infrastructure information-sharing mechanism might consider proper procedures for exchanging and protecting trade secrets and proprietary information. Trade secrets currently receive protection by Federal law from theft and unauthorized disclosure (see e.g., 18 U.S.C. § 1831 (economic espionage); 18 U.S.C. § 1832 (theft of trade secrets); 18 U.S.C. § 1905 (disclosure of trade secrets). Most states have also adopted the Uniform Trade Secrets Act which provides an additional layer of protection. These protections provide a framework for recovery from losses that may be incurred from unauthorized sharing of confidential information with competitors. Trade secrets may also receive protection when shared with the government. In order to structure a process that provides the maximum protection to sensitive corporate information, the information-sharing mechanism should consider whether it will receive trade secrets or other forms of proprietary private-sector information, whether adequate procedures exist to protect the status of that information, and whether current legal sanctions are adequate to the protect them from unauthorized disclosure.

- **Pro:** This approach elevates the importance of addressing trade secrets as a potential impediment to information sharing. It allows the issue to be studied within the context of the actual information-sharing process that develops.
- **Con:** It does not specifically and immediately address trade secrets as an impediment to information sharing.

PART THREE
MODELS FOR INFORMATION SHARING

Information Clearinghouse

An information clearinghouse may be public, private or mixed. It may collect information via voluntary or mandatory reporting. Its primary features are centralized collection and widespread dissemination of information. Information may also be accessed upon request rather than through a dissemination process. The Commercial Driver's License Information System is one model of an information clearinghouse.

- **Pro:** Widespread dissemination of information and availability upon request lower the potential liability issues associated with information sharing. Also, a clearinghouse with a large number of participants may raise fewer antitrust concerns (particularly if there is government involvement).
- **Con:** A clearinghouse may not provide the best protection for proprietary information. It may also raise privacy concerns if the information concerns individuals' medical, criminal, employment, or financial back-

grounds. Liability could be implicated with regard to those records. Antitrust issues and issues regarding government handling of sensitive information are also implicated. National security concerns may arise due to the widespread availability of information from the clearinghouse.

Advisory Committee/Commission

Advisory Committees and commissions are generally established to handle a specific and focused issue for a limited duration or on a more permanent basis. While advisory committees and commissions are not an exclusively governmental creation, they are employed on a widespread basis within the Federal government. Federal advisory committees often have private sector participants which usually will subject the committee to the Federal Advisory Committee Act which provides for open meetings and records. Advisory committees collect information related to their mission and may be specifically designed to further disseminate that information. But policy is generally their central focus.

- **Pro:** Federal advisory committees implicate fewer antitrust concerns than models that do not involve government creation or participation. An advisory committee is also more sheltered from liability for failure to disclose information because it is generally a limited exercise and seldom designed to handle "warning" type information. National security concerns are also diminished due to government involvement. The process of selecting participants and vetting participants allows screening for issues relating to "foreign persons" and clearances to be given to allow access to classified information.
- **Con:** Private sector advisory committees or commissions will have to deal with antitrust concerns. These are often small groups that represent companies which dominate an industry. Banding together to accomplish a particular mission without including other smaller members of the industry could give rise to an antitrust claim. It may also create some liability exposure. Antitrust does remain an issue if the body is government created but solely comprised of private sector participants. There will also be problems with protecting information from disclosure if the body is Federally created. FACA will require open records and meeting unless all members are full-time government employees or there is specific exemption.

Mandatory Provision of Information with Limited Exchange

Mandatory information-sharing models generally focus more on the collection of information than its dissemination. Usually, information is legally (by statute, subpoena, etc.) required to be provided to a specific body. Rarely is there any requirement that information be given in exchange or findings be reported anywhere but up the "chain of command." This model will almost always require some government action and most often involves government collection from the private sector and dissemination to government.

- **Pro:** Under a mandatory model there are few if any antitrust issues. Liability issues also are not a factor to the extent the information is not generally shared. Foreign ownership is also not a concern to the extent that the information is not disseminated outside the government.
- **Con:** The primary legal impediment encountered in such an option involves government protection of private sector information. This includes FOIA and trade secrets issues.

Voluntary Cooperation or Exchange of Information

Voluntary exchanges tend to be less formal than the other models considered. Information is collected through the voluntary provision of relevant information by willing participants. Dissemination may be limited to those participants or take more of a clearinghouse model where the information is generally accessible to a wider community. Voluntary information-sharing arrangements may be public, private or mixed.

- **Pro:** If there is government involvement in the process there will be few if any antitrust concerns. FOIA exemption b(4) extends special protection to private sector confidential information when voluntarily provided. Liability concerns are reduced in a model where information dissemination is limited to those who contribute or is generally accessible to interested parties.
- **Con:** Antitrust concerns will be implicated if this information-sharing mechanism is primarily or exclusively private. If it takes an informal shape, antitrust concerns will be heightened. Liability concerns may also be implicated by dissemination of information, if the groups to which it will be disseminated are not well-defined. This model, whether government or private, will have to handle trade secrets and privacy related information with care. If government is the collector and disseminator, it will have to address FOIA and foreign ownership issues.

Threat and Hazard Detection and Notification

A threat and hazard detection and notification model combines an information clearinghouse with a mandatory dissemination feature. The model involves the collection, analysis, and dissemination of information.

Dissemination plays a central role, though it may happen infrequently (i.e. only during emergencies). Dissemination also usually happens through predetermined channels and in this respect is more formal than some of the other models. Many threat detection and notification models occur within the Federal government. The Natural Oil and Hazard Substances Pollution warning mechanism is one such example.

- **Pro:** If government-sponsored, a threat detection and notification model will have no antitrust and few liability concerns. Liability concerns can be virtually eliminated by structuring the enabling legislation or executive order properly. Antitrust issues may also be limited in the private sector because such a model will likely have a limited focus that does not implicate price-fixing or market-splitting.
- **Con:** Liability will likely be a concern for the private sector and could be a concern for the government if special provisions are not made to limit it. Warning dissemination will raise special issues for the government with regard to classified information and foreign ownership.

PART FOUR
CONCLUSIONS

This paper postulates eight "principles" that ought to be considered in thinking about information sharing. The basic principles were derived from studying the five hypothetical information-sharing models that were themselves abstracted from existing information-sharing arrangements. These principles are summarized here as follows:

1. **Antitrust:** Crucial private sector participants have expressed reluctance to participate in the sharing of specific threat and vulnerability information because of impediments they perceive to arise from antitrust and unfair business practice laws. The Federal government might find it advantageous to offer limited assurances to the private sector that participation in information-sharing processes would not, at least under certain conditions, run afoul of such laws.

2. **Liability:** To limit potential government liability arising from its participation in an information sharing and threat warning program or process, the Federal government may wish to specifically enumerate its own duties and obligations arising out of such participation.

3. **Centralized Guidelines:** The Federal government might want to study whether centralized guidelines for sharing information with foreign corporations would serve the interests of national security and government efficiency.

4. **National Security:** Any mechanism established by the government to share information on threats to and vulnerabilities of the critical infrastruc-

tures (or to participate in such processes) must set guidelines for the sharing of such information with foreign corporations.

5. **Confidential Information:** Any mechanism established by the government to share information on threats to and vulnerabilities of the critical infrastructures (or to participate in such processes) should allow for appropriate protection of specific private-sector information. This might require, for example, inclusion of a b(3) FOIA exemption in enabling legislation.

6. **State Government Participation:** Because state "government in the sunshine" laws could impede participation by state officials by requiring disclosure of information that comes into their possession, a study should research and identify legal impediments to information sharing at the state level. A study group could propose solutions and draft model legislation to maximize state government participation in information sharing.

7. **Classified Information:** Any mechanism established by the government to share information on threats to and vulnerabilities of the critical infrastructures (or to participate in such processes) ought to take into account the need for classification of certain information, or certain bodies of aggregated information, and the impact that classification would have on the dissemination process.

8. **Trade Secrets and Proprietary Information:** Any mechanism established by the government to share information on threats to and vulnerabilities of the critical infrastructures (or to participate in such processes) should allow for appropriate protection of information containing trade secrets or other forms of proprietary information.

CHAPTER 13

FEDERAL GOVERNMENT MODEL PERFORMANCE

A "LEGAL FOUNDATIONS" STUDY

Report to the President's Commission on Critical Infrastructure Protection

1997

This report was submitted to the President's Commission on Critical Infrastructure Protection, and informed its deliberations and recommendations. This report represents the opinions and conclusions solely of its developers.

CONTENTS

PART ONE
INTRODUCTION

Issue

What are some of the methods available to the Federal government by which it can unilaterally change its own behavior in order to encourage the private sector, state or local governments to act consistently with infrastructure assurance objectives?[1]

Research Findings

- Performance programs and practices are currently being pursued by the Federal and state governments.
- Performance programs and practices are cost-effective, unilateral actions designed to improve government performance directly and to influence private sector behavior indirectly.
- A current trend in government is the movement toward outcome-based performance management, similar to business practices common in industry. Requiring the development and monitoring of performance measures to track progress in achieving goals allows the government to better meet policy objectives.
- While the government may not be able to directly require certain features through its procurement process because of the recent move to off-the-shelf products, it may nonetheless influence the marketplace by incorporating policy objectives in choosing among currently available products.
- Use of best practices is commonplace in the private sector. A clearinghouse for best practices within the government could draw on private sector expertise in order to improve government performance.
- Performance programs and practices are relatively inexpensive to implement and do not require reorganization of the government or the

---

1 Note on Methodology: SRA International performed research to identify successful government programs and best practices employed by these programs. These techniques and methods were then evaluated to determine their suitability for accomplishing infrastructure assurance objectives. The research process consisted of a literature search including searches of relevant Internet web sites such as those maintained by the National Performance Review (NPR) and National Academy of Public Administration (NAPA), and telephone interviews of government experts including individuals within the Office of Management and Budget (OMB) and the General Accounting Office (GAO). Summaries of model performance programs and a contact list containing the names and phone numbers of government experts contacted are available for review.

creation of a new government organization. Further study of performance-based options, however, may require a follow-on effort.

Assumptions

- The current trend towards use of performance-based programs within the Federal and state governments are likely to continue to grow and expand.
- Infrastructure assurance objectives can be achieved through performance-based methods.

Background: "Performance-Based" Categories

A number of performance-based practices common in industry and adapted by the government, were identified that have been successfully employed by government programs to achieve policy objectives. Some of the practices such as business process reengineering do not appear immediately applicable to infrastructure assurance, and others involve a level of government intrusion in the private sector that may prove unpopular. Figure 1 summarizes the practices and concepts determined to be potentially applicable for infrastructure assurance. (As indicated in Figure 1, not all of the practices are applicable to each of the critical infrastructure groups).

Figure 1: Current Government Practices and Possible Applications)

Practices & Possible Applications	Tele-com	Electrical Power	Gas & Oil	Banking & Finance	Transport	Water Supply	Emergncy Services	Govmnt Services
1. Performance Measurement								
a. Infrastructure Report Card	X	X	X	X	X	X	X	X
b. Publication of Infrastructure Reliability Data	X	X	?	?	X	?	?	?
c. Use ITMRA/ GPRA to Increase Federal IT Security	X							
d. High Risk Report for Infrastructure Assurance	X	X	X	X	X	X	X	X
2. Certification of Technologies								
a. Certification of Products Meeting Security Standards	X	?	?	?	?	?		
3. Procurement								
a. Infra. Assurance as Criterion for Fed. Procurements	X							
4. Identification & Dissemination of Best Practices								
a. Clearing-house for Best Practices in IT Security	X							X

Performance Measurement

Perhaps the most significant trend in government is the movement to outcome-based performance management. With the passage of legislation such as the Government Performance and Results Act (GPRA)[2] and the Information Technology Management Reform Act (ITMRA),[3] all agencies are accountable for achieving strategic goals. At the heart of this movement is performance measurement: the development and monitoring of performance measures to track progress in achieving strategic goals.[4] The GPRA, for example, requires Federal agencies to perform strategic planning and performance management. Specifically, it requires them to (1) define their mission, (2) establish strategic goals linked to the mission, and (3) develop and employ performance measures to track progress in achieving their strategic goals. When the system is fully in place, agencies, in their budget submissions, will have to forecast how funding for each of their programs will lead to specific improvements in the performance measures by fiscal year. In other words, they'll have to show OMB and Congress what is to be expected in return for funding their programs. Each year, agencies will also have to report to OMB and Congress whether the prior year's goals in terms of the performance measures were actually achieved and, if not, why. In this way, GPRA will make agencies accountable for achieving strategic goals and allow OMB and Congress to allocate funding based on past and expected performance.

The ITMRA operates in a similar manner, but only in the limited area of information technology. The Act requires that OMB review agency budget proposals and performance measurements for, among other things, adequate information security. The Chief Information Officer (CIO) for each agency is responsible for establishing performance measurements for information technology used by or acquired by his or her agency. The performance measures must evaluate how well the information technology supports agency programs. However, information security is not required to be one of these performance measures (even though it is included in the OMB review process). The ITMRA also links the performance measures to the government procurement process.[5]

---

2 Pub. L. No. 103-62 (1993).

3 Pub. L. No. 104-106 (1996).

4 The Government Accounting Office stressed the importance of performance measurement in its executive guide to *Effectively Implementing the Government Performance and Results Act*, stating,
"The second key step that successful results-oriented organizations studied take—after defining their missions and desired outcomes—is to measure their performance. Measuring performance allows these organizations to track the progress they are making toward their goals and gives managers crucial information on which to base their organizational and management decisions. Leading organizations recognize, as well, that performance measures can create powerful incentives to influence organizational and individual behavior."

5 *See* discussion *infra* p. A-5, Procurement.

This practice of performance measurement is also being successfully applied by state governments. The State of Oregon was recognized by the National Performance Review and Harvard's Kennedy School of Government for its *Oregon Benchmarks* initiative in which it established a number of performance measures, or benchmarks, linked to the state's strategic goals. The objective of the *Oregon Benchmarks* program was to translate the state's strategic goals into measurable objectives and create a report card for measuring progress in achieving the goals. The benchmarks have succeeded in providing a clear framework for engaging people both inside and outside the government in the pursuit of well-defined objectives. They are being used by the Oregon Legislature to develop and evaluate legislative proposals and have also helped to foster cooperation among government agencies and between the public and private sectors.

There are a number of possible applications of performance measurement to infrastructure assurance, and this practice appears to be applicable to all of the infrastructure areas. Following the Oregon example, an agency such as the National Security Council or the Federal Emergency Management Agency could be tasked to coordinate the development and annual preparation of an infrastructure assurance report card. The Federal agencies responsible for oversight of the various critical infrastructures would be required to develop the performance measures to be incorporated in the report card and provide the data as needed to the organization responsible for coordinating the preparation of the report. These agencies may already be capturing much of the data necessary to construct such a report card.

By developing an infrastructure assurance report card, the Federal government would be better able to communicate the status of infrastructure assurance, establish infrastructure assurance goals, and identify areas of high risk on which to focus its efforts and resources. Once valid performance data for infrastructure assurance have been collected, new strategies for improving infrastructure assurance may become apparent. For example, it might be possible to induce private sector owners of the infrastructures to improve assurance by publishing detailed company-specific data for various infrastructure assurance performance measures. As examples of this practice, the National Highway Traffic Safety Administration publishes comparative crash safety ratings for each car model, and the Department of Transportation publishes comparative airline on-time statistics. This approach is seen as a form of regulatory oversight.

An alternative to the development of an infrastructure assurance report card with performance measures for all of the infrastructure areas would be to apply performance measurement selectively within the existing GPRA framework. Under this application of performance measurement, infrastructure assurance would be made a strategic goal of the agencies with relationships to those infrastructure areas in which the threat or risk is significant. These agencies would then be required to develop specific infrastructure assurance goals and performance measures to be reported as part of their annual GPRA submissions to OMB.

The development of appropriate performance measures is a difficult task under any circumstances and one that takes time. In addition, there may be some unique problems in applying this practice to infrastructure assurance, particularly in the IT infrastructure area. One difficulty in collecting performance data for infrastructure assurance would be that incidents tend to be discrete events that one tries to prevent and react to rather than a continual activity that generates a stream of performance data. In some cases, such as information system incursions, incidents may occur without the knowledge of those operating the infrastructure. Another problem is that infrastructure owners in the private sector may be unwilling to release data for their infrastructures for fear of exposing themselves to attack. Nevertheless, it should be possible to overcome these measurement problems. For example, the Federal Aviation Administration has successfully tracked data for airline incidents for years. With regard to the problem of unobserved incidents, it may be possible to develop measures of incident prevention such as the percentage of agencies or Federal systems in compliance with IT security standards developed by the National Institute of Standards and Technology (NIST), OMB, or another agency. It might also be possible to conduct surveys of key IT infrastructure owners within the private sector to determine the percentage whose systems meet these security standards. At a minimum, the Federal government could produce an annual "high risk" report, such as those produced by the GAO and the National Transportation Safety Board, highlighting the major assurance risk areas and recent trends within the various infrastructure areas.

Certification Programs

Two model programs within the environmental area were identified involving certifications. In both of the programs it appears that government was able to accelerate the market's adoption of technologies that support policy objectives through certification of technologies or products that meet certain standards. In 1992 the EPA initiated the Energy STAR program which is designed to reduce air pollution by encouraging the development of energy-efficient office equipment. Under this program the EPA developed energy efficiency standards for office equipment. The program allows manufacturers who choose to participate in the program to affix the Energy STAR logo on equipment that meet these standards. EPA performs spot checks to determine that equipment displaying the Energy STAR logo does indeed meet the standards. According to the EPA, the Energy STAR program with its certification component has helped to promote the awareness of the availability of energy-saving technologies and accelerate industry's rollout of energy-efficient office equipment.

Based on the early success of the program, the White House issued Executive Order 12845[6] in 1993 directing all Federal agencies to purchase only computers, monitors, and printers that meet the EPA Energy STAR requirements. EPA has since expanded the Energy STAR program to other energy-consuming products such as light fixtures and heating, ventilation, and air conditioning units.

The California Environmental Protection Agency initiated an environmental technology certification program similar to the Energy STAR program in 1993. Unlike the Energy STAR program, however, the California EPA tests each of the products for which certification applications have been submitted to determine that they are effective in preventing or cleaning up pollution. The California program has certified about four dozen technologies and was a 1996 winner of the Innovations in American Government award.

In the standards area, the implementing body will have to carefully distinguish between performance standards and specific technical standards and consider the degree of resistance that each will receive from private industry. Generally, the private sector is less resistant to performance standards than specific standards. This practice of certifying effective technologies would appear to be applicable to at least the telecommunications infrastructure area and perhaps all of the infrastructure areas.

Procurement[7]

With the passage of the Federal Acquisition Streamlining Act of 1994 (FASA)[8] and the Federal Acquisition Reform Act of 1995 (FARA),[9] the Federal government is moving away from the practice of placing specific requirements on contractors and vendors. In fact, the simplified acquisition procedures for goods and services below $100,000 encourages government to purchase standard commercial products whenever possible. FASA and FARA instead encourage the government to use its old standards and requirements to choose the best commercially available product to meet the government's needs.

Furthermore, the Federal government probably does not have the market clout to directly and significantly influence the products developed by private sector firms through its purchasing decisions. For example, data obtained from GSA and the Information Technology Industry Council (ITI)

---

6 Executive Order No. 12845, *Requiring Agencies to Purchase Energy Efficient Computer Equipment* (April 21, 1993).

7 *See also* Appendix A, Summary of Procurement Trends and Legislation.

8 Pub. L. No. 103-355 (1994).

9 Pub. L. No. 104-106 (1995).

indicate that the Federal government accounts for only 1.5 to 3.5 percent of worldwide information technology purchases, and this percentage is expected to decrease in the future. However, the Federal government is still a large customer and can lead by example within the IT industry by incorporating information security considerations in its procurement practices. Moreover, the provision of vital government services represents a critical infrastructure area and government is increasingly dependent on information technology to provide services.

Therefore, it makes sense for the Federal government to use security as a criteria for its IT purchases if the threat of incursion is serious enough. This does not mean that the Federal government should include security-enhancing design specifications in its acquisitions; rather it should consider security provisions when choosing among currently available IT products.

There are a number of ways that this practice could be implemented. One indirect approach would be for the OMB, GAO, or CIO Council to develop guidelines for incorporating security safeguards in IT acquisitions. Among the more direct approaches, OMB could require agencies to include computer security plans in their budget submissions for proposed IT acquisition programs. OMB would then review these security plans as part of the budget process. The Administration should be very cautious, however, in recommending the use of the procurement process to achieve IT security goals. It must also be careful not to recommend an approach that will delay procurements or increase the cost and reduce the effectiveness of future government systems by placing unnecessary constraints on IT acquisitions. However, with the current re-competition of the FTS2000 contract, which provides telecommunications services for the majority of the Federal government, and many Federal agencies planning new IT applications to support their missions, there may be a significant opportunity to incorporate security safeguards within the government's IT infrastructure through the procurement process.

Identification and Dissemination of Best Practices

In recent years the practice of benchmarking and searching for best practices has become commonplace within private industry. The practice has also gained acceptance within the Federal government. For example, the ITMRA requires OMB to encourage the use of best practices in the acquisition of information technology.

Consistent with this concept, an agency, or agencies, be designated to serve as a clearinghouse within the Federal government for infrastructure assurance practices. This clearinghouse function should be performed in close conjunction with the private sector in order to leverage the expertise of private industry.

PART TWO
OPTIONS FOR MODEL PERFORMANCE

The options presented are based on performance-based categories discussed more fully in the section above. The option categories and the options themselves are not mutually exclusive and in many cases will be most effective when combined with other model performance programs.

Performance Measurement

The development of an appropriate set of infrastructure performance goals and measures appears to be an essential prerequisite for any infrastructure assurance program. Without them, it will be difficult to define the objectives and scope of the governments' infrastructure assurance efforts and virtually impossible to determine whether or not progress is being made. Without quantitative performance measures it will also be difficult to effectively communicate the need for increased infrastructure assurance. In fact, in the current GPRA environment, Congress and OMB will require the development of performance goals and measures for any infrastructure assurance program before it provides funding.

Develop And Prepare Annual High-Risk Report For Infrastructure Assurance

Under this option, a particular government organization or agency would coordinate the development and preparation of a report listing the areas within the nation's infrastructure facing the highest risk to cyber and physical threats. For each high-risk area, the report would describe the nature of the threats to the infrastructure and the risks they present. The preparation of this report would require the cooperation of the government agencies responsible for the various infrastructure areas. It might also be necessary to obtain information from private sector organizations and state and local governments to prepare the report. However, since the report would mainly be qualitative in nature, most of the necessary information would probably be available within the government.

- **Pro:** This approach would increase awareness of high-risk infrastructure areas and provide some indication of degree to which high risk areas were being addressed. It would also help to focus infrastructure improvement efforts. This approach could be developed very quickly and would be very inexpensive to prepare on a regular basis.
- **Con:** This approach would tend to result in a subjective listing. It runs the risk of not providing much substantiation for high-risk areas and would not be very useful for determining or communicating progress in improving infrastructure assurance.

"Infrastructure Assurance" Is Designated as a Strategic Goal for Federal Agencies to be Addressed within GPRA and ITMRA Frameworks

Under this option, infrastructure assurance is either appended to the mission or, more likely, assigned as a strategic goal by executive order to each of the Federal agencies with links to industries with critical infrastructures. If there are only a few high-risk infrastructures it may be necessary to assign infrastructure assurance as goal to only a few agencies. The effectiveness of this option would probably depend on the degree of specificity with which the Federal government could define the infrastructures to be protected, the nature of the threats, and its desired infrastructure assurance goals. Under GPRA, each of the agencies receiving the assignment would be required to develop specific strategic goals for infrastructure assurance, strategies for achieving the goals, and outcome-based performance measures to track progress in achieving the goals. OMB could then prepare an annual, consolidated status report on infrastructure assurance based on the performance data and narrative information provided by agencies in their annual budget submissions and performance reports. Revisions to the ITMRA to include security in information technology management considerations would also link infrastructure assurance objectives to the developing government management process.

- **Pro:** This approach would institutionalize infrastructure assurance as a mission of the Federal government by directly linking it to the budget process. It would also force agencies assigned the mission of protecting infrastructures to devise outcome-based infrastructure assurance strategies and programs.
- **Con:** This approach would place an additional performance reporting requirement on government agencies at a time when they are just beginning to implement performance management. It might fail to develop coordinated efforts among Federal agencies, and it could encourage development of unnecessarily costly infrastructure assurance programs.

Establish an Interagency Task Force to Investigate the Feasibility of Developing an Infrastructure Assurance "Report Card"

The development of a concise, quantitative, annual "report card" for infrastructure assurance (similar in format to that developed by the State of Oregon), would provide the Federal government with a powerful tool for raising awareness of the need for infrastructure assurance and communicating the status of infrastructure assurance. It would also help policy makers establish infrastructure assurance goals and identify areas of high risk on which to focus its efforts and resources. Under this option, the Administration could recommend the formation of a follow-on task force to determine whether it is possible to develop such a summary report for infrastructure assurance and, if so, which government organizations should be responsible for preparing the report and what data and resources might be needed.

- **Pro:** This approach recognizes the difficulty in developing appropriate performance measures. The report would be a powerful tool for communicating infrastructure assurance goals and for tracking progress in achieving them. The report would also help to focus and justify subsequent government infrastructure assurance initiatives.
- **Con:** A follow-on study does not represent immediate action to protect critical infrastructures. The report would not make any government agency accountable for achieving infrastructure assurance goals.

Establish An Interagency Task Force To Develop An Infrastructure Assurance Report Card

In the alternative, an interagency task force could be formed to develop the report, or a particular agency could be assigned the responsibility of coordinating development with the assistance of other agencies as needed. It probably would not be necessary to include members of the private sector directly on the report development team, but this might be a good strategy for a number of reasons. If a considerable amount of the data required for the report will have to be obtained from the private sector or state and local governments, involvement of these stakeholders in the development of the report may make them more willing to provide the necessary data.

Involvement of external stakeholders in the development of the report would also help to ensure that performance measures included in the report are valid. Finally, the involvement of external stakeholders would help in gaining widespread buy-in of the results presented in the report.

- **Pro:** This report would be powerful tool for communicating infrastructure assurance goals and for tracking progress in achieving them. It would also help to focus and justify subsequent government infrastructure assurance initiatives.
- **Con:** The report would probably not be capable of making any government agency accountable for achieving infrastructure assurance goals.

Publication of Infrastructure Reliability Data

This approach potentially represents an inexpensive strategy that may be quite effective for certain infrastructure areas.

Investigate Applicability Of Publishing Infrastructure Reliability Data

A study could be performed to determine where among the infrastructure areas this practice might be applicable and effective, and the extent to which it can be accomplished by existing government structures within the framework of existing data collection efforts. This course may also be a logical outgrowth of an effort to develop an infrastructure assurance report card as described above.

- **Pro:** This option would allow time to develop valid measures of infrastructure reliability and security. It would also provide time to gauge stakeholder reaction to the concept.
- **Con:** A general recommendation is less likely to be implemented and a recommendation to perform a follow-on study implies that more time will pass before specific action is taken.

Conduct a Pilot Study on the Feasibility of Publishing Comparative Telecommunications or Electric Utility Reliability Data

The Administration could direct the Federal Communications Commission (FCC), the Network Reliability and Interoperability Council (NRIC), the Federal Energy Regulatory Commission (FERC), or the North American Electric Reliability Council (NERC) to perform a pilot study of the feasibility of publishing comparative data of the service reliability among either the telecommunications or electric service providers. The study would also address the degree to which reliability assurance might also achieve desirable infrastructure assurance objectives. If the results of the pilot study indicate that the practice is likely to effective in promoting infrastructure assurance objectives among telecommunications or electric service providers, then the government can implement the practice for the telecommunications or electric service industry and investigate the feasibility of applying the practice to others.

- **Pro:** This option appears to be good test of this practice, since telecommunications and electric industries are being deregulated to spur competition, and consumers are likely to be concerned about the continuing reliability of these services. A pilot study conducted by an existing government or government-sponsored industry council would be relatively inexpensive and likely to be more cost-effective than broad study of applicability of concept across all critical infrastructure areas performed by new interagency task force. A more focused study by organizations designed promote infrastructure reliability and with access to reliability data may lead to a faster implementation of the concept. Since NRIC and NERC include significant industry representation, having either of these organizations perform a pilot study would provide immediate stakeholder reaction to the concept of publishing comparative reliability data and help gain buy-in for the concept.
- **Con:** Increased infrastructure reliability may not necessarily translate to increased infrastructure protection or security. Current reliability may be so high in these industries that consumers base their purchasing decisions on factors other than reliability, such as price. Given the use of

shared networks and other facilities in these industries, individual providers may not be able to adequately influence reliability figures.

Begin Publication Of Specific Infrastructure Reliability Data

Under this option, the government would begin publishing detailed data comparing infrastructure reliability and security among private sector owners within particular industries. In order for this to be effective, it would have to be specified which agencies should publish what types of reliability and security data.

- **Pro:** This option would result in highly visible actions in the near future that would help to promote awareness of the issue of infrastructure assurance.
- **Con:** Without knowing what kind of data are available for each of the infrastructure areas it may be premature to recommend publishing comparative data. There is also the risk that by rushing to implement this practice the government could publish data that present an inaccurate comparison of the infrastructure reliability or security among the various private sector owners. This might expose the government to some kind of legal liability or at least poison relations between the government and the private sector organizations it needs to partner with to enhance infrastructure assurance. Rushing to publication might also result in measures that service providers can "game," again leading to invalid comparisons.

Identification And Dissemination Of Best Practices

Recognition of IT security problem is increasing as is the number of products and procedures designed to enhance IT security. Failure to establish a government clearinghouse for best practices in IT security will delay the adoption of such practices within the Federal government and, perhaps, the private sector. This approach also ignores the potentially significant need in other infrastructure areas, where assurance needs are less well recognized, to identify and disseminate best practices in infrastructure assurance. Finally, the identification of standards underlying best practices appears to be a prerequisite for employing any certification practice.

NIST, Commerce, NSA or Some other Agency be Assigned to Serve as Clearinghouse for Best Practices in IT Security

Under this option, the tasked agency would work with private industry to determine the best practices in IT security including those technologies, products, and procedures that offer the best mitigation and assurance from cyber threats as well as those that provide the best recovery capability from incursions. This expands the current requirement that government CIO's collect and share best practices. The agency would be expected to maintain and disseminate information concerning these best practices to appropriate public or private organizations upon request. These best practices could form the guidelines to be used for auditing the security of Federal information systems. They might also become the basis for establishing security standards for IT products and services that can inform a certification process.

- **Pro:** This option is consistent with the direction of recent legislation. NIST appears to already have a substantial knowledge of IT security and in, FedCIRC, is already disseminating information to other government agencies. This would also test the concept of a Federal clearinghouse for best practices in infrastructure assurance before attempting to expand the concept to other infrastructure areas.
- **Con:** There may already be, or soon be, a private sector organization that government and private sector organizations can look to for guidance on best practices and products in IT security. This approach ignores potentially significant need in other infrastructure areas to identify and disseminate best practices in infrastructure assurance.

NIST, Commerce, NSA or Some other Agency be Assigned to Serve as Clearinghouse for Best Practices in IT Security and the Federal Government would Investigate the Feasibility/Advisability of Establishing Clearinghouses for Infrastructure Assurance in other Infrastructure Areas

Under this option, a specific approach to establish NIST, Commerce, NSA or another government agency as a clearinghouse for best practices in IT security within the Federal government would be combined with a more general recommendation to study the applicability of the clearinghouse concept to the other critical infrastructure areas.

- **Pro:** This approach would address the immediate need in the IT infrastructure area as well as that of the other critical infrastructure areas.
- **Con:** In order to ensure that the study of the applicability of the clearinghouse concept to other, non-IT infrastructure areas is performed, a follow-on task force would have to be formed or a specific agency would have to be assigned the task of performing the study. This assignment would compete with and might dilute the efforts of any follow-on task force assembled to implement infrastructure assurance efforts. In addition, the task force or agency performing the study would not have the benefit of the lessons learned from the establishment of a clearinghouse for best practices in IT security.

The Federal Government Encourages and Supports the Establishment of a Private Sector Clearinghouse for Best Practices in IT Security

Under this option, NIST would work with private industry to establish a clearinghouse for best practices in IT security within an IT industry association such as ITI or EIA. The private sector industry association serving as the clearinghouse would be expected to maintain and disseminate information concerning these best practices to any public or private organization that requested it. (The fact that ITI formed the Information Security Exploratory Committee (ISEC) at the request of the President is an indication that this option is feasible). However, in order to induce an industry association to accept the larger responsibility of establishing and maintaining a clearinghouse for best practices in IT security, it may be necessary for the government to provide some funding as well as other forms of support.

- **Pro:** This approach would avoid potential competition between government and any private sector organizations considering establishing a clearinghouse for best practices in IT security. It would probably be more cost-effective than establishing a government clearinghouse. A private-sector clearinghouse may be more effective in obtaining information concerning IT security from private sector firms than government. It may tend to establish an infrastructure protection "beachhead" within the private sector that could help government monitor security of IT infrastructure area and encourage development of new IT security products and practices.
- **Con:** This approach ignores a potentially significant need in other infrastructure areas to identify and disseminate best practices in infrastructure security. Private sector industry associations may be unwilling to accept a role as clearinghouse. If the need for IT security is so great, the private sector would likely to establish clearinghouse without a push from government.

Procurement

Establish a Task Force to Determine how Infrastructure Assurance Considerations might be Incorporated into Federal Acquisitions

A task force could review the IT acquisition guidelines to be developed by OMB to determine whether they adequately address infrastructure assurance concerns. To be effective, this option (like a number of the others) requires well-defined infrastructure assurance goals and a clear understanding of what kind of infrastructures, especially within the Federal government, need to be protected and from what kinds of threats.

- **Pro:** This approach appears to be a reasonable first step in attempting to use the procurement process to promote infrastructure assurance. It would help to determine if infrastructure assurance is currently being adequately considered in acquisition programs.
- **Con:** This approach takes a broad perspective across all infrastructure areas where a more specific focus on one or two high risk infrastructure areas, such as telecommunications, may provide faster and more focused results. The government may miss significant opportunities to

build-in security within its own infrastructure while it is evaluating the applicability of using procurement to improve infrastructure assurance.

Form a Task Force to Review the Adequacy of Infrastructure Assurance Provisions in Large Pending Federal Procurements

The focus of this option would be to form a "SWAT" team to ensure that major, near-term Federal acquisitions such as FTS2001 (follow-on to FTS2000) adequately provide for infrastructure assurance. This option is especially appropriate if current procurement procedures do not adequately consider and provide for infrastructure assurance needs and if a number of specific acquisition programs that should be reviewed can be readily identified.

- **Pro:** This approach involves immediate action to improve assurance of government-owned infrastructures.
- **Con:** The findings are likely to *raise* costs of underlying acquisitions. This approach does nothing to institutionalize infrastructure assurance considerations in Federal acquisition process. Agencies are likely to react negatively to the recommendation to conduct an additional audit of their near-term acquisition programs. Without well-defined assurance goals, this approach might delay or compromise acquisitions without providing significant improvements in infrastructure assurance.

The President Issues an Executive order Requiring Agencies to Procure only IT Products and Services that Meet NIST, Commerce, NSA or other Agency Security Standards

Under this option, the Federal government would apply the EPA Energy STAR model to IT acquisition. To be viable, this option would require that NIST, Commerce, NSA or another agency establish security standards for IT products and services and develop certification methods for those products and practices that meet the standards.

- **Pro:** This option would promote conformance with established IT security standards throughout the Federal government.
- **Con:** Such a recommendation may be viewed as being premature until the NIST, Commerce or NSA certification process is in place. It might place too much weight on security considerations during IT acquisition process, and would not address other critical infrastructure areas.

Certification Programs

NIST, Commerce, NSA or Another Agency Works with IT Industry to Evaluate the Advisability and Feasibility of Certifying Security Protection of IT Products and Services

Under this option, a Federal agency works with an IT industry association such as the Information Technology Industry Council (ITI) or the Electronics Industry Association (EIA), or forms a joint government/IT industry council, to determine whether IT security could be improved through the certification of IT products and services that meet certain security standards. Among other things, the association group or council would determine the scope of IT products and services to be certified, the feasibility of establishing security standards for these products and services, and define procedures for certifying products and services.

- **Pro:** This approach would enable the government to determine whether the practice of certification is likely to be effective prior to implementation, and would help to define the proper scope of an IT certification program as well as the appropriate role of government. It would also provide early industry reaction to the concept.
- **Con:** It does not address other infrastructure areas besides IT. A recommendation for follow-on study implies more time will pass before specific action is taken, and would delay certification program for IT products and services which may be ready for implementation.

NIST, Commerce, NSA or Another Agency Works with Industry to Establish Security Standards for IT Products and Services and Develop A Private-Sector Certification Process

The Federal government appears to be already moving toward establishing security standards for IT products and services. Under this option, the Federal government would attempt to accelerate and build on this movement by applying something along the lines of an EPA Energy STAR model to the IT infrastructure area. In addition, some Federal agency or agencies could work with IT industry groups to establish security standards for IT products and services and develop a process within the private sector for certifying these products and services. Ultimately, an industry association, university center, or FFRDC could assume responsibility for managing the certification process. As in the Energy STAR program, the certifying industry association would provide "Security STAR" labels or a seal to producers to affix to their IT products or service brochures to indicate they are compliant with the IT security standards.

- **Pro:** This approach would increase the market awareness of IT infrastructure security technology by allowing producers of IT products and services to advertise that their products comply with specific industry security standards. The program could involve minimal government expense since industry would administer the largest portions of the certification process. This approach also sets the stage for eventually requiring government agencies to buy only certified IT products and services. Collaboration between government and private sector in setting standards, and private sector management of the certification process would help in gaining industry buy-in for the program.

- **Con:** This approach does not address other infrastructure areas besides IT. An industry association may encounter conflicts of interest in attempting to perform impartial certification of products and services because of its dependence on corporate membership.

NIST, Commerce, NSA or Another Agency Establishes Security Standards for IT Products and Services, and Certify those Products and Services that meet the Standards

This is the same as the earlier option, except that a Federal agency would take full responsibility for establishing IT security standards and performing or contracting for tests to determine if IT products and services meet the security standards.

- **Pro:** This approach would increase market awareness of IT infrastructure security technology, and set the stage for requiring government agencies to buy only certified IT products and services. The public may have more confidence in certification if the tests are performed by the Federal government.
- **Con:** This approach does not address other infrastructure areas besides IT. Private industry may resent and resist a program in which the Federal government establishes standards and performs certification. It is a potentially costly endeavor to certify all current IT products, especially in the constantly-changing IT market, and there may be a liability issue as well if government-developed standards or government-tested IT products prove faulty.

NIST, Commerce, NSA or another Agency Certifies IT Products and Services, and a Follow-On Task Force is Established to Evaluate the Advisability/Feasibility of Certifying Assurance Products and Services for other Infrastructure Areas

This option assumes that the IT certification program will be successful in promoting IT infrastructure assurance.

- **Pro:** This option initiates IT certification program and positions the Federal government to quickly apply practice to other infrastructure areas.

- **Con:** If the IT certification program fails, time and resources invested to study the application of certification to other infrastructure areas will probably have been wasted.

PART THREE
CONCLUSIONS

The possible permutations of options itemized in the previous section can be reasonably boiled down into the following conclusions as to what would constitute an optimum mix for best ensuring the success of infrastructure assurance initiatives. By virtue of the various competing equities involved, a "blended" approach to various options offers the best combination of political feasibility and the highest probability of achieving infrastructure assurance objectives.

Performance Measurement

The production of a "high risk" report for infrastructure assurance is a laudable goal for a Federal government infrastructure assurance organization. Such a report could make use of information assembled in furtherance of such an organization's responsibility to "assess the national risk." Once sufficient baseline data has been accumulated, the report might take on a more normative character. Furthermore, specifying infrastructure assurance as a strategic goal or category of specific goals under the GPRA or ITMRA frameworks would be very helpful. There appears to be a need for information security considerations to be specifically included in the management framework for information technology acquisition and use as set out in the ITMRA and other pieces of legislation and regulations. There are, however, differences in the respective goals, structure and relative specificity of the GPRA and ITMRA.

Agencies should identify existing infrastructure assurance-related responsibilities and include them in strategic plans submitted in compliance with the GPRA. Agencies should incorporate any new infrastructure assurance-related taskings into strategic plans under the GPRA, thus facilitating coordinated budgetary performance review. Currently, agencies are required by the GPRA to prepare a five year plan including "major functions and operations" of the agency. If an agency does not consider infrastructure assurance to be a "major function or operation," it may not be included in the strategic plan. The failure to include such a function in the five-year plan allows it be left out of the annual performance plan which serves as the foundation for budgetary review. In order to assure performance measures are put in place and reviewed as part of the budget process, infrastructure assurance functions should be explicitly included in agency strategic plans.

- **Pro:** This approach encourages infrastructure assurance to be included in strategic planning, thus facilitating coordinated budget review.
- **Con:** The current rate of implementation of the GPRA has been slow, and this option could promote delay unless used in conjunction with other performance measures.

Congress could also consider the propriety of a modest revision to the ITMRA to encourage agencies to establish security-based performance measures in addition to other performance measures required by the Act (see proposed language in Appendix B).

- **Pro:** This approach promotes agency development of performance measures in a minimally burdensome way, and in a manner likely to lead to the development of government wide standards and best practices.
- **Con:** The effectiveness of implementing ITMRA requirements through agency CIOs remains uncertain.

Publication of Infrastructure Assurance Data

An immediate pilot study ought to be conducted to investigate the feasibility and likely effectiveness of publishing comparative telecommunications or electric utility reliability data in order to accomplish infrastructure assurance objectives.

Publication of performance data (e.g. by the FAA for on-time arrival/departure), is a possible means of increasing general sensitivity to and awareness of reliability, security or other infrastructure-related performance of infrastructure services. In light of deregulation and restructuring of major infrastructures such as telecommunications and electric power, additional information, provided by a neutral source, may allow consumers to make educated choices in selecting between available infrastructure services.

Insofar as publication may be one possible means of promoting infrastructure assurance objectives, the Administration could direct the FCC's Network Reliability and Interoperability Council (NRIC) to initiate a pilot study as to the feasibility of publishing comparative infrastructure assurance-related data for the telecommunications industry. The NRIC could conduct a pilot study of the likely impact and feasibility of such publication.

Such a study that could address, among other issues: (1) the types of infrastructure assurance-related data that could be published; (2) to what degree the practice of publishing infrastructure assurance-related statistics would likely achieve infrastructure assurance objectives; (3) whether effective publication could be accomplished using data already made available to the Federal government, or whether additional data collection might be necessary; and (4) how the practice of publication might influence industry practices, consumer choices, and the liability climate.

- **Pro:** The approach promotes the practice and initiates further study of its application. A careful study will allow complex issues such as liability to be further explored and resolved, and use of a pilot allows the practice to be studied thoroughly before being considered for extension to other infrastructures.
- **Con:** The use a pilot in one industry only may slow the spread of the practice to other areas where it may be needed, particularly by consumers, in the near future.

Identification and Dissemination of Best Practices

Additional research has revealed that under Executive Order 13011, the Government Information Technology Services Board (GITSB) is responsible for collecting and disseminating information on IT best practices for the government. In order to implement this responsibility, GITSB allocated funds to NIST's Computer Security Resource Clearinghouse (CSRC). The CSRC is a website with information available on a wide range of information technology and security issues. The information available originates primarily with NIST, but includes some private sector information of significant impact. The CSRC information is available to both government and private sector. Additional informal avenues are available to government CIOs to share information about IT practices. These include the CIO Council and a GSA inter-agency newsletter on information technology management. Implementation of information security recommendations made by the National Performance Review are still underway.

Existing Federal government efforts to identify and disseminate best practices in information technology could be expanded to promote greater private sector participation and that the resulting efforts be considered as potential models for best practices clearinghouses for other critical infrastructures. Sharing of best practices as mandated by E.O. 13011 could be further expanded to draw on the expertise of other government agencies (e.g NSA) and the private sector. Additional resources should be used to expand awareness and use of the information so collected. It may prove desirable for the CRSC to take a more active role in disseminating information to government IT managers. In concert with the expansion of the CSRC efforts, a study should also be undertaken to determine whether the structure, funding, participants, and management of the IT best practices effort could be applied to other infrastructure assurance areas. This study could be undertaken by an interagency task force.

- **Pro:** This approach allows the Federal government to highlight best practices as a model performance practice and to endorse the practice in a number of areas while making a specific recommendation for the fortification of current efforts in IT. It also enhances the private sector's role in sharing best practices information with the government while allowing them to benefit through dissemination channels. The option requires only minimal resources to implement and existing channels may be used. Limited success in the IT arena can be achieved before moving to other infrastructure assurance areas.

- **Con:** The current Federal government IT environment has a number of players and can be confusing. This approach does little to simplify the environment.

Procurement

The Administration could identify an interagency task force to (1) identify large pending procurements related to infrastructure assurance issues; (2) study the degree to which large pending procurements adequately address infrastructure assurance objectives, (3) make recommendations as to how such procurements might be adapted, in the short term, without unduly adding to their cost; and (4) based on lessons learned through review of individual procurements, make recommendations as to how the Federal acquisition process might be revised to address infrastructure assurance. The impending re-competition of several long term Federal government contracts (including, for example, FTS 2000) provides an important opportunity for the government to assess the degree to which these contracts reflect infrastructure assurance concerns. An interagency task force should be convened to address the issues raised in those large procurements identified as needing review. The task force could use what is learned from this study to develop recommendations for revising the procurement *process* to better take into account infrastructure assurance objectives (specifically taking into account, for example, the current security criteria in place and the procedures for obtaining waivers). The task force might also consider the feasibility of requiring procurements to meet security standards set out by other government agencies (NIST, NSA, DOE, etc.).

- **Pro:** With such an approach, the Federal government acts quickly to promote infrastructure assurance objectives in large pending procurements, yet allows for informed study of the procurement process using modest resources.

- **Con:** The current procurement environment, marked by the movement to commercially available/off-the-shelf products, may constrain potential recommendations of such a body in all but the truly large-scale procurements.

Certification Programs

Energy STAR is a certification program for energy efficient computer hardware and office equipment administered by the EPA. The Administration could consider programs like the EnergySTAR certification program as a potential model to guide infrastructure assurance efforts.

Energy STAR is enforced by Memoranda of Understanding between the company seeking to use the logo and the EPA. The MOUs dictate the appropriate uses of the Energy STAR logo and allow companies to self-test or employ a third party to test the products for compliance with the EPA standards. EPA spot checks products for compliance with standards and to ensure that only companies with an MOU are using the logo. In the event a company is not in compliance, EPA will generally attempt to resolve the issue informally. If that fails, EPA will require that the company either meet the standards or remove the label.

The Energy STAR program could provide an effective model for government programs aimed at the certification of IT products and services. Consideration of such a program might also encourage development of similar programs in other infrastructure areas to promote assurance objectives. The Energy STAR program may be held up as an efficient and appropriate use of government authority and resources, and advocates its broad consideration as a potential model for other government programs aimed at promoting infrastructure assurance through government and private sector adoption and implementation of certification standards.

- **Pro:** This approach would highlight the success of the Federal government initiative while providing a useful model for a certification process. It carries the additional benefit of low costs of implementation and enforcement and high visibility.
- **Con:** Merely acknowledging the effectiveness or appropriateness of a model does not, in itself, significantly further the effort.

Standards

Before meaningful certification programs can be established, standards must be in place to guide the certification process. There are many means by which the government can develop standards, however, in advocating a performance-based approach to certification, and a similar approach can be taken with regard to standards development as well. Research was performed into the various mechanisms by which the Federal government can influence or engage in the standards development process. The results of that research are attached as a supplemental report. (*See* Appendix C.)

APPENDIX A
SUMMARY OF PROCUREMENT TRENDS AND LEGISLATION[10]

Background

The Department of Defense (DoD) is the world's largest buyer of goods and services. In Fiscal Year 1996, DoD completed over 8.7 million procurement actions, of which more than 8.4 million were for purchases under $25,000. These purchasing actions involved the efforts of more than 20,000 contracting personnel in over 1,300 contracting offices worldwide. Virtually all procurement actions under $100,000 require significant time and effort to define needs, identify funding, define sources and secure contracts. The acquisition reforms recently enacted are intended to streamline the acquisition process for goods and services valued under $100,000.

Two major reforms in DoD acquisitions were the increase of the small purchase threshold from $25,000 to $100,000 and the institution of the International Merchants Purchase Authorization Card (IMPAC). By increasing the small purchase threshold to $100,000, DoD contracting officers are not required to issue notices in the Commerce Business Daily (CBD) and may solicit from a short list of vendors. With the IMPAC card, buyers holding the card may purchase goods and services in the same way as a customer holding a VISA credit card, reducing the amount of paperwork and time required by DoD buyers.

Regulatory Reform

The last three years have witnessed radical reforms in government procurement regulations. Three key reform acts have resulted in a vastly different landscape for government buyers. Below are the three Federal acts, with a summary of the changes which affect acquisitions under $100,000.

Federal Acquisition Streamlining Act of 1994 (FASA)

The Federal Acquisition Streamlining Act of 1994 (FASA) (Public Law 101-365) made a number of changes in the way goods and services at, or below, $100,000 are acquired. The Act replaces the $25,000 threshold with a new "Simplified Acquisition Threshold" (SAT) of $100,000 once an agency (or procuring activity within the agency) has achieved certain electronic commerce (FACNET) capabilities, is using them and certifies that they have met the criteria. Until that time, the threshold is only increased to $50,000.

---

10 This appendix was prepared by Mr. Paul Harder of SRA International.

Other key changes made to procurement of goods and services below $100,000 are:

- Quotes will be solicited in writing 15 days after placing a notice in the Commerce Business Daily (CBD). The notice may be used: (1) to ask for expressions of interest from vendors so that after the 15-day waiting period they may be requested, in writing, to submit a quotation, or (2) to request written quotations, to be submitted within a reasonable time after the 15-day waiting period. In addition, the CBD notice will contain: (1) an accurate description of the product(s) to be acquired, (2) a description of the procedures to be used to select and make the award, (3) a statement reserving the procurement for small business, and (4) a statement of the period of time expected for the contracting officer to solicit/evaluate vendor quotes, and issue the purchase order. All quotes received from a small business will be considered.
- FASA reserves new contracts, above $2,500 but under the simplified acquisition threshold, for small business. It specifically authorizes continued set-asides of all contracts under the threshold for minority small businesses. However, FASA excludes purchases of $2,500 or less (Micro-Purchases) from the small business reservation, to make it possible for agency officials to make simplified purchases and credit card purchases. These purchases do not have to be competitive and are exempt from the Buy American Act.
- FASA eliminated the applicability of several FAR solicitation clauses to SAT transactions and established a $100,000 threshold for fifteen (15) different statutes that establish paperwork and recordkeeping requirements not applicable in the commercial sector.

FASA retains the requirement that a notice of any procurement over $25,000 be published in the Commerce Business Daily 15 days prior to the issuance of a solicitation. After the issuance of this notice, however, simplified acquisitions could follow any procedures described in the notice — for example, by shortening the period for submission of offers.

FASA phases out the requirement to publish advance notice of purchases below $100,000 when electronic commerce procedures and systems (interim FACNET) are in place. It also provides that all procurements of $250,000 or less need not be synopsized in the CBD if a "full FACNET" system is being used for the procurement.

Federal Acquisition Reform Act of 1995 (FARA)

FARA revises the standard for "full and open competition" for procurement of goods and services to provide for open access to bidding that is "consistent with the need to efficiently fulfill the Government's requirements." The Act also revises requirements for the use of other than competitive procedures, allowing such procedures only when the use of competitive procedures is not feasible or appropriate. Standards for making this determination are to be set forth in the Federal Acquisition Regulation (FAR).

FARA provides that the FAR shall ensure that the requirement to obtain full and open competition is implemented in a manner that is consistent with the need to fulfill the Government's requirements efficiently. However, this provision makes no change to the requirement for full and open competition, nor to the definition of full and open competition. Therefore, it is unclear how this requirement will be implemented.

FARA prohibits sole source procurements of commercial items valued above the simplified acquisition threshold "unless the need to do so is justified in writing and approved in accordance with the [FAR]." Section 4202 of FARA permits the use of simplified acquisition procedures for procurements expected to exceed the simplified acquisition threshold, but not to exceed $5 million, when the contracting officer reasonably expects, based on the nature of the property or services and market research, that offers will include only commercial items.

Regarding Commercial-Off-The-Shelf (COTS) products, FARA provides that the FAR shall include a list of provisions that are inapplicable to contracts for the procurement of commercially available off-the-shelf items. The House-Senate Conference Report states that the list would include each provision of law that, in the opinion of the Administrator of the Office of Federal Procurement Policy (OFPP), imposes government-unique policies, procedures, requirements, or restrictions, except any which the Administrator determines it would be in the best interests of the United States to include in contracts for commercially available off-the-shelf items. The list would not include generally applicable provisions of law (i.e., those not uniquely applicable to government contractors) nor several categories of statutes, such as any that provide for civil or criminal penalties.

FARA also expanded the definition of "commercial services" to include services sold based on "market" as well as catalog prices. The new definition reads: "Services offered and sold competitively, in substantial quantities, in the commercial marketplace based on established catalog [or market] prices for the specific tasks performed and under standard commercial terms and conditions." This revision allows service firms to qualify as commercial services by charging market prices, which are defined as prices that are established in the ordinary course of trade between buyers and sellers free to bargain and that can be substantiated from sources independent of the offeror.

Finally, FARA amends FASA by eliminating the $50,000 cap on use of simplified acquisition procedures until an agency receives interim FACNET certification (i.e., all activities may use simplified acquisition procedures for procurements up to $100,000). Further, it provides that the threshold will revert back to $50,000 after 31 DEC 1999 if an agency does not have full FACNET certification.

The Information Technology Management Reform Act of 1996 (Section E of the FY96 Defense Authorization Act) also known as the Clinger-Cohen Act

The Information Technology Management Reform Act of 1996 (ITMRA), radically changes the way information technology (IT) goods and services are procured within the Federal Government. ITMRA repeals the Brooks Act, thereby eliminating GSA's role in the oversight of IT acquisitions and the requirement for agencies to get a Delegation of Procurement Authority (DPA) from GSA.

Agencies are given the direct authority to procure IT with a focus on capital planning and investment control and performance and results-based management. Agencies are required to designate a Chief Information Officer (Section 5125). OMB is given an enhanced role in IT management and oversight including evaluations of agency IT programs and investments. National security systems are exempted from some of the Act's requirements.

Procedures required for IT acquisitions are destined to change due to a requirement that the FAR Council ensure that the process for acquisition of IT is simplified, clear, and understandable and specifically addresses the management of risk, incremental acquisitions, and the need to incorporate commercial IT in a timely manner. Also, ITMRA directs agency heads to use "modular contracting" to the maximum extent practicable for the acquisition of major IT systems. The Act describes modular contracting as an approach under which an "agency's need for a system is satisfied in successive acquisitions of interoperable increments. Each increment complies with common or commercially accepted standards . . . so the increments are compatible with other increments . . . comprising the system." The FAR must provide that a contract for an increment of an IT system should be awarded within 180 days after the solicitation is issued, and if it cannot be, the increment should be considered for cancellation.

The critical portion of ITMRA with respect to small purchases is the requirement that by January 1, 1998, FACNET (or, if use of FACNET is not practical, another automated system) must provide government-wide on-line computer access to information on products and services that are available for ordering under the GSA multiple award schedules. This is precisely the requirement which will be addressed herein.

Significant Policy Changes

The past three years have also resulted in significant changes in the policies surrounding defense acquisitions. This change reflects the Defense Department's interest in becoming less of a driver of technology, specifying what goods and services should look like and becoming more of a buyer of technology, looking for the best commercially available unit to meet the government's requirements.

The Defense Department's change is reflected in their approach to specifications. During the past three years, DoD has eliminated over one hundred military specifications and standards, replacing them with equivalent commercial specifications and standards. By doing this, DoD opens the doors to new commercial vendors to provide goods to meet the needs of the warfighter without having to re-tool or re-manufacture goods to meet a specific military requirement. Instead, the same product which meets the commercial market's standards may also meet the military's standards.

Another change is the renewed emphasis on using existing contracts rather than entering into new contracts. DoD has ordered its personnel to maximize use of the GSA schedules for goods and services and to minimize the use of new contracts where possible. This shift away from making contracts whenever a need arose to consulting a pre-negotiated contract before starting a new contract is a shift in the procurement mindset.

Further enabling this shift has been the creation of electronic catalogs. Through the Internet, government buyers have been able to procure goods and services through electronic catalogs like GSAdvantage! and the DLA Buyer website. Further, Dr. Steve Kelman, Administrator of the Office of Federal Procurement Policy issued on March 14, 1997 a policy statement on Electronic Catalogs. "Electronic Catalogs" are defined by OFPP as a Web-based electronic ordering system which involves: (1) a contract with pre-established business arrangements with industry; (2) a means for the customer to identify and order goods and services, either from within an agency (intra-agency) or by more than one agency (inter-agency); and (3) sufficient information (updated to reflect changes) for the customer to compare the items offered by performance, price and delivery.

Dr. Kelman recognized that electronic catalogs were the way of the future for federal acquisitions. He stated that:

> "Electronic catalogs have become a prime method for us to take advantage of the operational efficiencies offered by evolving electronic commerce technologies. I encourage you, in your efforts to provide greater value to the taxpayer, to aggressively promote the use of Federal electronic catalogs. As our use of this evolving technology increases, however, there is a corresponding need for greater interagency coordination to maximize the effective use of electronic catalogs."

Electronic catalogs offer two advantages. They require fewer resources to make repetitive purchases and they offer opportunities for agencies to pursue more effective purchasing strategies. They can help the federal government leverage its buying power through volume purchasing. Alternatively, they enable government customers to make "spot" purchases with on-line comparisons of the price, features, and performance of similar products and services.

APPENDIX B
INFORMATION TECHNOLOGY MANAGEMENT REFORM ACT

(Proposed revisions as marked)

Sec. 5123 Performance and Results-Based Management

In fulfilling the responsibilities under section 3506(h) of title 44, United States Code, the head of an executive agency shall—

(1) establish goals for improving the efficiency and effectiveness, **and security** of agency operations and, as appropriate, the delivery of services to the public through the effective use of information technology;

(2) prepare an annual report, to be included in the executive agency's budget submission to Congress, on the progress in achieving the goals;

(3) ensure that performance measurements are prescribed for information technology used by or to be acquired for, the executive agency and that the performance measurements measure **the level of security of the information technology, and** how well the information technology supports programs of the executive agency; . . .

* * *

(4) ensure that the information security policies, procedures, and practices of the executive agency are adequate.

APPENDIX C
FEDERAL ROLE IN DEVELOPMENT OF STANDARDS[11]

This discussion examines ways the Federal government can provide incentives and promote the standards settings process without necessarily requiring or mandating the setting or use of such standards. The Legal Team

---

11 This appendix was prepared by Elizabeth A. Banker, Assistant General Counsel to the President's Commission on Critical Infrastructure Protection (PCCIP). Background research was compiled by Mr. Ted Drake of SRA International.

conducted research on available methods for creating such incentives. These efforts have been fruitful and may merit further consideration.

Background

The U.S. standardization system is quite complex compared to those of other countries. In the United States alone, there are 700 organizations that take part in the standardization process. Both government and non-government organizations have an almost equal hand in the system. Of the 93,000 standards in the United States, the Federal government developed 44,000 and non-government organizations developed 49,000. Major players include Federal government agencies, scientific and professional organizations, trade associations, standards developing organizations, and developers of informal standards.

The leading standards developers in the government include the Department of Defense (DoD) and the General Services Administration. Although 80 government organizations have developed standards, these two organizations alone account for over 80 percent of Federally developed standards and almost 40 percent of all U.S. developed standards. Most government standards are developed for either internal or regulatory use, although there are a few cases where the government coordinates the development of voluntary standards for the private sector.

In the private sector, over 620 organizations take part in the standards development process. The American National Standards Institute (ANSI) coordinates 175 of these organizations, promoting the development of voluntary, consensus-developed standards. Other large non-government standards developers include the American Society for Testing and Materials, the U.S. Pharmacopeial Convention, the Society of Automotive Engineers International, and the Aerospace Industries Association.

Figure 1. Number of U.S. Standards by Type of Organization

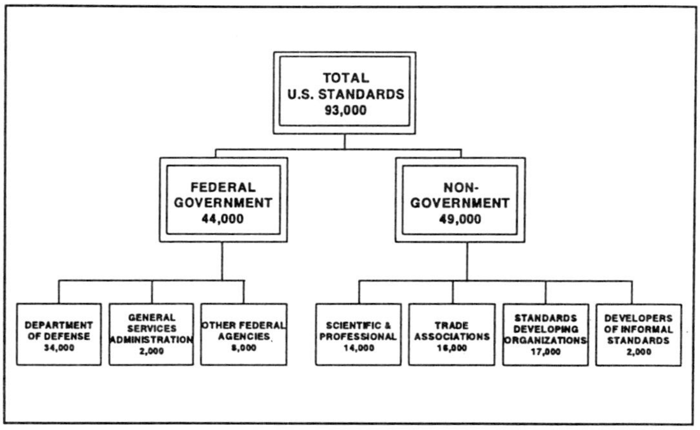

The following sections will discuss the standards development process from a government perspective, focusing on methods that the government can use to facilitate standards development. We will look at industry evaluation and adoption of standards and obstacles to the development of effective standards. We will discuss relevant legislation as well as the government's role in the standards development process.

Models for Standards Setting

Figure 2. provides a framework for analyzing the standards development process from a government perspective. The Federal government has several options, involving varying degrees of intervention, for developing a standard. Once the government has defined a need for a standard, it has five options ranging from letting the market develop the standard to specifying compliance with a standard in government regulation. Figure 2. illustrates these options and the associated outcomes of each. As the government increases its intervention in the standardization process, it also increases the likelihood that the industry will adopt the proposed standards. However, with more intervention, there are additional costs that will be placed on the affected industry, and the longer it will take.

Figure 2. Models for Standards Setting

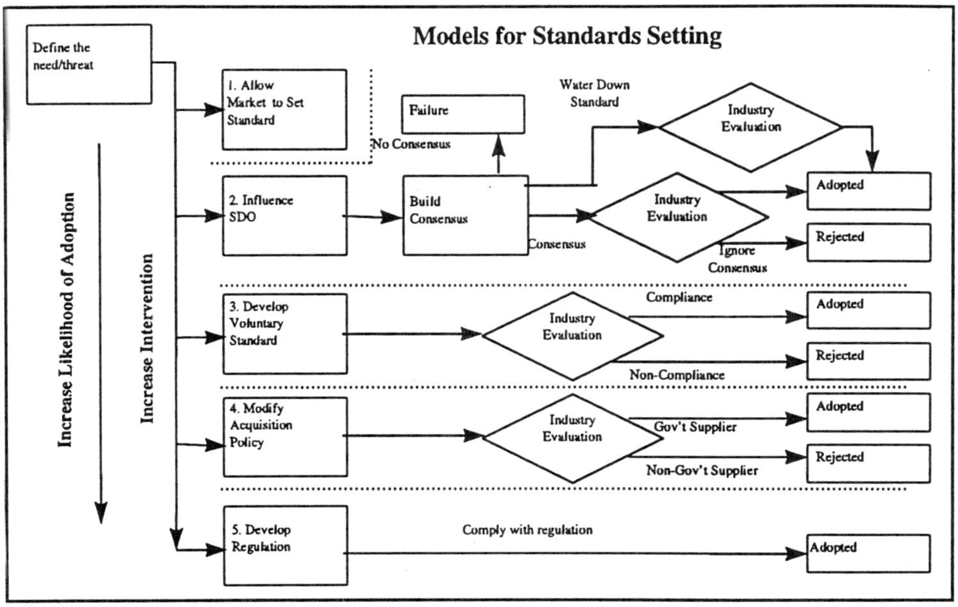

Allow The Market To Set Standards

In many cases, the market will develop and implement standards on its own. There are two key players in this area, standards developers and standards development coordinators. Developers include industry organizations, professional associations, and standards development organizations. Coordination programs facilitate both the development of "needs" for standards, as well as the actual development and implementation of standards among the various standards developers.

ANSI is the leader in the coordination and facilitation of standards within the private sector. Founded in 1918 by five engineering societies and three government agencies, ANSI remains a private non-profit organization. The organization's goal is to enhance the global competitiveness of U.S. business and the American quality of life by promoting and facilitating voluntary consensus standards. ANSI accredits 175 standards developing organizations that follow specific procedures regarding the development of standards. In addition, it reviews standards that are submitted by accredited organizations. If the standards satisfy certain requirements, they are labeled as American National Standards. The Information Infrastructure Standards Panel (IISP), a high-level arm of ANSI, promotes and accelerates the cross-industry coordination of national and global standards for the deployment of the information infrastructure. The panel tries to identify and obtain sponsors for needed standards.

Influence Standards Developing Organizations

There is significant opportunity for the Federal government to influence private sector standards development. Most standards development organizations are completely open to external involvement. ANSI, in particular, has many government members participating on its boards and committees. Both NIST and DoD actually track employee participation in these committees, and manage this participation to ensure that there are no "holes" in coverage.

The National Institute for Standards and Technology (NIST) plays the role as the Federal government's chief coordinator for the development and maintenance of standards. The Interagency Committee on Standards Policy (ICSP), within NIST, brings together representatives from each Federal executive agency to promote effective and consistent standards policies in furtherance of U.S. domestic and foreign goals. NIST has over three hundred employees who participate on the technical committees of various standards developing organizations. In many cases, NIST works directly with these organizations to develop standards requested by government agencies.

The Department of Defense (DoD), as the largest standards developer in the United States, develops its own requirements as well as works with other standards organizations to develop new standards. DoD participates as an "equal partner" with the private sector and other government agencies on the technical committees of non-government standards bodies. The Department's goal is to ensure the proper consideration of DoD requirements, enhance the technical knowledge of DoD personnel, and allow the DoD to contribute its considerable technical capacity. The Department of Defense monitors its influence through the SD-11 database, which identifies and tracks the many DoD personnel who participate on these technical committees. This allows the Department to manage its resources by determining where participation is adequate, and where it is weak and requires increased support.

This option is of particular interest because all standards are developed through a consensus process, which requires the involvement of all interests including the government. This adds significant weight to the standard when it goes to the evaluation and adoption process. Since all interests are involved, it is much more likely that the standard will be adopted by the affected industry and adhered to in the long run.

Develop Voluntary Standards

In some cases, the Federal government has actually developed voluntary standards for use in the private sector. This usually occurs in markets or industries where private sector organizations have failed to provide sufficient coverage with regard to standards development. The government has assisted in the development of standards where liability issues have kept other private organizations away. For example, NIST coordinates the development and maintenance of voluntary product standards for softwood lumber and construction. Private organizations have shied away from developing standards in these two areas because of various liability issues.

In addition to the direct development of voluntary standards, the government can offer incentives to organizations to develop standards in a particular area. In the past, these incentives have been in the form of grants and subsidies. For example, pursuant to the Goals 2000: Educate America Act, 20 U.S.C. § 5932 *et seq.*, the Federal government established the National Skills Standards Board (NSSB). Even though Congress created the NSSB, it is not a government agency. It is a group of business, education, labor and civic leaders dedicated to encouraging the development of voluntary skill standards.

To accomplish its goal, the NSSB "awards grants to voluntary partnerships for the development of [the] skill standards." 20 U.S.C. § 5934(g). The NSSB determines the career fields where skill standards are needed and then accepts proposals from interested groups to set the necessary standards. Funds are awarded to associations or coalitions that demonstrate an ability to convene key stakeholders in specific industry sectors in order to set the necessary standards. The NSSB then approves the standards proposed by the grantees.

The Goals 2000: Educate American Act was passed after 10 years of looking at what was wrong with the United States education system. Some believed that our education system was in a perilous condition and that the Federal government had not done enough to improve it. (cite Mr. Biden & Mr. Chafee, Senate 2/8/94) Presently, the NSSB is thought to be an organization that is trying to keep ahead of a rapidly changing environment.

Modify Acquisition Policy

Many agencies use acquisition policy as a way to influence standards development. This is particularly effective when the government has a large market share for the product being supplied. Acquisition policies result in supplier compliance, but may also have an effect on non-suppliers who are interested in supplying products to the government. On the downside, in those areas where the government has a minimal market share, acquisition policies may not have any effect outside of the government.

Develop Regulation

Developing and instituting a regulation is always an option, but there is a significant downside. Lack of involvement of the private sector in the development process may result in lackluster adherence to the regulation. In addition, regulations may place significant costs on the affected industries. In the area of information technology, where conditions change rapidly, regulation would be hard pressed to keep up and remain appropriate.

Industry Evaluation

The most important step in the voluntary standards development process is to obtain industry approval for the proposed standard. This applies to every standards development model, with the exception of government regulations. There are several factors that increase the likelihood that a standard will be accepted. Clearly, the more benefits that accrue from a standard, the more likely it is to be adopted. Standards that promote compatibility, efficiency, and reliability are much more likely to be implemented by industry. A standard that increases the market share of a particular industry sector or serves to reduce an industry's liability in a particular area tend to be more successful as well.

Product standards are more likely to be accepted when the "users" of the product have a hand in the standard's development rather than the suppliers or producers. When suppliers are involved in the development process, they tend to advocate much more generic standards so as not to make any particular supplier's product obsolete. Users, when involved in the process, develop more specific standards especially with regard to reliability and compatibility.

The aerospace industry is a good example. A typical aircraft, whether military or commercial, contains an extensive array of parts and components. An exceptional number of these parts (60 to 70 percent) are defined by standards developed by the aerospace industry. Aerospace customers are very interested in parts reliability and compatibility among different types of planes, and, in part, drive the development of many of these standards.

In addition, the "users" of these parts (airplane manufacturers) are extensively involved in the development of aerospace standards, and promote the development of well-defined aviation standards. As a result, the industry has a significant number of distinct standards for airplane parts.

In comparison, the automobile industry is much less organized in the standards arena. Suppliers have a much greater role in standards development, while automobile consumers have little role. This results in a situation opposite to that of the aerospace industry. The industry uses a much smaller number of standardized parts, and the standards that they do have are much more generic. This is evidenced by the multitudes of proprietary parts that currently exist in the automobile industry today.

Obstacles Which Prevent the Efficient Development of Standards

Several obstacles exist that can prevent standards from being established relatively efficiently by the public and private sectors. Organizational issues can distort the process. These include decentralization, market incentives, and antitrust issues.

The decentralization of the U.S. voluntary standardization system is one of these obstacles. Compared with other countries, the United States has no central authority for standards development. The American National Standards Institute (ANSI) is the primary coordinator in the private sector, but it only has a minor hold on the standards market. In total, there are over 600 private sector organizations involved in standards development. Approximately 175 of these organizations are ANSI accredited. While ANSI accredited organizations account for over 75 percent (36,000) of non-government standards, only a third (11,500) of these standards are actually certified as American National Standards. There are so many organizations involved in standards development that it makes it difficult to coordinate the development of a standard. Competing organizations often develop different standards for similar products. In the building and construction sector, there are more than 12,000 standards alone. This creates the problem of deciding which standards to implement in an organization.

Compared to the United States, Canada has a much more centralized system for developing standards. The primary standards body is the Standards Council of Canada, a body of the Canadian government. The Standards Council of Canada is a federal Crown Corporation with the mandate to foster and promote voluntary standardization for the benefit of industry, consumers, and the economy. The council has the legal mandate to accredit organizations performing standards development functions. As a result of this coordination, Canada has significantly fewer standards and standards developing organizations than the United States.

Market incentives may hinder the efficient development of standards. In the private sector, the development of standards is supply-driven. Most standards development organizations fund their operations through membership fees and the sale of standards, creating an incentive to develop standards for income rather than need. As a result, a relative glut of standards has emerged and the quality and specificity of these standards has decreased. The United States has three times the number of standards as Germany, the second largest standards developer in the world. In addition, many of these standards are unused because they are either obsolete or too generic for their purpose. It is estimated that 25 to 30 percent of our national standards, both government and industry, apply to obsolete technology. Often, standards are redundant or overlap one another.

Antitrust issues may hinder the full consideration of certain issues in the private sector. Industry organizations and standards development organizations must be particularly careful in working within the antitrust regulations in the development of product standards. Most organizations have strict guidelines as to what participants may or may not discuss in the course of deliberations. In some cases, this may prevent the complete consideration of certain standards.

Relevant Legislation/Government Role

There are two recent pieces of legislation that have served to modify the Federal government's role in the standards arena. OMB Circular A-119 set policy regarding the government's procurement and regulatory activities. The Circular states that the government should rely on voluntary standards, both domestic and international, wherever feasible and consistent with the law. In addition, it requested that government agencies participate in and coordinate their participation in voluntary standards bodies. As a result of this legislation, many Federal agencies have started to adopt voluntary standards and include them in regulatory actions. The DoD and NIST have a significant number of employees now participating in standards development organizations. As noted above, agencies like DoD are starting to track their participation to manage their influence in these organizations.

The National Technology Transfer and Advancement Act sets several requirements for NIST. The Act requires NIST to compare standards used in the private sector with standards adopted by the Federal government and coordinate the use of private sector standards by Federal agencies. NIST is to coordinate public and private standards activities to eliminate unnecessary duplication and complexity. This act basically gives NIST the role of standards coordinator for the Federal government. NIST has the mandate to be the primary facilitator of standards in the United States. In response to this, NIST has increased its influence in various standards development organizations, but has much work to do to integrate public and private standards development.

Analysis

Standards may currently be lacking in important areas of the critical infrastructures. Normally, the market will respond to such a void and fill it through the numerous private standards organizations working today within the United States. In some cases, the market may not identify the same needs for standards as does the Federal government. It may be appropriate for the Federal government to take action to fill that void. Five options to fill the void have been identified. However, only two of those options are truly consistent with an overall approach favoring market forces to work with only minimal coaxing through incentives and other indirect means.

Given recent legislation and the value of consensus, the most appropriate way for the Federal government to promote standards is to allow relevant markets to set the standards, and to help standards-developing organizations identify needed standards and incentivize the process. Both OMB Circular A-119 and the Federal Technology Transfer and Advancement Act have shifted the government's focus from the development of required government standards to the use of non-government voluntary standards. It is also important to understand the market factors that influence the evolution and adoption of standards. Industry organization and the costs/benefits of a standard can drive or inhibit industry adoption of a standard. The Federal government has several opportunities to influence the standards development system, but must clearly define the need for a standard as well as address any market factors that could influence industry acceptance.

The NSSB is one example of such an approach to standards. The Federal government, through the Congress, responded to an apparent void in the education and training leading to the job market because of undefined skill requirements for positions which should be uniform throughout the country. The mechanism the Congress designed allows private sector representatives which wholly comprise and run the NSSB to identify where the voids are and to chose those most capable of creating the standards to fill the void. The only role for the Federal government is one of funding.

The Federal government could also fund efforts of existing standard setting bodies to pursue specific projects. For example, through NIST or another agency, the government could sponsor a group like ANSI's standard setting process in a specific area where the Federal agency felt there was a void. This type of approach allows the government more control over the direction of the process, however, because of Federal involvement in many of the standard setting bodies' work and the Federal allocation of funds, private sector groups may be reluctant to participate because of the degree of Federal control.

The best way to take advantage of the numerous benefits of private standard setting processes it likely a combination of the two approaches.

Conclusion

The Administration should consider putting forth recommendations in the standards area for the following purposes: (1) promoting standards setting in areas where the market is not sufficient to spur development of private sector standards; (2) alleviating the burden on already taxed Federal bodies with standard setting responsibilities; and (3) providing additional support for existing Federal standards bodies.

In order to further these ends, the following steps should be considered:

- The Federal government should find ways to use Federal funds and other incentives to promote creation of standards, by the private sector or by government, in productive areas where the market alone has not yet prompted the need for such standards (including the development of government standards).
- Congress should consider the creation of a government-funded private sector board, modeled on the NSSB, to identify areas where standards are needed and to make grants available to deserving standard setting bodies for projects related to infrastructure assurance objectives, including information technology.

CHAPTER 14

APPROACHES TO CYBER INTRUSION RESPONSE

A "LEGAL FOUNDATIONS" STUDY

Report to the President's Commission on Critical Infrastructure Protection

1997

This report was submitted to the President's Commission on Critical Infrastructure Protection, and informed its deliberations and recommendations. This report represents the opinions and conclusions solely of its developers.

CONTENTS

PART ONE
INTRODUCTION

Research Issues

Are there viable alternatives to traditional criminal law enforcement responses (i.e., investigations, prosecutions and sanctions) that would serve as an additional deterrent to unauthorized computer intrusions?

Research Findings

- Many intrusion incidents go undetected.
- When detected, intrusion incidents are underreported to law enforcement.
- Reasons for failure to report include fear of negative publicity, fear of bestowing undue advantage to competitors, or preference for a civil remedy.
- Of the incidents reported, only a relative few result in fully successful criminal prosecutions.
- Failure of detection and under-reporting of intrusion incidents significantly dilutes potential deterrent effects of a criminal enforcement scheme.
- The limited ability of law enforcement to successfully prosecute all reported incidents significantly dilutes potential deterrent effects of a criminal enforcement scheme.

Assumptions

- The number of unauthorized intrusions will continue to grow, owing to a steady increase in the numbers of people with capability and the increased availability of intrusion tools.
- Even assuming perfect reporting of intrusion incidents, law enforcement will not be able to provide a sufficiently robust response to achieve maximum deterrence.
- Additional deterrence will be needed to thwart a growing threat.
- Deterring computer crime will reduce the cyber-based threat to critical infrastructures.

Background

Although figures vary, only a small percentage of successful computer intrusions are detected. Only a small percentage of those detected are reported to law enforcement. According to the 1997 Computer Crime and

Security Study (CSI Spring 1997), only 17.87 percent of surveyed companies that experienced computer intrusions over the last 12 months reported those incidents to law enforcement. This figure is essentially unchanged from 1996 (16 percent). Among the reasons cited for not reporting intrusions to law enforcement are a fear of negative publicity, restoring advantage to competitors, or preference for a civil remedy. Whether in the interest of maintaining the confidence of their customer base or retaining control of their systems, personnel and resources, the private sector has, by and large, opted to incur the costs of intrusion incidents rather than respond through currently available law enforcement avenues.

In a very real sense, private sector victims are forced to choose between notifying law enforcement and losing public trust because of constitutional considerations that require public airing of criminal allegations.[1] (Assuming as a given that the constitutional rights afforded a criminal defendant are unlikely to be radically transformed in the foreseeable future–this conflict will not resolve itself any time soon.) Add to this the intense media interest in high technology crime, and private sector reluctance to report known criminal incidents becomes increasingly understandable.

Victims are placed in the awkward position of having to choose between cooperation and maintaining their anonymity. They must weigh, on one hand, the benefits derived through full cooperation with law enforcement (which, in most cases, amounts only to the deterrent value derived through a successful criminal prosecution, as few criminal defendants have had the ability to make restitution for damage done), against the inordinate expense incurred in supporting an investigation and the reputational harm that can result from public acknowledgment of security vulnerabilities.[2] Maintenance of public confidence as a primary business objective often militates in favor of non-reporting. It is reasonable to expect a similar calculus to apply for the owners and operators of the critical infrastructures, given the high value that they place on maintaining public trust and confidence in the reliability and continuity of their infrastructure services. This may be especially true in highly competitive industries such as commercial airlines, banking, and long-distance telecommunications, or in infrastructure sectors that are just starting to move toward competition, such as electric power.

In addition, several writers have acknowledged obstacles to the successful investigation and prosecution of computer-related offenses.[3] First there are

---

1 *See* Darryl C. Wilson, *Viewing Computer Crime: Where Does the Systems Error Really Exist?*, 11 Computer/Law J. 265, 284 (1991).

2 *Id.*

3 *See, e.g.*, Michael P. Dierks, *Computer Network Abuse*, 6 Harv. J.L. & Tech. 307 (1993); James A. Fagin, *Computer Crime: A Technology Gap*, 15 Int'l J. Comp. & Applied Crim.

legal obstacles. The current legal structure for prosecuting computer crimes including intrusions is fraught with difficulties ranging from proving intent to establishing jurisdiction and it may be many years until the actual laws, law enforcement personnel, and judicial systems operate in such a way that they will maximize deterrence to criminal behavior.[4]

Even assuming an ideal legal climate, practical difficulties arising from the resource-intensive nature of computer crime investigations will continue to hinder the achievement of maximum deterrence through a traditional law enforcement response. Investigations are extraordinarily resource-intensive. In terms of personnel alone, while an average "low-tech" state or local investigator may handle a normal load of 40-50 cases per month, a "high-tech" investigator can only handle 3 or 4 per month.[5] Equipment is expensive, and must be frequently improved or replaced to keep pace with the technology used by intruders. Incidental investigative costs (such as travel) remain high, owing to the multi-jurisdictional nature of the offenses. And these costs are frequently borne by more than one investigative agency that may be actively involved in investigating the same series of intrusion incidents.

Congress has been agreeable to improving laws relating to computer crime, as evidenced by amendments to the Computer Fraud and Abuse Act and the passage of the Economic Espionage Act of 1996. The Department of Justice is currently working on proposed legislation to further streamline the national and international law enforcement response.

Congress has now taken the first step toward the creation of a meaningful federal *civil* remedy for victims of computer crime. The 1994 amendments to the Computer Fraud and Abuse Act created a limited civil cause of action for victims to use against violators to obtain compensatory damages (limited to economic damages) or injunctive relief. The 1996 amendments to Section 1030 now make civil remedies available for any violation of Section 1030, including any of its 7 subsections (a)(1) through (a)(7). Economic damages are available for violations that cause a loss of $5,000 or more. No limitation on damages is imposed for violations that modify or impair medical treatment, examination or diagnosis, that cause or threaten physical injury, or that threaten public health or safety. This new cause of action is rarely exercised however. To date, the civil remedy has been

---

Justice 285 (1991); B.J. George, *Contemporary Legislation Governing Computer Crimes,* 21 Crim. L. Bull. 389 (1985).

4 *See* Robert L. Dunne, *Deterring Unauthorized Access to Computers: Controlling Behavior in Cyberspace through a Contract Law Paradigm,* 35 Jurimetrics J. 1, 8-9 (1994).

5 Ingrid Becker, *Cybercrime: Cops Can't Keep Up with Technobandits,* 15-Jun Cal. Law. 47, 91 (1995) (quoting Bill Spernow, SEARCH Group, Sacremento California).

invoked only once–unsuccessfully, by a pro se plaintiff against his former employer–in an unpublished opinion[6]

The Administration, in its *Framework for Global Electronic Commerce,* has defined the appropriate role for government as "encourag[ing] industry self-regulation wherever appropriate and support[ing] the efforts of private sector organizations to develop mechanisms to facilitate the successful operation of the Internet."[7] With regard specifically to electronic commerce, the Administration has registered a preference for "a non-regulatory, market-oriented approach to electronic commerce, one that facilitates the emergence of a transparent and predictable legal environment to support global business and commerce." Fostering a more well defined and robust civil liability climate appears to be consistent with this general guidance. Toward this end, the President and Vice President have embraced efforts by a number of groups to adapt the Uniform Commercial Code and other laws to cyberspace, and the adoption of uniform legislation by all states. In the *Framework,* the President and the Vice-President set forth both broad and specific principles to guide growth of commerce on the Internet. Some of these principles may be a helpful frame of reference when thinking about alternatives to the traditional law enforcement response.

Principles for Commerce on the Internet

- **Principles for the Global Information Infrastructure:**
  - the private sector should lead;
  - governments should recognize the unique qualities of the Internet.
- **Principles for Rules Governing Electronic Commerce:**
  - parties should be free to order the contractual relationship between themselves as they see fit;
  - rules should be technology neutral (i.e., the rules should neither require nor assume a particular technology) and forward looking (i.e., the rules should not hinder the use or development of technologies in the future); and
  - existing rules should be modified and new rules should be adopted only as necessary or substantially desirable to support the use of electronic technologies.

It may prove productive to consider whether there may be viable alternatives to the traditional criminal law enforcement response that may counter

---

6 *See* Letscher v. Swiss Bank Corp., No. 94 Civ. 8277 (LBS), 1996 WL 183019 (S.D.N.Y. April 16, 1996) (unpublished).

7 Http://www.whitehouse.gov/WH/New/Commerce/read.html (July 18, 1997).

or avoid some of the perceived shortfalls and difficulties noted in this paper. The following options are specifically designed to supplement the existing law enforcement response and will work equally well in conjunction with an enhanced law enforcement response, to the extent options are considered under the PCCIP supplemental report *Adequacy of Criminal Law and Procedure (Cyber)*. Instead of further regulating reporting of intrusions, there may be alternatives that will be more effective in garnering private sector support, participation, and, that will ultimately result in enhanced deterrence. Civil and administrative sanctions and proceedings, for example, can occur in a less public environment in comparison to criminal prosecutions, but with a proportionate loss of deterrent effect. However private sector responses are structured, issues regarding liability, maintenance of the rule of law and privacy should be addressed and balanced.

PART TWO
ADDITIONAL APPROACHES TO CYBER INTRUSION RESPONSE

Emphasizing Increased Security Measures by Private Industry

A strategy aimed predominantly at enhancing governmental response is by its very nature, both government-centric and reactive. There are benefits from the private sector reserving for itself a greater degree of defensive and responsive capability. Inherent in this idea is for Congress to help the private sector concentrate funding on security measures that will reduce intrusions from small-time hackers, thereby allowing more limited law enforcement resources to be spent apprehending sophisticated and dangerous computer criminals.

- **Pro:** An emphasis on increased security may prove more cost-effective for government by alleviating or externalizing some of the costs associated with the investigation and prosecution of recreational hackers. A security-based approach to minor incidents can prevent damage which reactive measures like prosecutions and civil actions may not be able to repair.
- **Con:** This option may convey to the public an impression of law enforcement complacency to computer crime, which could ultimately reduce rather than increase deterrence. Government may not be able to contribute the resources necessary to effectively raise the private sector's level of security given changes in technology over time.

Creating an Expert Study Group to Explore Alternative Legal Regimes

It is difficult to arrive at a comprehensive solution that strikes delicate balances between public and private response, and between criminal, tort, and contractual liability. The White House or Congress may consider forming an expert study group to explore supplemental response mechanisms.

- **Pro:** This approach acknowledges that a fundamental shift in controlling legal regimes must necessarily involve careful consideration of a broad range of legal alternatives, each of which must appropriately address societal concerns such as liability exposure, maintenance of the rule of law, and privacy.
- **Con:** This approach invariably delays resolution of the problem. A study group may be tempted to "wait and see" what courts do under prevailing legal regimes before developing positions.

Exploration, Expansion and Use of Private Contractual or Tort Remedies

The Administration can expressly recognize the value of contractual or tort remedies as applied to various "cyber relationships," and can urge Congress to enable the private sector to implement avenues through which it can pursue alternative remedies. One possibility might be to encourage Internet hosts and users to enter into binding contractual arrangements specifically setting forth one another's rights and responsibilities. The Administration might also describe and suggest the use of tort-based remedies such as trespass and invasion of privacy, or perhaps even the creation of new civil rights of action for "cyber torts." Either method or some combination will require further study to determine long-term effects of such a system as well as the proper modes of implementation.

Private, Contract-Based Approach

Under a contract-based approach, contractual agreements between Internet users would provide a basis by which to claim damages for breach of contract as a result of unauthorized access. In order to form a network of contractual relationships, users would sign agreements stating acceptable conduct in order to obtain access privileges from Internet providers. Providers would link to other sites using similar agreements. Users would have to register aliases with their providers or hosts. Providers themselves would be the first line of enforcement, as system operators would monitor systems for inappropriate actions and identify users acting in violation of their contractual agreements. Providers and sites could take action against an intruder, including notifying them of the inappropriate nature of their conduct; requiring them to submit to monitoring as a condition of continued access; or suspension or withdrawal of access privileges.[8]

---

8 This scheme is based on a model suggested in Robert L. Dunne, *Deterring Unauthorized Access to Computers: Controlling Behavior in Cyberspace through a Contract Law Paradigm,* 35 JURIMETRICS J. 1-15 (1994).

- **Pro:** This approach would allow "policing" of the Internet to be done by those in the best position to monitor activities–the system operators. The scheme would be entirely consensual; all monitoring is with knowledge and consent of users. Wide acceptance of a contractually based "code of conduct" may reduce other problems related to unsolicited e-mail, indecency, or privacy violations.
- **Con:** A network of consensual relationships between authorized users would take time to develop and will be difficult to implement. Laws may have to be changed to allow for expanded monitoring by system administrators in order to enforce the system. Privacy advocates and Internet users may be reluctant to sign on to a system of involving increased monitoring. Sanctions may not be severe enough to provide adequate deterrence to a hacker who poses a serious threat. The availability of alternate channels to the Internet may nullify effectiveness and deterrent effect.

Private, Tort-Based Approach

Under an enhanced tort-based approach, substantial remedies in tort for unauthorized access to or use of computers and computer systems would provide financial deterrents to potential hackers. Such an approach is likely to take shape, in the states, out of existing causes of action for trespass or violations of privacy, or out of new civil actions specifically tailored to cyber intrusions and the potential damages they cause. As noted above, although a federal civil cause of action now exists, it is not being used. Congress may want further consideration of or modification to the civil right of action set forth at 18 U.S.C. Section 1030(g) to make it more widely available as an alternative and effective remedy.

- **Pro:** This approach allows the victim an alternate route if criminal prosecution is declined or unsuccessful. The lower standard of proof in civil matters may ease burdens on investigators and plaintiffs allowing greater success than in criminal proceedings. Monetary damages would allow a victim to be financially compensated for the harm done to their systems or businesses as a result of the unauthorized access. Injunctive relief could be used to establish limited conditions of system use for known violators.
- **Con:** The difficulty of identifying the proper defendant and collecting evidence may prevent the victim from going forward with a civil action. Current civil actions and damage awards may not provide a sufficient incentive for a victim to go forward with a costly civil action, necessitating some expansion of remedies or other change in the law (e.g., treble damage provisions, etc.). For this option to be successful in terms of deterrence, numerous cases will have to go forward with substantial damage awards which may create a burden for the court system. Pub-

licity surrounding large damage awards may not give the victim the confidentiality that they desire.

Exploring Administrative Remedies

There may be a need for an effective "middle ground" response to supplement existing private or law enforcement response mechanisms. Such an approach may take the form of administrative remedies and enforcement mechanisms to adjudicate broad "invasions of privacy" or "trespass" claims arising from unauthorized intrusions. This enforcement regime, the equivalent of "speeding tickets in cyberspace," could include injunctive remedies and substantial fines, as well as procedures that include the use of non-disclosure provisions to promote confidentiality and avoid publicity inherent in criminal actions.

In such a regime, a body could establish "rules of the road" and an administrative enforcement mechanism would be put in place to adjudicate violations. Rules could be set by a group after appropriate public comment period, and could be designed to prohibit unauthorized access, spamming, viruses, and privacy violations. Enforcement could be carried out by local, state and federal law enforcement personnel, who would issue on-line citations for violations observed. Victims or observers of prohibited activity could file complaints with an administrative hearing board. Sanctions could include citations with fines, targeted courses on rules, suspension of privileges, etc. Law enforcement can refer to the board cases that are not appropriate for criminal prosecution.

- **Pro:** This approach allows for enforcement and application of penalties without publicity associated with criminal investigations. It allows adjudication of computer incidents without the expense of standing up full-scale criminal investigations or engaging in a lengthy civil trial.
- **Con:** It would demand a complex regulatory enforcement mechanisms to oversee investigation and adjudication of disputes. This relatively heavy-handed, governmental type of solution may still fail to achieve widespread acceptance within the private sector. It will require considerable study and resources before implementation is possible. It would be heavily reliant on law enforcement resources to "patrol" the Internet for violations and issue citations, contrary to existing law enforcement policies and public expectations.

Federal Licensing of Private Parties Engaged in the Business of Investigating Computer Intrusion Incidents

The Federal government could issue professional licenses to qualified individuals engaged in the growing business of investigating computer intrusions. Under such a scheme, the FCC or other federal regulatory body would issue licenses to qualified private investigators to track and identify unauthorized intruders. The licensing body would dictate requirements for eligibility and renewal of the license, public liability, and standards of conduct.

This novel idea would have Congress look at the feasibility and advisability of Federally licensing computer security experts to track and identify unauthorized intruders. The proposal could be based extensively on the state model for licensing private investigators. The hallmarks of such licensing schemes are stringent educational requirements, training courses and on-the-job training requirements, including examinations and continuing education requirements; liability requirements in the form of bonds and/or public liability insurance; and administrative oversight of misconduct.[9]

Each of the elements of a licensing scheme would have to be carefully considered and delineated to ensure coverage is neither too broad nor too narrow and to ensure that the licensing scheme provides benefits to all interested parties–potential licensees, customers, and the government. The licensing scheme will have to be carefully placed and designed to avoid an appearance that recipients of licenses are acting as federal agents and are subject to the same legal constraints and liability.[10] In fact, by using state laws for private investigators as a model, it may be necessary to insulate the investigator-client relationship through a statutory privilege. While these vary from state to state in terms of coverage, most states do protect clients from unauthorized disclosure of investigatory information.

Such a proposal raises related issues as to the availability of *legal* investigatory techniques in the area of computer security. Under the current statutes governing interception of electronic communications consent is an exception to the general prohibition on interception. It may be premature to determine whether consent is an adequate legal basis for any investigator to perform the necessary steps to trace the source of an intrusion. To answer this question more must be known about the types of techniques available to private investigators and the use and effectiveness of their potential customers' "banner" warnings and notifications. While some amendment to current laws may be necessary and appropriate to facilitate such investigations, and for other purposes, it is not clear at this time that the current legal framework is an impediment to such a licensing scheme.

---

9 For a proposal whereby a Federal licensing scheme for computer security expert investigators is described in more detail, *see* Stevan A. Mitchell & Elizabeth A. Banker, *Private Intrusion Response,* 11 Harv. J. L. & Tech. 699-732 (1998). (Note: This report has been updated since the publication of the PCCIP report to refer to the "Private Intrusion Response" article.)

10 For this reason, the licensing scheme should be administered by a non-law enforcement agency. It may be appropriately housed within the FCC as it is fairly congruent with its current statutory mission.

The benefits of elevating cyber-investigation to a legitimate, sanctioned profession will include enhanced deterrence of computer-based attacks on critical infrastructures. It would likely increase successful investigations of intrusions by creating an avenue for the private sector to investigate incidents with the degree of confidentiality and control that they desire, while allowing the government to externalize costs of petty computer intrusions. The government can then concentrate resources on computer crime investigations with implications for national security. Ultimately, both avenues may lead to more successful prosecutions of cyber criminals. In the meantime, a licensing scheme would create the needed standards of professionalism and protections of privacy that are currently lacking from what is a growing, but seldom discussed line of business. The license may contribute to the open and successful growth of such businesses, providing a defined liability climate and a government stamp of approval for investigator credentials. The cadre of investigators licensed through such a program will provide valuable services not only to the private sector, but also to the government through limited reporting (e.g., investigation statistics–how many per month, types of systems, successful tracing of intruders) and a promise of availability to assist in national cyber-emergencies.

- **Pro:** This approach might enhance deterrence while promoting responsible cyber-investigation. It might help clarify the liability climate for practitioners and for those potentially damaged. Provisions could be made to assure private-sector confidentiality and control. It would allow government to externalize the costs of minor investigations and concentrate its limited resources.
- **Con:** This approach calls for an administrative mechanism for oversight. Further study will be needed into the details before implementation will be possible.

Expanding Statutory "Right To Monitor and Detect" for Broad Range of Users and Service Providers

Currently, 18 U.S.C. Section 2511(2)(a)(i) permits electronic service providers to intercept, disclose, or use communications when it is necessary incident to the rendition of service or to protect rights or property of the service provider. Wire service providers are limited to observing or random monitoring for mechanical or service quality control checks. The Administration could recognize a need for this statute to specifically authorize system owners and operators to monitor their systems and compile evidence of intrusions, their source, and any illegal activities while in their systems for use in civil suits or by law enforcement.

- **Pro:** This approach would more clearly set forth rights and responsibilities of owners and users that is lacking in current federal legislative scheme, which is controlled by laws originally devised to govern telephonic but not computer-based communications. It may work in conjunction with other alternatives to traditional law enforcement.

- **Con:** Enforcement exceptions may, of necessity, be so broad as to authorize a wide range of activities currently prohibited by exiting legislation. This approach would likely substantially reduce expectations of privacy in otherwise unprotected communications.

Developing an International Civil Enforcement Mechanism for Computer Crime Modeled on the WTO - TRIPS Intellectual Property Enforcement Mechanism

Congress and the Administration could also actively expand the international availability of civil remedies for computer crime violations through organizations such as the World Trade Organization. These efforts could be modeled after existing mechanisms that provide for international civil enforcement of intellectual property violations.

The World Trade Organization's (WTO) mechanism for enforcing intellectual property rights infractions is a potential model for international computer crime legal development–including civil remedy enforcement. Implemented by the WTO, a successor to the GATT, the Trade-Related Aspects of Intellectual Property Rights (TRIPS) requires all 131 signatory countries to create and maintain minimum standards for protecting intellectual property rights; TRIPS additionally requires that all countries create legal mechanisms for adjudicating civil disputes. TRIPS sets a timetable for developing:

- law enforcement investigative procedures;
- procedural rules that are not onerous or overly complicated; and
- enforcement proceedings, whether administrative or judicial.

TRIPS permits signatory countries to develop legal and law enforcement mechanisms within the parameters of existing legal structures and institutions. TRIPS establishes a timetable for all countries to meet minimum standards. The more advanced countries offer technical assistance and training to assist lesser developed and developing signatory members. For a more detailed treatment of this issue, refer to Appendix A.

- **Pro:** The WTO-TRIPS model advances international communication and may be available to set timetables for cooperation over complex computer-related intrusions and related investigations. The model provides a framework for defining and implementing methods and goals currently used in the United States, Japan, and most European countries. The WTO TRIPS model may be used to offer developing and lesser-developed countries incentives to implement legal mechanisms for investigating and enforcing civil judgments for computer-related intrusions. WTO-TRIPS could channel information and training in law enforcement, computer forensics, and administrative and judicial procedures to member countries. International organizations are not focusing on lesser developed or developing countries. The WTO-TRIPS model mandates National Treatment and Most-Favored-Nation benefits for law enforcement investigations; additionally, WTO-TRIPS mandates similar treatment for judicial proceedings and civil penalty enforcement.

- **Con:** Potentially tenuous connection between world trade and computer abuse may cause skepticism among member countries. Implementing the agenda will require a tremendous amount of time and effort. Numerous other international organizations are addressing issues associated with computer-related crime, including the P8/G7, EU, OECD, and the United Nations. There is little if no support for an additional international organization to focus on computer crime enforcement.

PART THREE
CONCLUSIONS

Traditionally, law enforcement investigations and prosecutions have been seen as the principal source of deterrence for criminal activity. However, based on what appears to be a growing trend toward private "security" services ($6 billion in 1996), and the unique nature of computer intrusions, there some day may be a need to supplement law enforcement deterrence. While there are many possible supplements to a traditional law enforcement response, all of the options considered have one thing in common–they place the control of the investigation and prosecution and their costs in the hands of the victim.

APPENDIX A
WTO INTELLECTUAL PROPERTY ENFORCEMENT MECHANISM

Existing international organizations and agreements offer mechanisms for enforcing civil penalties associated with computer-related crimes. These agreements also present an excellent model for providing training and technical assistance to countries to adopt and enforce substantive and procedural standards for computer-related crimes.

One particular model might be the agreements negotiated and implemented by the General Agreement on Tariffs and Trade (the "GATT"), and its successor organization, the World Trade Organization ("WTO").[11]

The TRIPS agreement is recognized by all 131 signatory countries and mandates legal protection for intellectual property rights and enforcement mechanisms.[12] TRIPS includes highly-negotiated and extensive provisions for enforcing a final decision—including a dispute resolution mechanism. TRIPS also creates responsibilities for more advanced countries to offer technical assistance and training to lesser developed signatory countries.

Flexibility

TRIPS is a minimum standards agreement, which allows members to provide more extensive protection. The agreement also permits signatories to determine the appropriate method of implementation in the context of their own legal system and practices.

Enforcing Judgments

TRIPS allows for differences in various legal regimes, but mandates specific *minimum standards* and *general principles* that all signatory enforcement regimes must incorporate, including civil and administrative remedies, special requirements for criminal sanctions, and procedures and remedies available so right holders may effectively enforce their rights.[13]

The following summarizes TRIPS' specific obligations for enforcing civil judgments:

---

11 The WTO is an inter-governmental organization resulting from the GATT Uruguay Round negotiations (1986-1994). The WTO adopted the GATT agreement in addition to other agreements from the Uruguay Round talks – including the General Agreement in Trade and Services ("GATS") and, more importantly, the Agreement on Trade-Related Aspects of Intellectual Property ("TRIPS").

12 The WTO's purpose is to facilitate trade flow as smoothly as possible in a system based on rules, to settle trade disputes between governments, and to organize trade negotiations. In May, 1997 WTO membership included 131 countries. For more background information, please refer to the WTO's Web Site at http://www.wto.org/.

13 The general enforcement provisions are in Part III of TRIPS.

General Obligations

TRIPS mandates clear and transparent general requirements, and include:

- rights to expeditious process of claims and remedies;
- construction of legal and administrative deterrents to breaches of law;
- rights to enforcement procedures that are not costly or complicated;
- rights to judicial and administrative appeal;
- obligations to create mechanisms to facilitate legal transparency.

Civil and Administrative Procedures and Remedies

TRIPS outlines in great detail the minimum standards and basic features for civil judicial procedures. Particulars include minimum standards for application of:

- basic rules of evidence;
- due process (*e.g.*, written notice, expeditious remedy, and complaint with sufficient detail);
- court or administrative body's right to seek production of evidence;
- court or administrative body's right to punish an uncooperative party;
- safeguards for abuse of judicial or administrative power; and
- application of criminal sanctions for certain willful violations.

National Treatment and Most Favored Nation Treatment

All WTO Members must provide *National Treatment* and *Most-Favored-Nation-Treatment ("MFN")* to all other Signatories under TRIPS.[14] National Treatment means that a Member country must afford to foreign nationals (from Member countries) all protections and access to enforcement mechanisms no less favorable than it accords to its own nationals. MFN awards "any advantage, favor, privilege and immunity granted by a Member to the nationals of any other country."[15] This would include:

- law enforcement arrangements;
- judicial assistance in capturing perpetrators; and
- administrative or judicial assistance in enforcing civil penalties.

With regard to judicial and law enforcement assistance, both National Treatment and MFN could facilitate investigations of computer crimes. Further, these legal regimes would create incentives for all countries to

---

14 TRIPS Agreement, Articles 3 (National Treatment) 4 (MFN Treatment).

15 TRIPS, Article 4.

develop legal mechanisms to enforce civil penalties arising from computer-related crimes.

Dispute Settlement

The TRIPS agreement makes disputes over obligations subject to the WTO's dispute settlement procedures.[16] The Dispute Settlement Body ("DSB") has exclusive jurisdiction to implement the dispute settlement process. TRIPS sets forth a process to form DSB panels for mediation and arbitration. TRIPS also allows for appeals to the WTO from DSB panel final decisions. The DSB enforces all decisions and maintains surveillance over the decision implementation process.

Transitional Arrangements - Obligations Created

TRIPS gives all WTO members transitional periods so that they can meet their obligations. Transitional periods depend on the level of signatory-country legal and economic development[17] In all cases, however, signatories are not permitted to "back-slide." All lesser-developed countries are expected to implement TRIPS within set periods of time.[18]

TRIPS additionally imposes affirmative obligations on developed countries to train and assist lesser-developed countries to meet time deadlines to implement various requirements.[19] Such technical assistance includes assistance in preparing laws and regulations, systems to prevent judicial and administrative abuse of power, and training for officials involved in enforcement of legal authorities and mechanisms.

TRIPS also includes a general agreement by all members to cooperate in general enforcement of TRIPS. This includes greater communication, establishment of contact points, and promotion of exchanges in cooperation and communication.[20]

Summary of a Dispute within the WTO - TRIPS Legal Regime

The Dispute

1. Dispute arises; party may seek redress within Member country legal system. Under WTO-TRIPS, all Member countries must afford National Treatment ("no less favorable than it accords its own nationals. . . ") and

---

16 TRIPS, Article 64.

17 TRIPS, Articles 65-66.

18 For example, so-called developing countries have five years to implement TRIPS.

19 TRIPS, Article 67.

20 TRIPS, Article 69.

Most-Favored-Nation-Treatment (". . . any advantage favor, privilege or immunity granted by a Member to the nationals of any other country . . . ").

2. Violating Country must afford to other Member country judicial and law enforcement assistance consistent with benefits for its own nationals.

3. WTO will expect parties to attempt settlement bilaterally;

4. If this fails, WTO, in strict accordance with TRIPS, invites parties to seek assistance from the WTO Director-General - acting in an *ex officio* capacity seeks to mediate dispute;

5. If consultations fail to arrive at a solution after 60 days, complainant may ask the Dispute Settlement Board to establish a "Panel"; the establishment of a Panel is almost automatic;

6. The Panel's procedures are set forth clearly in TRIPS and require the Panel to assist the DSB in making its decision in light of the existing documents;

7. The Panel, after hearing the matter, drafts a Report; the DSB must then "adopt" the Report;

8. Once adopted, either party may appeal to the Appellate Body, which can then modify the Report;

Implementation and Enforcement of Remedy

1. According to TRIPS, parties are expected to adhere promptly to all DRB/Appellate Body decisions.

2. TRIPS instructs parties to report, within 30 days of the decision, their intentions with regard to the decision;

3. If impractical to comply immediately, DSB provides a "reasonable period of time" to comply;

4. After this period, if the party has still not complied, the losing party is instructed to negotiate on a mutually accepted compensation;

5. If the parties cannot come to agreement, the DSB may suspend certain concessions that the losing party may enjoy under TRIPS.

6. DSB monitors compliance.